XXIII International Symposium on the Jahn-Teller Effect 2016

Vibronic Coupling and Electron-Phonon Interactions in Molecules and Crystals

Journal of Physics: Conference Series Volume 833

Tartu, Estonia
27 August - 1 September 2016

Editors:

Vladimir Hizhnyakov
Isaac Bersuker
Pablo Garcia-Fernandez

ISBN: 978-1-5108-4076-8
ISSN: 1742-6588

Printed from e-media with permission by:

Curran Associates, Inc.
57 Morehouse Lane
Red Hook, NY 12571

Printed by Curran Associates, Inc. (2017)

For permission requests, please contact the Institute of Physics at the address below.

Institute of Physics
Dirac House, Temple Back
Bristol BS1 6BE UK

Phone: 44 1 17 929 7481
Fax: 44 1 17 920 0979

techtracking@iop.org

Additional copies of this publication are available from:

Curran Associates, Inc.
57 Morehouse Lane
Red Hook, NY 12571 USA
Phone: 845-758-0400
Fax: 845-758-2633
Email: curran@proceedings.com
Web: www.proceedings.com

Table of contents

Volume 833

Vibronic Coupling and Electron-Phonon Interactions in Molecules and Crystals: XXIII International Symposium on the Jahn-Teller Effect

27 August to 1 September 2016, Tartu, Estonia

Accepted papers received: 3 April 2017
Published online: 19 May 2017

Preface

Papers

Jahn-Teller effect

Pseudo Jahn-Teller effect

Superconductivity in Jahn-Teller systems

Vibronic interactions

IOP Conf. Series: Journal of Physics: Conf. Series **833** (2017) 011001 doi:10.1088/1742-6596/833/1/011001

Foreword

This issue of *J. Phys. Conf. Series* contains the proceedings of the 23rd International Symposium on the Jahn-Teller Effect with the main title "Vibronic coupling and electron-phonon interaction in molecules and crystals", which took place in Tartu (Estonia), August 27-September 1, 2016, bringing together over 70 participants from various fields of physics and chemistry. More information on the symposium program can be found at https://ttl.ut.ee/jt16/.
The series of mostly biannual Jahn-Teller symposia started in 1976. It addresses topics involving the coupling between electronic and nuclear motions in molecules and solids in conditions of electronic degeneracy and pseudodegeneracy. The Jahn-Teller effect, pseudo Jahn-Teller effect, and related vibronic coupling and electron-phonon interactions are responsible for structural instabilities and non-trivial dynamics in polyatomic systems, including pseudo-rotations and non-adiabatic effects around conical intersections and seams, and play a key role in the explanation of seemingly diverse phenomena, such as spontaneous symmetry breakings, structural phase transitions, colossal magnetoresistance, exotic molecular magnetism, superconductivity in fullerides, giant permittivity and flexoelectricity, and chemical reactions after thermo- or photoexcitation. The meeting brings together theoreticians and experimentalists working in molecular physics and chemistry with researches working in solid state physics.

A characteristic of these symposia is that they are devoted to discussion of a wide range of physical and chemical phenomena which, at first sight, are different in scope and nature, but in fact they are strongly entangled by vibronic coupling, and their joint presentation and discussion is deemed to mutually enrich the presenters from different fields with stimulating novel ideas.

The 23rd Symposium included the following topics:

- Fundamental theory, computation, and experimental observation of vibronic coupling Jahn-Teller (JT), pseudo JT (PJT), and Renner-Teller effects
- Vibronic coupling effects in structural, reactivity, and spectroscopy properties of molecules and clusters, fullerenes and fullerides, graphene, silicene, and related systems
- Conical intersections, geometric phase effects, and nonadiabatic dynamics on intersecting potential energy surfaces; implications in photophysical and decay processes
- Local and cooperative JT and PJT effects and electron-phonon interactions in solids: impurity centers, ferroelectrics, multiferroics, topological insulators, superconductors, spin-orbital and spin-crossover effects
- Further applications of the JT and PJT effects in materials science: spintronics, magnetic-dipolar bistability, molecular magnets, permittivity, flexoelectricity, and electrostriction.

The Proceedings contain contributions that reflect all the above topics.

We gladly acknowledge support from institutional research funding IUT2-27 of the Estonian Ministry of Education and Research, and from European Union through the European Regional Development Fund (project EU49695).

Vladimir Hizhnyakov, University of Tartu, Estonia
Isaac Bersuker, University of Texas at Austin, USA
Pablo García-Fernández, University of Cantabria, Spain

IOP Conf. Series: Journal of Physics: Conf. Series **833** (2017) 011002 doi:10.1088/1742-6596/833/1/011002

Peer review statement

All papers published in this volume of *Journal of Physics: Conference Series* have been peer reviewed through processes administered by the proceedings Editors. Reviews were conducted by expert referees to the professional and scientific standards expected of a proceedings journal published by IOP Publishing.

IOP Conf. Series: Journal of Physics: Conf. Series **833** (2017) 012001 doi:10.1088/1742-6596/833/1/012001

The Jahn-Teller and pseudo Jahn-Teller effect in materials science

I B Bersuker

Institute for Theoretical Chemistry, The University of Texas at Austin, Austin, TX 78712, USA

E-mail: bersuker@cm.utexas.edu

Abstract. Defining ***materials science*** as *a translation from microscopic foundations of matter to macroscopic performance in applications,* this mostly review paper is devoted to the special features of matter with Jahn-Teller (JT) and pseudo JT (PJT) centers that makes it outstanding in the search of novel properties and applications as novel materials. There are three kinds of problems in this respect. The first is related to the *difficulties in the use of computer simulation* of observable properties (a methodology widely employed in materials science) in application to systems with dynamic JT and PJT effects. The second is due to the specifics of such systems in *interaction with external perturbations*, which contribute with a strong orientational effect that enhances the observable properties by orders of magnitude. The third kind of problems relevant to materials science is in the *use of the JT and PJT effects in revealing the origin of structural properties* of a relevant polyatomic system and working out methods to influence them in designing novel materials.
The paper formulates these problems and provides for examples that demonstrate their importance. For the second group of problems, illustrative examples include flexoelectricity, permittivity, and electrostriction in systems with dynamical JT and PJT centers, and it is shown that the latter enhance the observable properties by several orders of magnitude. Also related to this group of problems are the magnetic-dielectric bistabilities in molecular systems and solids induced by JT and PJT effect, illustrated on molecules like CuF_3 and crystals of the type $LiCuO_2$. The third group of problems is shown by the already multiply used application of the PJTE to reveal the origin of distortions of planar configuration in 2D (graphene-like) and quasi-2D materials and to work out methods of restoration of their planar configurations by means of external perturbations.

1. An open problem for materials simulation: density functional theory (DFT) does not take into account local JT and PJT dynamics

One of the most important features of polyatomic systems with a JT effect (JTE) or pseudo JTE (PJTE) is the very strong vibronic coupling between the electronic cloud and nuclear motions that does not allow to fully separate their coordinates in a the adiabatic approximation (see [1-3] and references therein). For a visible example, consider a polyatomic formation in a degenerate or pseudodegenerate state for which the adiabatic potential energy surface (APES) has several or an infinite number of equivalent minima induced by the JTE or PJTE. The stationary states of such a

IOP Conf. Series: Journal of Physics: Conf. Series **833** (2017) 012001 doi:10.1088/1742-6596/833/1/012001

system can be presented as a linear combination of the states in the minima, each of which is a product of *local* electronic and nuclear-vibrational functions, and there is no way to separate their electronic and nuclear coordinates in the total wavefunction. For systems without such dynamic JT or PJT centers the modern computer simulation methods relay heavily on electronic structure calculations based on density-functional theory (DFT) (see, e.g., [4-7]). However, the existing DFT methods do not include the nuclear motions of the disordered distortions produced by the centers with independent dynamic JTE or PJTE (before the phase transition to the cooperative JT phase). In a paper of 1997 [8] we have outlined the difficulties of the DFT in application to degenerate and pseudodegenerate states. The main assumption of DFT is that the electron-nuclear V'(q, Q) and nuclear-nuclear V"(Q) interactions, V(q, Q) = V'(q, Q) + V"(Q), where q and Q stand for all the electron and nuclear coordinates, respectively, may be regarded as an "external potential" to the electronic subsystem of the polyatomic system. Under this assumption, the Hohenberg-Kohn (HK) theorem [4] establishes a one-to-one correspondence between the electronic density σ of the ground state and the potential V, and hence between σ and the energies and wavefunctions of the Hamiltonian H, since the latter is determined by V and the number of electrons which enter σ by normalization. In the original HK theorem the ground state of the system is assumed to be non-degenerate. This limitation has been formally removed by the Levy-Lieb constraint search formulation [6] in which the density σ is that of one of the degenerate states, provided the main assumption remains valid.

However, it seems unlikely that the Schrodinger wavefunction contains waste information. This means that the description by multidimensional functions can hardly be reduced to the 3-D case without loss of information about the system. As stated by N. March (private communication) this loss of information is intentional: if properly managed, it simplifies the calculations without loss of accuracy for some particular results. Then *one must know explicitly what kind of information is lost and how it influences the DFT and the limits of its applicability*. In both degenerate and pseudodegenerate cases with sufficiently strong vibronic coupling, any DFT formulation of the problem, at least in the form of the DFT known so far, is impossible because, as mentioned above, in these cases the wavefunction is non-multiplicative with regard to the electronic and nuclear subsystems. In such situations the Born-Oppenheimer adiabatic approximation is violated, the assumption of electrons moving in the field of a fixed configuration of the nuclei does not hold, and the basic assumption of DFT that the electron-nuclear and nuclear-nuclear interactions V(q,Q) can be regarded as an external perturbation to the electronic density is inapplicable [8].

Essential improvements in the DFT and applications to systems without JT and PJT centers were achieved in the last two decades (see [9-12] and references therein), but the degeneracy and pseudodegeneracy problem resulting in nonadiabacity and inseparability of the electronic and nuclear motions was not properly countered and not solved. For example, in one of the latest contributions to the extension of the DFT that is aimed at improving materials simulation [10] the electronic density and related properties of solids are considered as a function of the (fixed) nuclear framework, in a way that any changes in the latter as a whole (phase transitions, ferroelectricity, etc.) can be taken into account in the electronic density. However, again, the electronic and nuclear motions are separated in this approach, and the nonadiabacity in the local JTE and PJTE independent (non-cooperative!) dynamic distortions is fully ignored (the same problem remains in Ref. [11] in which the nuclear coordinates are incorporated in a DFT scheme). In the Levy-Lieb constraint search formulation [6] all the f electronic densities of the f-fold degenerate state can be defined, in principle, for *fixed nuclear configurations*. Distinct from the nondegenerate state for f=1, these electronic densities in the ensemble density matrix have no independent physical meaning, they are not observable, and separately they do not determine the charge distribution and energy spectrum of the system. Note that the same is true for the wavefunctions and eigenvalues of a degenerate state obtained for fixed nuclei. Again, the point is here in the *inseparability* of the electronic and nuclear coordinates. In another recent improvement of DFT calculation [12], which combines the basic DFT with elements of Coupled Cluster methods, more accurate numerical results of electronic structure calculations with fixed nuclei were achieved, but the nuclear motion problem and nonadiabacity are even not mentioned. They are not mentioned also in the otherwise inclusive review article [9].

In the next Section we show where and why is the inseparable nuclear dynamics important in description of the properties of matter relevant to materials science.

2. In interaction with external perturbations dynamic JT and PJT effects enhance observable properties *by several orders of magnitude*

Interactions of polyatomic systems with external perturbations are well studied and documented in many original publications and textbooks. For dielectric systems, very important are their properties in external electric fields and under strain, in particular, permittivity, flexoelectricity, electrostriction, and structural phase transitions under pressure. Many theoretical investigations are devoted to these properties (see, e.g., [13] and references therein). *These works are based on the theoretical fundament of the adiabatic approximation*, which separates electronic and nuclear motions, assuming that nonadiabatic effects in such systems are negligible. As stated above, this assumption does not hold in some specific, but very important situations of the JTE and PJTE that lead to spontaneous symmetry breaking (SSB) [1-3]. In the presence of these effects the electronic and nuclear motions are not fully separated, the electronic states are mixed by nuclear displacements, and hence the adiabatic approximation is violated. Another misperception is that in the absence of electronic degeneracy the PJTE may occur only when the energy gap Δ to the exited state is small. In fact, the criterion of occurrence of this effect includes three parameters (see below), among which the energy gap Δ between the interacting electronic states may be very large (there are cases of PJTE instability at Δ~14 eV [14]). Since the JTE and PJTE are essentially of local origin, they strongly depend on the local electronic (atomic) structure and vibronic coupling constants. The majority of papers on dielectric properties do not take into account these nonadiabatic vibronic coupling effects. Note that the PJT mixing of electronic states with differing symmetries makes them dependent on nuclear displacements with appropriate symmetry, and hence it is essentially different from orbital hybridization, which takes place without nuclear displacements.

In polyatomic systems with JT and PJT centers, in the absence of external perturbations and/or strong interaction between the centers the distortions at each center are independent from the others, and they locally resonate between the equivalent minima by means of tunnelling or over-the-barrier transitions. In such dynamic systems the quantum-mechanically averaged expectation values of observables are that of unbroken high symmetry. In the presence of lower symmetry perturbations the possibility to reveal the SSB in such systems is time and frequency dependent, and hence directly related to the methods of observation [3]. Denote the (temperature dependent) “lifetime” of the system in the minimum of the APES by $\tau=2\pi/\omega$, where $\hbar\omega=\delta_0$ is the splitting of the vibrational levels in the minima due to the tunneling between them, and the “time of measurement” of the specific method of observation by $\tau'=2\pi/\omega_p$, where ω_p is the characteristic frequency of the perturbation influence. Distorted configurations can be revealed only if $\tau'<\tau$, while for $\tau'>\tau$ the system is seen averaged over the equivalent minima with unbroken high symmetry.

Consider the properties of dielectric polyatomic systems with dynamic JT or PJT centers (finite τ values) under external perturbations with either $\tau'\to\infty$ (constant electric field or strain), or $\tau'>\tau$ (low-frequency field or strain), for which the SSB is not revealed explicitly: the system is seen as averaged, totally symmetric with regard to such perturbations. Dielectric polyatomic formations with a JTE or PJTE in any high-symmetry environment without strong cooperative interactions (in solids, at temperatures above structural phase transitions) are systems with finite τ values. In these cases *the JTE and PJTE enhance observable properties by several orders of magnitude* [15, 16]. This statement is illustrated below by estimates of polarization, permittivity, and striction in electric fields, and flexoelectricity under strain gradients.

The statement of several orders of magnitude larger properties induced by the dynamic JTE or PJTE is very strong and requires background explanation. For clarity, consider an illustrative example,

IOP Conf. Series: Journal of Physics: Conf. Series **833** (2017) 012001 doi:10.1088/1742-6596/833/1/012001

dielectric polarization of a liquid with dipolar molecules, e.g., water. According to the classical Debye theory the permittivity ε in this case follows the well-known equation:

$$\varepsilon=\varepsilon_0[1+\frac{\chi_d}{1+(\omega_p\tau_d)2}+\frac{\chi_o}{1+(\omega_p\tau_o)2}] \tag{1}$$

where χ_d and χ_o are the displacive and orientational dielectric susceptibilities, respectively, τ_d and τ_o are the relaxation times in the corresponding processes, ω_p is the perturbation frequency, and ε_0 is the permittivity of vacuum. The orientational contribution to the polarization is always much larger than the displacive one (in water $\chi_o = 72$, $\chi_d = 5$). This is quite understandable: it is much easier to rotate a ready-made dipole than to polarize a stable polyatomic formation. *The enhanced properties in systems with dynamic JTE or PJTE are due to the orientational contribution, which is not there without these effects.* In systems with multiminimum APES streaming from the SSB in the high-symmetry configuration, any external perturbation of lower symmetry makes the minima non-equivalent and stops the tunnelings between some of them, locking the system in a broken symmetry configuration. This process is similar to (but easier than) rotating (orienting) a rigid dipole along the external perturbation (as in the liquid water above). In JTE cases, three equivalent dipolar distorted configurations occur in the JTE in systems with trigonal symmetry in degenerate E states (in the $E\otimes e$ problem), and four equivalent dipolar distortions take place in tetrahedral systems with a degenerate T terms in the $T\otimes t_2$ problem [1-3]. Among the vast PJTE cases [2], systems with the $(A_{1g}+T_{1u})\otimes t_{1u}$ problem were explored in application to metal centers in perovskites.

Consider the PJTE in the paraelectric phase of $BaTiO_3$ [17, 18]. Very briefly, in cluster language for TiO_6^{8-}, the PJTE emerges from the (nonadiabatic) mixing of the ground state A_{1g} formed by the fully occupied t_{1u} combinations of oxygen atomic $2p$ states, with the excited state T_{1u} (produced by one-electron ($t_{1u}\rightarrow t_{2g}$) excitations to the three empty $3d(t_{2g})$ states of the Ti^{4+} ion) by the polar t_{1u} type normal coordinates Q_x, Q_x, and Q_x. The numerical results below were obtained recently [18] by the Green's functions method taking into account the interaction of the Ti ion with the whole crystal via its electronic and vibrational bands. With the linear vibronic coupling terms included, the solution of the secular equation of this $(A_{1g} + T_{1u}) \otimes t_{1u}$ problem yields the following APES [17, 18] ($Q^2 = Q_x^2 + Q_y^2 + Q_z^2$):

$$U(Q)=\tfrac{1}{2}K_0Q^2-2\left[\sqrt{\Delta^2+2F^2\left(Q^2-Q_x^2\right)}+\sqrt{\Delta^2+2F^2\left(Q^2-Q_y^2\right)}+\sqrt{\Delta^2+2F^2\left(Q^2-Q_z^2\right)}\right] \tag{2}$$

where 2Δ is the energy gap between the mixing electronic states (appropriately band-averaged in the Green's functions approach [18]), K_0 is the primary force constant for the Q displacements (stiffness of the crystal without the vibronic coupling), and F is the vibronic coupling constant (H is the Hamiltonian),

$$F = \langle 2p_z(0)\left|\left(\frac{\partial H}{\partial Q_x}\right)_0\right|3d_{xz}(\mathrm{Ti})\rangle \tag{3}$$

The three-dimensional APES (2) has a specific form. Under the condition

$$\Delta < 8F^2/K_0 \tag{4}$$

the surface (2) has a maximum (meaning instability) when the Ti ion is in the center of the octahedron, eight equivalent minima placed along the four trigonal axes in which the Ti ion is displaced toward three oxygen ions (away from the other three); higher-in-energy 12 equivalent saddle points along the six C_{2v} axes at which the Ti ion is displaced toward two oxygen ions (at the top of the lowest barrier between two near-neighbor minima); and next six higher-in-energy equivalent saddle points at which the Ti ion is displaced to one of the oxygen ions along the fourfold axes [17]. For $BaTiO_3$ the vibronic coupling parameters were estimated using the Green's functions description [18] and two experimentally determined structural constants: the energy gap 2Δ = 2.8 eV and the vibrational frequency at the bottom of the trigonal minimum $\hbar\omega_E = 193 cm^{-1}$. With these data all the other essential parameters were estimated (table 1), including K_0, F, the positions of the minima $Q_x= Q_y=Q_z=Q_0$ and first saddle points $Q_x= Q_y= q_0$, $Q_z=0$, their PJTE stabilization energies, and the tunnelling splitting δ_0.

Table 1. Numerical values of the PJTE vibronic coupling and APES parameters of the Ti active centers in the $BaTiO_3$ crystal [18].

K_0	2Δ	F	$\hbar\omega_E$	Q_0	q_0	$E_{JT}[111]$	$E_{JT}[110]$	δ_0
55eV/Å^2	2.8 eV	3.42eV/Å	193 cm^{-1}	0.14 Å	0.16 Å	-1250 cm^{-1}	-1130cm^{-1}	34cm^{-1}

The four phases of $BaTiO_3$ emerge directly from this APES by gradually populating the states in the minima with temperature and stepwise overcoming its barriers [17, 19]. Of particular interest here is the *prediction of disorder in the orthorhombic and tetragonal ferroelectric phases and in the cubic paraelectric phase*. It was confirmed later together with a variety of other predictions of the theory that have no explanation in displacive theories (see in [17, 18]). The basic experimental conclusions is that in all the phases the Ti ion is *instantly* displaced in the trigonal direction in disagreement with displacive theories in which the metal off-center displacement occurs as a result of the phase transition to the ferroelectric phases. This was confirmed directly by a variety of experimental methods, noticeable diffuse X-ray scattering [20], EXAFS [21], ESR [22], NMR [23], neutron scattering [24], light refraction [25], etc. Not only is the Ti ion displaced along the [111] type direction in the paraelectric phase where the averaged symmetry is cubic, but it is as well displaced in this trigonal direction in the tetragonal phase where the crystal symmetry and the macroscopic polarization are tetragonal [24].

The order of magnitude of the tunnelling frequency was estimated in ESR experiments with probing ions by substituting Ti^{4+} with the Mn^{4+} ion which produces a very similar APES [22] (Mn^{4+} is a "ferroelectric" ion [26]) yielding $10^{-9} > \tau > 10^{-10}$. It is also seen indirectly by EXAFS [21] and NMR [23] experiments: in the former $\tau' \sim 10^{-15}$, and the Ti ion is seen in the trigonal minima in all the phases of $BaTiO_3$, while in NMR experiments $\tau' \sim 10^{-8}$, and only an averaged picture is revealed. The vibronic coupling theory was extended recently to formulate the necessary condition of coexisting magnetic and ferroelectric (multiferroicity) properties in ABO_3 perovskite crystals with B as a transition metal ion in a d^n configuration, n=1, 2, …10 [26], that found already confirmation in experiments [27, 28] and ab initio calculations.

In the paraelectric phase of $BaTiO_3$ the Ti ions are displaced in the trigonal minima, but in the macroscopic picture the crystal is cubic and centrosymmetric because of their noted tunnelings and over-the-barrier transitions. Perturbations of lower symmetry make the PJTE minima non-equivalent, thus terminating a part or all the tunnelings. If the perturbation is a homogeneous electric field $\boldsymbol{E}$ directed along the trigonal axis, it has the same direction for all the unit cells leading to the polarization of the crystal in this direction. We assume that the polar distortions become fully oriented along the electric field $\boldsymbol{E}$ when the energy difference between two opposite minima of the APES Δ_E reaches the order of magnitude of the tunnelling splitting, $\Delta_E = k\delta_0$, where k is a numerical coefficient, $k \leq 1$. Since $\Delta_E \cong 2d_0E$, where d_0 is the dipole moment of the unit cell when the Ti ions ion is locked in one of the minima, in the electric field $E_m \cong k\delta_0/2d_0$ all the Ti ions are displaced along the same trigonal direction, and the polarization of the crystal in this direction is $P=d_0/a^3$, where a is the unit cell constant. Hence for the orientational dielectric susceptibility χ_o in the relation $P_o=\chi_o E$ up to $E=E_m$ we get:

$$\chi_o = 2d_0^2/ka^3\delta_0 \qquad (5)$$

The data for the $BaTiO_3$ crystal in table 1 allow us to estimate the numerical value of the susceptibility and permittivity. Under the conditions $\tau' \to \infty$ or $\tau' > \tau$, outlined above, $\varepsilon/\varepsilon_0 \approx 1 + \chi_o + \chi_d$, and in all the cases below $\chi_o >> \chi_d$. We have: $d_0 = |e|Z^*\sqrt{3}Q_0$, where $Z^*=7.8$ is the effective (Born) charge of the Ti ion [18], and in the limits of the coefficient $k=0.1 \leftrightarrow 1.0$ we get for the orientational dielectric susceptibility of the $BaTiO_3$ crystal χ_o=0.45×10^3↔ 0.45×10^4. This agrees well by orders of magnitude with experimental measurements of permittivity in this system (see, e.g., [29, 30]). On the other hand, ignoring the PJTE, the polarization of the unit cell P_d is limited by its stiffness K_0 with respect to the polar distortions along the above normal coordinate Q. Hence the electric force $2|e|Z^*E$ producing the dipole moment $|e|Z^*Q$ is compensated by the elastic force K_0Q. Thus $2|e|Z^*E= K_0Q$ and $P_d= |e|Z^*Q/a^3=\chi_dE$, which leads us to the expression: $\chi_d=2(eZ^*)^2/a^3K_0$. Using the same numerical values as above and the K_0 value from table 1 we get the following estimate: χ_d=1.22. This yields a typical value expected for permittivity in systems without orientational contribution. Compared with the value χ_o=0.45×10^3↔ 0.45×10^4, obtained above, we see that *by taking into account the PJTE the permittivity of $BaTiO_3$ increases by more than three orders of magnitude.*

In fact the whole permittivity in the field E up to the orientational saturation at $E=E_m$ is of PJTE origin, acquiring a less-significant displacive-only addition in higher fields. The value of E_m can be estimated from the above expression $E_m \cong k\delta_0/2d_0$. For $BaTiO_3$ $E_m \cong k\times 1.15\times 10^7$ V/m. It follows that up to this voltage the permittivity is strongly enhanced by the orientational contribution of the dynamic PJTE, reaching the value of polarization $P= d_0/a^3 \cong 0.47$ Cm^{-2} which coincides approximately with that of the ferroelectric phase. Very similar to $BaTiO_3$ are the expected properties of $KNbO_3$ perovskites; the NbO_6 centers have a qualitatively similar to the TiO_6 PJTE polar instability producing local dipole moments ($BaTiO_3$ and $KNbO_3$ were the first two crystals in which the order-disorder nature of the phase transitions predicted in [17] was observed by X-ray diffuse scattering [20]; see also [31]).

Since the tunnelling splitting is temperature dependent, the permittivity will be temperature dependent too, especially near the phase transition to the tetragonal phase. In a more elaborate investigation of the induced by the PJTE orientational contribution to the permittivity of perovskite ABO_3 crystals [32], its temperature dependence was revealed, and it was also shown that the permittivity in this case has a *cubic anisotropy* with maximum values along the trigonal axes. This anisotropy is unexpected in cubic crystals without the PJTE.

As mentioned above, polyatomic formations with trigonal symmetry may have multiminimum APES with dipolar distortions in the minima, and such dielectric systems in gas, liquid and solid state are vast. For instance, there is a whole class of molecular systems with a typical hidden PJTE [33] producing equivalent distorted configurations with nonzero dipole moments, and hence an expected giant permittivity in the regime of the dynamic effect. As an example, in CuF_3 the resonating dipole moment d_0=1.67 D [33], and this molecule is notable also by its possessing bistability properties, namely, coexistence of the dipolar equilibrium configuration with d_0=1.67 D, but no magnetic moment, μ=0 (electronic ^1A state), with a slightly higher in energy undistorted D_{3h} one in the electronic state ^3A that has no dipole moment, d_0=0, but a nonzero magnetic moment μ=1.82 μ_B [33] (see Section 3 below). Another class of molecular systems and solids with expected enhanced permittivity is presented by tetrahedral molecular formations and solid states with tetrahedral AX_4 or X_4 unit in an electronic T state, in which the JTE creates four equivalent polar configurations with one long and three shorter A—X bonds [1] (examples see, e.g., in [1], Chs. 7, 8). Note, however, that the JTE distortions and hence all expected observable properties based on their dynamic nature are, in general, smaller in tetrahedral systems as compared with octahedral systems or with some PJTE systems.

In line with permittivity is *flexoelectricity* where the polarization of a centrosymmetric dielectric crystal is induced by a strain gradient instead of the electric field. All theoretical estimates (without involving JTE or PJTE) show that the flexoelectric coefficient f (the component of the tensor of proportionality between the polarization P and strain gradient G in one-dimensional models, $P=fG$), is rather small, of the order of nCm^{-1}[34-36], whereas the experimental measurements for some systems yield by *several orders of magnitude larger values* [37-40]. The biggest flexoelectric coefficient f occurs in the paraelectric phase of $BaTiO_3$, and it is, in general, approximately proportional to the permittivity of the system. This is quite understandable: both dielectric and flexoelectric polarization in the presence of dynamic JTE or PJTE are produced by the same orientational contribution of the local dipolar distortions trapped in the direction of the external perturbation.

For the paraelectric $BaTiO_3$ with the strain gradient G along the trigonal direction, similar to the considered above dielectric polarization, we assume that the tunnelling between the dipolar minima of the APES is suppressed when the energy difference between the opposite minima induced by the strain gradient $\Delta_s=2gG$ (where $g=\mathrm{d}\Lambda/\mathrm{d}G$ with Λ as the strain energy) is of the order of magnitude of the tunnelling splitting, $\Delta_s=\mathrm{k}\delta_0$ with $\mathrm{k}\leq 1$. For the strain gradient we have: $G=(\mathrm{d}S/\mathrm{d}x)\approx\Delta S/a$, where S is the unit cell strain, $\Delta S=-Q/a$, where Q is the introduced above normal coordinate of the dipolar distortion of the unit cell along the (opposite to strain) trigonal direction , and $G=-Q/a^2$. Then for the strain energy we have: $\Lambda=(1/2)K_0Q^2=(1/2)K_0a^4G^2$, where K_0 is the stiffness in the Q direction, and $g=\mathrm{d}\Lambda/\mathrm{d}G=K_0a^4G$. Putting together these formulas we find that the strain gradient $|G_m|=(\mathrm{k}\delta_0/2K_0a^4)^{1/2}$ restores the dipole moment of the unit cell $-d_0$, and hence the polarization of the crystal $P=-d_0/a^3$ along the trigonal direction. It follows that the flexoelectric coefficient $f=P/G_m$ is:

$$f= -(d_0/a)(2K_0/\mathrm{k}\delta_0)^{1/2} \tag{6}$$

With the same numerical values of the parameters, employed above, we get the following estimate: $f = -(0.43\cdot 10^{-6}\div 0.43\cdot 10^{-5})$ Cm^{-1}. Thus instead of (and in addition to) producing a polar charge redistribution in a rigid cubic crystal of the order of nCm^{-1} the strain gradient in the paraelectric phase of a PJTE ferroelectric just recovers the virtual local dipole moments (suppresses their dynamical averaging) inducing the ferroelectric polarization of the order of $\mu\mathrm{Cm}^{-1}$. This removes the main controversy in the theory of flexoelectricity explaining the origin of the 3-4 orders of magnitude larger flexoelectric coefficient f in paraelectric $BaTiO_3$ as compared with systems without the PJTE. The enhanced flexoelectricity takes place for G values below G_m, at which point the dynamic PJTE is quenched and the orientational contribution to the polarization is maximal; for larger values of G the additions to f are much smaller. The numerical data above yield $|G_m|=(\mathrm{k}\delta_0/2K_0a^4)^{1/2}\sim 10^6\mathrm{m}^{-1}$.

As the PJTE and JTE are of local origin, the enhanced flexoelectric effect takes place everywhere these effects produce local dipolar distortions. The mentioned above similarity between $KNbO_3$ and $BaTiO_3$ explains the origin of the enhanced flexoelectricity in perovskite systems that contain a significant amount of NbO_6 centers, e.g., $Pb(Mg_{1/3}Nb_{2/3})O_3$ [40] and $(Bi_{1.5}Zn_{0.5})(Zn_{0.5}Nb_{1.5})O_7/Ag$ [41]. The different flexoelectric properties of the two at-face similar crystals, paraelectric $BaTiO_3$ and $SrTiO_3$, find also explanation: the criterion of PJTE polar instability (4) holds well for the Ti center in $BaTiO_3$ (table 1) [18], but it does not hold for the same ion in $SrTiO_3$. The reason of this is straightforward: the size of the Sr^{2+} ion that controls the unit cell dimensions is smaller, and hence, the interatomic Ti–O distance is smaller in the latter case, and this, in turn, leads to larger stiffness and smaller vibronic coupling constant value, both deteriorating the condition of instability [17, 18]. For the same reason, $SrTiO_3$ becomes ferroelectric under negative pressure that increases the Ti–O distance, $Ba_xSr_{1-x}TiO_3$ becomes ferroelectric when x>0.45 [42], and the nonferroelectric $CaMnO_3$ becomes ferroelectric when Ca is substituted by Ba [43] (as mentioned above, according to the PJTE theory Mn4+ is a "ferroelectric" ion [26]). Although not ferroelectric, the $SrTiO_3$ crystal is a virtual ferroelectric: the PJTE in its Ti centers is not strong enough to obey the condition of dipolar instability [17, 18], but it is very near to it, so the criterion of the PJTE instability may be induced by the strain gradient. Indeed, the latter is softening the crystal in at least one direction thus lowering its stiffness, hence allowing dipolar distortion (polarization) in this direction. This explains (qualitatively) the order of magnitude larger flexoelectric coefficient in this crystal [44] as compared with expected in the theory without the PJTE. The assumptions in some publications (e.g., [39, 40]) of possible polar formations in the bulk crystal that are responsible for the enhanced flexoelectricity, in a way resonate (are compatible) with these results. Obviously, many other crystals with dipolar centers of PJTE or JTE origin including organic and biological systems may have similar enhanced flexoelectric properties.

One of the consequences of the orientational origin of the polarization in the enhanced flexoelectricity is in its frequency dependence in case of dynamical strain gradient, e.g., in sonic or ultrasonic perturbations. The mechanism of relaxation of distortions does not depend directly on the perturbation that produces the distortion. This means that the Debye formula (1) for relaxation in dielectrics is applicable also to flexoelectric polarization. As the orientational contribution to the latter is much larger than the displacive one, we can assume that the frequency dependence of the coefficient of flexoelectric polarization follows Eq. (6), which is thus a characteristic equation of frequency dependent flexoelectric piezo-effect. For the paraelectric $BaTiO_3$ system, in the regions where the tunneling frequency is high compared with ultrasound frequencies, the system may serve as a flexoelectric ultrasound transducer.

If the electric field along the trigonal axis of $BaTiO_3$ suppresses the tunnelling between the dipolar minima of the APES trapping all the units of the system in the same kind of trigonal minimum along the whole crystal, the latter will be not only polarized, but also strained in this direction. Similar to permittivity and flexoelectricity, the change in the dimensions of the unit cell in the electric field E (the electrostriction) can be estimated for the $BaTiO_3$ crystal. In fact the electrostriction problem is closely related to the inverse of flexoelectricity: instead of polarization P induced by strain gradient G in flexoelectricity, electrostriction is the strain S (not strain gradient) induced by the polarization. In the one-dimensional model with the electric field and strain along the same trigonal axis of the crystal we have: $S = \gamma P^2$, where γ is the coefficient of electrostriction. More convenient is to measure the strain as a function of the E, $S=\eta E^2$. At E_m~$(10^6—10^7)$ Vm^{-1} in the trigonal direction the full orientational ordering of the local unit cell dipoles is reached resulting in the strain $S=Q/a \cong\sqrt{3}Q_0/4\cong0.05$. Without the orientational contribution of the dynamic PJTE, from the data above we have: $2|e|Z^*E=K_0Q$ which at $E=E_m$ yields the strain $S \sim 10^{-5}$, a typical value for a regular ionic crystal. Again, as for permittivity and flexoelectricity, we obtain a giant electrostriction of about three orders of magnitude bigger than that in similar systems without the dynamic PJTE, and this effect is in accord with experimental measurements [45, 46].

3. The JT and PJT effects produce a novel class of molecular and solid state materials with correlated magnetic and dielectric bistability

Molecular systems and solids with two coexisting stable and close-in-energy configurations that exhibit markedly different magnetic and structural properties with controllable conversion between them (bistability) are very attractive for application in electronics as high-density information storage materials and sensor devices [47-52]. So far, the most studied systems with magnetic bistability involve transition metal spin crossover complexes (TMSCC) that are virtually possible in d^4-d^7 transition metal compounds with octahedral symmetry and d^3-d^7 in tetrahedral or cubic environment. In these systems the low-spin (LS) to high-spin (HS) transition can be induced by temperature or pressure changes or by light irradiation [51, 52]. The latter case is best suited for information storage devices. However, in spite of almost three decades of study, the lifetime of the metastable state populated after irradiation before its decay to the ground state in these materials is typically [51, 52] a few nanoseconds which is too short for practical application. The decay of the metastable state is strongly dependent on the spin-orbital coupling and the energy barrier between the HS and LS states involving the structural changes in the conformational transformation [51], which in TMSCC is characterized by a metal-ligand distance change of ~0.2Å. The TMSCC are thus restricted in chemical variety and parameters that influence their properties.

On the other hand, it was shown recently [33] that in molecular systems and solids with at least one threefold (or higher) symmetry axis and a half filled degenerate valence shell (valence orbital configurations e^2 or t^3, where e and t are symmetrized molecular orbitals or analogous states in solids), a similar to the spin-crossover (but richer) bistability effect takes place. According to Hund's rule the two (in e^2) or three (in t^3) electrons should have parallel spins in the ground state (figure 1a), but in these electronic configurations low-lying low-spin excited states are strongly involved in the formation of the stable configurations via the pseudo Jahn-Teller effect (PJTE) [33]. In many cases the latter produces an additional low-spin state accompanied by structural distortions, as illustrated in Figure 1b. In fact the PJTE mechanism explains why and how orbital disproportionation and additional covalency lead to the formation of the second stable state [33]. The two equilibrium configurations in this class of polyatomic systems are close in energy, have different spin values (magnetic bistability) and geometries (structural bistability). Both bistabilities are strongly correlated and are switchable by external perturbations.

The general results obtained in [33] were applied to specific systems, and it was shown that the above properties may be indeed observable [53]. The PJTE induced bistabilities have not only much better parameters and lifetimes (by orders of magnitude) than the TMSCC magnetic bistability, but are accompanied by symmetry breaking with changes from non-polar to dipolar or quadrupolar structures (dielectric bistability) (Figure 1). In other words, switching from one magnetic state to the other one in these systems simultaneously changes their dielectric properties. In addition, some of these systems in the solid-state display also strong magnetoelastic and ferroelastic properties, and all these effects can be tuned by means of external perturbations. The vast variety of systems with such properties which have in common a special feature of electronic structure, half-filled degenerate valence shells, form a class of materials which can be termed *half-filled–shell (HFS) materials* (HFSM). As the number of molecular and solid state systems with e^2 and t^3 electronic configurations is practically infinite, we thereby suggests an inexhaustible source of materials with bistability properties. Such systems may occur in all aggregate states of matter formed by organic and inorganic, main group and transition metal compounds in all varieties of crystal structures and impurity centers in crystals, as well as in clusters, quantum dots, and thin films.

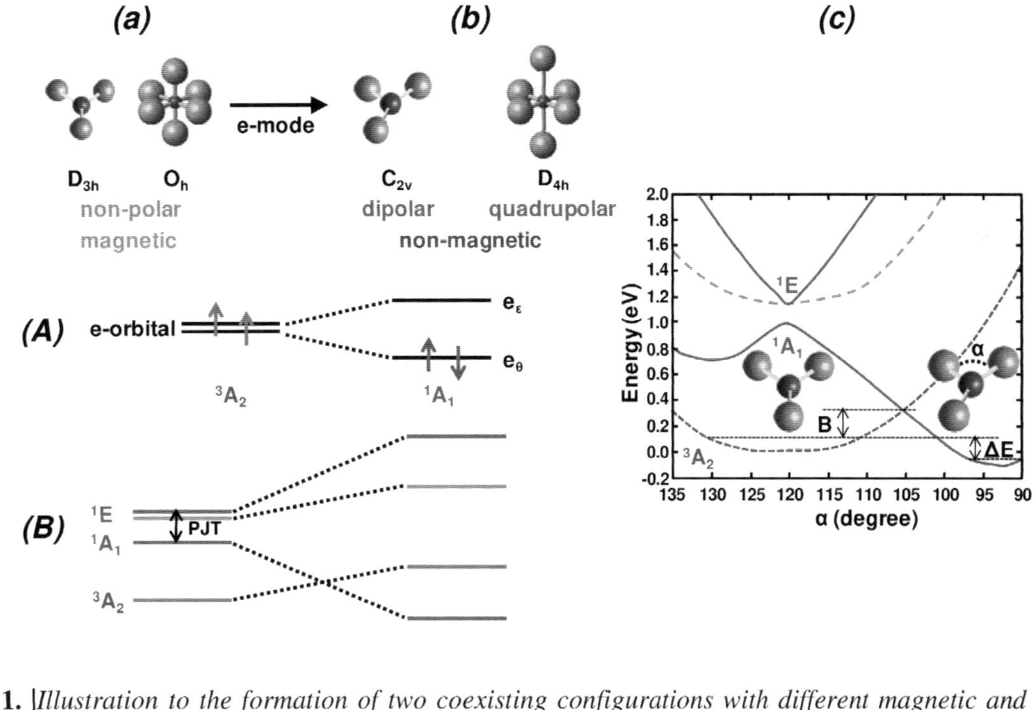

Figure 1. |*Illustration to the formation of two coexisting configurations with different magnetic and structural properties*: (*a*) In a typical energy level scheme of the electronic e^2 configuration (*A*) the ground state $^3A_2$ is formed by the two electrons with parallel spins, each of them occupying one *e* orbital, resulting in the $e_\theta\uparrow e_\varepsilon\uparrow$configuration which is totally symmetric and does not distort the system; the pictures show examples of planar triangular (e.g.,CuF_3) and octahedral systems (e.g., the CuO_6 units in the $LiCuO_2$ and $NaCuO_2$ crystals considered below) with the e^2 configuration which have the high-symmetry geometry in this state; the energy terms of this configuration are shown in (*B*); (b) In the first excited singlet state $^1A_1$ the electron configuration $(1/\sqrt{2})(e_\theta\uparrow e_\theta\downarrow + e_\varepsilon\uparrow e_\varepsilon\downarrow)$ is totally

symmetric too, but due to the PJT interaction with the higher $^1E_1$ state it undergoes orbital disproportionation resulting in either $e_\theta\uparrow e_\theta\downarrow$ or $e_\varepsilon\uparrow e_\varepsilon\downarrow$, both having a lower symmetry of charge distribution that distorts the systems in the *e* mode direction, as shown in the pictures and energy levels (*B*); (*c*) Ab initio calculated energy levels of CuF_3 as a function of the angle α (*e* mode distortion) fully confirm the theoretical predictions in (*a*) and (*b*). At $\alpha=120^o$ the ground state is a spin-triplet $^3A_2$ with a magnetic moment of 1.62 μ_B and planar, regular-triangular geometry, but due to the strong pseudo Jahn-Teller effect between the lowest two excited spin-singlet states $^1A_1$ and $^1E_1$ the molecule acquires a global minimum at $\alpha\approx 93^o$ in which it is distorted with a dipole moment of 1.67 D, but no magnetic moment (the values of B and ΔE are given in table 2).

The common properties of this class of materials and their origin can be illustrated using the mentioned above CuF_3 molecule as an example. In the high-symmetry configuration this molecule is planar regular triangular with the D_{3h} symmetry and the electron configuration e^2 (so it is a representative of HFSM). Figure 1*c* shows the results of ab initio calculations of its ground and excited electronic states as a function of the angle α(FCuF). We see that when the molecule is an equilateral triangle ($\alpha=120^o$) the ground state is $^3A_2$ with the lowest excited states $^1A_1$ and ^1E, which is *typical for all electronic e^2 configurations* (note, however, that the order of occurrence of the two singlet states may be inverted, in octahedral systems the ^1E level is lower than $^1A_1$, but this does not alter the general conclusions). The spin-triplet ground state $^3A_2$ in the D_{3h} geometry is stable, but there is a strong PJT coupling between the two excited states $^1A_1$ and ^1E by *e* type distortions: the lower state $^1A_1$ goes down, crosses the $^3A_2$ state, and produces a global minimum in which the initial geometry D_{3h} is reduced to C_{2v} and the electronic state is a spin-singlet $^1A_1$. The origin of this PJTE is quite understandable [33, 53]: when the system is distorted the non-magnetic structure with the two electrons paired in one orbital is energetically more convenient than the totally symmetric magnetic distribution with two electrons in two orbitals with parallel spins because of additional covalency gained by distortion (figure 1b). As a result the molecule CuF_3 has two stable states: the triplet (high-spin, HS) state has a magnetic moment of μ_1=1.82 μ_B and a zero dipole moment p_1=0, whereas the low-spin (LS) state has a zero magnetic moment μ_2=0 and a strong dipole moment of p_2=1.67 D. The HS state is higher in energy by ΔE~0.2 eV and it has an energy barrier of B~0.33 eV to the point of spin crossover with the LS state (energy values are read off the zero vibrations).

We have thus two coexisting bistabilities, magnetic and dielectric (dipolar), which are strongly correlated: you may have either a HS state with a zero dipole moment, or a zero spin state with nonzero dipole moment. This means that by magnetic switching from LS μ_2=0 to HS μ_1=1.82 MB the system simultaneously undergoes a dipolar switching from p_2=1.67 D to p_1=0, and *vice versa*. Such correlations between magnetic and polar properties are typical for multiferroics. Hence the CuF_3 molecule may be termed as a *single-molecule multiferroic.* An ensemble of such molecules, e.g., organized on a thin film is a representative of HFSM. This picture of bistabilities demonstrated for the example of the CuF_3 molecule is qualitatively the same for all the systems with e^2 configurations, but with varying parameter values. Table 2 shows the parameters of instability calculated ab initio for several molecular systems with e^2 and t^3 electronic configurations.

S*olid state HFSM* may be even more important in applications because of their larger variety and better parameters. They may include virtually all the cubic symmetry coordination compounds, ionic crystals, and impurity centers in crystals containing transition metal ions with the electronic configuration e^2 (d^8 ions in octahedral coordination and d^2 ions in tetrahedral or cubic environment), and t^3 (d^3 in octahedral and d^7 in tetrahedral coordination). Remarkably, these transition metal systems *are not TMSCC*. Experimentally [54, 55], there is evidence of magnetic bistability in $Ni^{2+}(d^8)$ complexes where a reversible equilibrium between octahedral magnetic and square-planar diamagnetic complexes has been observed. Electronic structure calculations for two solids from the HFSM class, the crystals $LiCuO_2$ and $NaCuO_2$ were performed in Ref. [53]. The structure of these crystals is formed by alternating layers of vertex-sharing $Cu^{3+}(O^{2-})_6$ octahedrons and Li^+ or Na^+ ions, respectively (see figure 2). The Cu^{3+} ion has a $(d^8)e^2$ electronic configuration (in general *e* denotes molecular orbital or symmetrized band states).

Table 2. HS-LS distortions, totally symmetric Q_1 and *e* type Q_2 and Q_3, energy difference ΔE, barrier β (read off the zero-point energy), and lifetimes τ at low temperatures (T=5K) calculated by two methods for several HFSM (the CuO_6 system is taken as a cluster in the $LiCuO_2$ crystal).

System	Q_1 (Å)	Q_2 (Å)	Q_3 (Å)	ΔE (eV)	B (eV)	β (cm^{-1})	τ (s) method 1	τ (s) method 2
Si_3	0.34	0.06	0.00	0.17	0.12	340	$12\cdot10^{-8}$	$14\cdot10^{-8}$
CuF_3	0.61	0.20	0.00	0.21	0.33	1040	$29\cdot10^{-6}$	$91\cdot10^{-6}$
Si_4	0.82	0.22	0.79	1.08	0.72	166	$30\cdot10^{-1}$	$84\cdot10^{-3}$
Ge_4	0.79	0.05	0.81	0.82	0.81	1940	$26\cdot10^{3}$	$35\cdot10^{-1}$
C_4H_4	0.56	0.06	0.0	0.78	0.13	1.2	$37\cdot10^{-7}$	$44\cdot10^{-7}$
CuO_6	1.24	0.46	0.00	0.23	0.17	890	$67\cdot10^{13}$	$14\cdot10^{16}$

As predicted by the general theory and shown above for molecular cases, the HS configuration of these crystals is undistorted, while the LS state is diamagnetic with a very strong elongation of the CuO_6 octahedrons (figure 2). In Ref [53], Density Functional Theory (DFT) calculations with the hybrid B1WC functional [56] was used to optimize the $LiCuO_2$ and $NaCuO_2$ crystal structures in both the HS (ferromagnetic and antiferromagnetic) and LS (diamagnetic) states. A summary of the numerical results is presented in table 3. As expected, the ferromagnetic and antiferromagnetic phases have very similar geometries with regular octahedral coordination of the oxygen atoms around Cu, whereas the diamagnetic phase is strongly distorted to almost square-planar coordination (in $LiCuO_2$ the metal-ligand distance in the magnetic phase is 2.01Å, while in the diamagnetic phase the equatorial and axial ligands are at 1.86Å and 2.76Å, respectively). In $NaCuO_2$ the same pattern is found, although the lattice parameter and distortions are by about 10% larger than in $LiCuO_2$. The results obtained for the geometry of the diamagnetic phase of these crystals agree well (within 2%) with the experimental data [57,58]; the interatomic distance changes are here 3-4 times larger than in TMSCC.

Table 3. Experimental (in parenthesis) and calculated by DFT B1WC lattice parameters A, B, C (in Å), lattice angle γ (degrees), interatomic equatorial dCu-O_{eq} and axial dCu-O_{ax} distances in the diamagnetic (DM), ferromagnetic (FM), and antiferromagnetic (AF) phases of $LiCuO_2$ and $NaCuO_2$ and their relative energies ΔE (in cm^{-1}).

System	Phase	A	B	C	γ	$dCuO_{eq}$	$dCuO_{ax}$	ΔE
$LiCuO_2$	DM	5.61	2.742	5.539	119.9	1.858	2.768	20
		(5.733))	(2.718))	(5.622)	(120.7)	(1.836)	(2.718)	
$LiCuO_2$	FM	5.032	2.910	5.032	109.4	2.009	2.012	230
$LiCuO_2$	AF	5.021	2.905	5.021	108.6	2.008	2.012	0
$NaCuO_2$	DM	6.265	2.785	6.032	120.3	1.862	3.013	0
		(6.358)	(2.750)	(6.059)	(120.5)	(1.846)	(3.029)	
$NaCuO_2$	FM	5.203	3.004	5.493	108.3	2.034	2.030	700
$NaCuO_2$	AF	5.207	3.006	5.493	108.3	2.021	2.032	1420

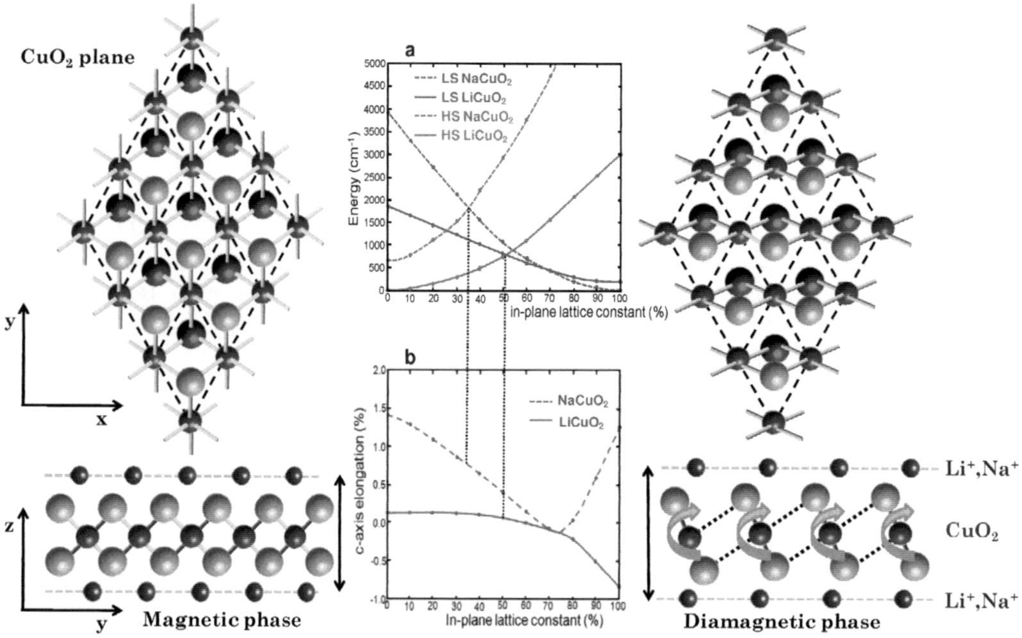

Figure 2. *Giant magnetoelastic effect in $LiCuO_2$ and $NaCuO_2$ crystals.* The vertex-sharing CuO_6 octahedrons forming the CuO_2 layers in the two coexisting magnetic-undistorted (left) and non-magnetic-distorted (right) phases are shown in the upper pictures with Cu in red and O in light-green (front) and dark-green (rear); in the lower part the tilting of the CuO_2 units to accommodate the octahedral elongations induced by the PJTE is illustrated (Li and Na ions are shown in blue). In the graphic insertions (*a*) the total energy of the magnetic (red) and non-magnetic (blue) phases of the two crystals is plotted as a function of lattice strain (coordinate of distortion) termed as 0 in the undistorted magnetic phase and 100% in the distorted non-magnetic phase, and (b) the c-axis lattice elongation in the diamagnetic phase in % to that of the magnetic phase as a function of the same strain is shown. We see that at ~35% of strain in $NaCuO_2$ and ~50% in $LiCuO_2$ the two phases become approximately degenerate with a lattice relaxation of ~0.8% in the former and 0.2% in the latter systems in a magnetic phase transition.

The magnetic to diamagnetic phase transition has a strong repercussion on the crystal structure along the c-axis enlarging the unit cell by ~10% (≈0.5Å) but the geometrical difference between the two bistability states does not translate in a large energy difference between them. Indeed, as shown in table 3 and figure 2 the energy difference between the phases is only of a few hundred cm^{-1}. In the case of $LiCuO_2$ the calculations yield that the magnetic phase (at the minimum of the APES) is lower by 150 cm^{-1} than the diamagnetic one, but with the vibrational zero-point-energy correction the two phases are almost degenerate. In $NaCuO_2$ the non-magnetic phase is more stable. Due to these small energy differences between the phases it should be relatively easy to trigger a phase transition between them by external perturbation. Indeed, for $LiCuO_2$ a phase transition [59] is observed at T=180K. For $NaCuO_2$ the ab initio calculations [53] show that a diamagnetic to magnetic phase transition is achievable at a pressure of ~6 GPa.

In view of the close energy values of the two phases, the possibility was explored [53] to further tune the system by means of epitaxial strain, so that they become almost degenerate and hence more vulnerable to external perturbations. The in-plane geometry of the unit cell of these crystals is almost hexagonal (figure 2), so their growth on a particular hexagonal substrate could constrain the in-plane lattice constant to a particular value. Geometry optimizations with the in-plane lattice parameters fixed to a value between that of the magnetic phase and the diamagnetic phase while the rest of the parameters were let free to vary for $LiCuO_2$ and $NaCuO_2$ (figure 2*b*) show that at some intermediate strains (≈50% for $LiCuO_2$, ≈35% for $NaCuO_2$) the magnetic and non-magnetic phases become degenerate. At this point a diamagnetic to magnetic phase transition can easily be triggered by a magnetic field, resulting in a strong distortion of the crystal along the c-axis; as seen from figure 2*b*, the lattice parameter of $LiCuO_2$ (≈5Å) changes by ~0.2% which amounts to ~ 1pm. Such changes in lattice parameters are typical in ferroelectrics, but very seldom found in magnetic transitions. In fact such distortions are termed in literature as *giant magnetoelastic effects* (see, e.g., [60]). Moreover, a similar change in the lattice parameter of $NaCuO_2$ is ~0.8% (figure 2b), *almost 4 times larger than those of* $LiCuO_2$. In the strained systems this change in lattice parameter is realized by a tilting of the CuO_2 (almost) square planar units in the diamagnetic phase toward the formation of CuO_6 octahedrons by the near neighbour rows in the magnetic phase (figure 2).

Note that, as mentioned above, in the whole class of HFSM the distortion of the high-symmetry (magnetic) configuration toward the diamagnetic one by the PJTE in two excited states is accompanied by *orbital disproportionation* [33] due to which the two electrons in two e orbitals with parallel spins (and symmetric charge distribution) prefer to occupy one orbital with anti-parallel spins lowering the symmetry of the system, but gaining additional covalent metal-ligand bonding. This explains also the insulating properties of the diamagnetic $NaCuO_2$ or $LiCuO_2$ crystals that are subject to discussion in the literature [61]: the orbital disproportionation splits the two-fold degenerate ($3z^2$-r^2 and x^2-y^2 of Cu) E level (band) and fully occupies the lower one (figure 1).

One of the most important characteristics of the bistability with regard to possible applications is the lifetime τ (relaxation time) of the induced higher energy state (in cases in table 1 the spin-triplet state) with respect to its thermal relaxation to the lower (spin singlet) state. This is less a problem for solids where the cooperative interactions stabilize well the two states in the bistability. For molecular systems or small aggregates the relaxation lifetime depends on several factors, mainly on the spin-orbital interaction, the barrier height and width between the two minima of the APES, and the dissipation of energy in interaction with the environment. Direct calculation of the lifetimes τ is not trivial; in Ref 53 [62, 63] we employed two approximations to estimate the τ values for a series molecular systems and for (isolated) octahedral units in $LiCuO_2$ (taken as examples of the HFSM class). The results are shown in table 2, while figure 3 illustrates the temperature dependence of some of them. As expected, the distortions in all HFSM are dominated by the symmetry-breaking coordinate (Q_1 in table 2), while in normal TMSCC they are dominated by the totally symmetric coordinate Q_2, and in general they are larger or much larger in the former case than in the latter. This feature increases essentially the lifetime of HFSM as compared with TMSCC because of the strong dependence of the overlap of the vibrational functions of the two states, which decreases exponentially with the distortion. A decrease in τ is also achieved in organic and other compounds of non-heavy elements due to their smaller spin-orbital coupling constant.

In table 2 Si_3 has the shortest lifetime (comparable to that of most TMSCC) which increases in systems with larger distortions, and reaches macroscopic values in Si_4, Ge_4 and $LiCuO_2$ at low temperatures (but in Ge_4 there are low-lying triplet excited states that may reduce the lifetime). In the crystals $LiCuO_2$ and $NaCuO_2$ the estimates of the lifetime of the HS configuration are carried out for a separate cluster unit CuO_6; it is extremely large at low temperatures (due to the large distortions in the diamagnetic phase) and remains macroscopic at room temperature. Since the cooperative interactions between the complexes in the condensed phases obviously increase the lifetimes, the values in table 2 and in figure 3 show the lower limit of the expected lifetimes. This means that the crystals $LiCuO_2$ and $NaCuO_2$ are quite feasible for practical use as bistability materials, and their large lifetime of both states in this bistability indicates that they can be used in the form of clusters, quantum dots, or thin films.

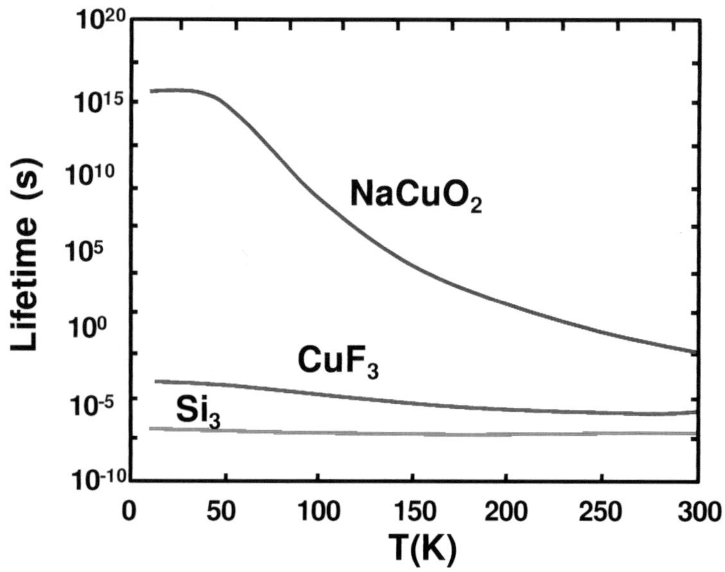

Figure 3. *Temperature dependence of lifetimes* estimated for the higher in energy high-spin states of Si_3, CuF_3 and an isolated octahedral center in $NaCuO_2$ by means of model calculations.

With practically inexhaustible numbers and with the shown above examples of their feasibility for practical use polyatomic systems with electronic configurations e^2 and t^3 may be regarded as *a "database" of novel materials with bistabilities for information storage and sensor devices.*

4. The PJT effect is instrumental in design of compounds with required properties: "planarization" of puckered 2D (graphene-like) or quasi-2D compounds

Two-dimensional (2D) or quasi-2D materials have unique physical properties due to a variety of their planar-surface-specific physical and chemical properties. The best-known example of this kind is graphene [64, 65]. Similar properties are sought for in silicene [66], boron nitride [67], zinc oxide [68], phosphorene [69], gallium nitride [70], as well as in other attractive subjects like 2D transition metal dichalcogenids and oxides [71], and there is an unlimited number of other organic and inorganic 2D and quasi-2D compounds with expected similar properties.

One of the main structural features that makes these compounds special in applications is their planarity or quasi-planarity. On the other hand, it was shown that all deviations from the planar configurations of polyatomic systems in nondegenerate states in the form of puckering (or buckling) are due to, and only to the pseudo Jahn-Teller effect (PJTE) [1-3]. In degenerate states, both the JTE and PJTE may be active, which for extended systems results in Peierls distortions (see in [1, 2]). In application to a series of cyclic and heterocyclic compounds the PJTE explains the origin of their out-of-plan distortions, puckering (see [3] and references therein; puckering as a PJTE was first formulated in [72]). Explaining the origin and mechanism of such distortions leads us directly to another important problem related to the possible use of these compounds in applications: if the system under consideration is out-of-plan distorted, how can we restore the planar configuration with minor other changes?

Starting with the high-symmetry (planar) configuration of 2D systems, the PJTE reveals the active excited states that produces the instability of the ground state, its symmetry and energy, and the numerical condition of instability [1-3]. Therefore we can try to influence the system by means of external perturbations that violate the PJTE condition of instability of the ground state mainly by changing the active excited states. In principle, there are two possibilities to do this: (1) increasing just the energy gap between the electronic states mixed by the PJT vibronic coupling, e.g. by weak coordination with outside systems, and (2) changing the symmetry of the active excited states in a

IOP Conf. Series: Journal of Physics: Conf. Series **833** (2017) 012001 doi:10.1088/1742-6596/833/1/012001

redox process by removing or adding electrons (which may change all the parameters of the PJTE including the energy gaps). The possibility (1) was first demonstrated in the mechanism of hemoglobin oxygenation [73] in which the PJTE out-of-plane displacement of the iron atom from the porphyrin ring returns into the plane by oxygenation due to the increase of the energy gap to the active excited state, thus suppressing the PJTE. Examples of realization of the mechanism (2) were demonstrated recently [74, 75]. For instance, by coordination of two chloride anions to nonplanar (distorted by the PJTE) chair structures of Si_6X_{12} (X = Cl, Br) to form dianions of the type $[Si_6Cl_{14}]^{2-}$ in a "triple-decker-sandwich"-like structure, they were proved to have a planar Si_6 ring, in which the PJTE is suppressed because the orbitals of the active excited states of the ring become occupied by the excessive Cl^- electrons [74] (more examples are shown below). It is assumed that such external perturbations preserve the stability and main ring properties of the system, but the quenching of the PJTE can be achieved also by other external perturbations, e.g., by spectroscopic excitations, and substitutions (see below).

The PJTE has been involved in exploration of the origin of out-of-plane (puckering) distortions and possible ways of restoration of the planar configuration of quite a number of small and large virtual 2D systems [73-82]. For illustration and as an example consider the mechanism of distortion and restoration of the planar configuration of a series of tetra-heterocyclic 1,2-diazetes, $C_2N_2E_4$, E= H, F, Cl, Br [80]. These compounds are well known and studied before in their ground state [83]. According to the general PJTE theory, the study of possible equilibrium geometries of any molecular system in nondegenerate electronic states starts with a reference configuration of the highest possible symmetry for which all the first derivatives of the APES with respect to the nuclear displacements are zero (see, e.g. [1, 3]). In case of the $C_2N_2E_4$ molecules the planar nuclear configuration of the C_{2v} symmetry is the starting one. The eight-atom systems $C_2N_2E_4$ have eighteen vibrational degrees of freedom. In the planar configuration of C_{2v} symmetry they transform according to the irreducible representations $7A_1 + 3A_2 + 2B_1 + 6B_2$. Five symmetrized displacements, three of a_2 type and two of b_1 type, correspond to the out-of-plane distortions that transforms the planar C_{2v} nuclear configuration to the puckered structures of C_2 or C_s symmetry, respectively (figure 4).

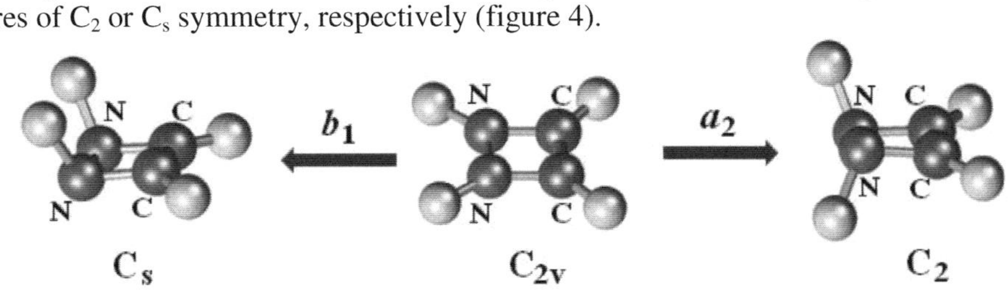

Figure 4. Illustration to the $C_{2v} \rightarrow C_2$ and $C_{2v} \rightarrow C_s$ puckering transformations of the $C_2N_2E_4$ systems resulting from the $(^1A_1 + {}^1A_2)\otimes a_2$ and $(^1A_1 + {}^1B_1)\otimes b_1$ PJTE problems, respectively.

To reveal the vibronic origin of the out-of-plane distortions of these molecules the three-mode $(^1A_1 + {}^1A_2)\otimes(a_2 + a_2 + a_2)$ and the two-mode $(^1A_1 + {}^1B_1)\otimes(b_1 + b_1)$ PJTE problems should be solved. Direct solution of these problems requires the construction of the 3x3 and the 2x2 secular matrices and calculation of a large number of vibronic coupling constants. Fortunately, the problem can be simplified by employing the results of electronic structure calculations and vibrational frequency analysis of the systems in the high-symmetry C_{2v} configuration. Indeed, in all these cases calculations show the presence of two imaginary frequencies: the one with a larger absolute value corresponds to the a_2 type nuclear displacements, and the second one describes displacements of the b_1 type.

In other words, the high-symmetry C_{2v} nuclear configurations of all the considered molecules are unstable with respect to both types of the out-of-plane distortions. The calculated normal coordinates of the a_2 and b_1 types, which are linear combinations of corresponding symmetrized displacements, comply with the lines of the steepest descent in energy toward the two minima on the APES of the molecules. The first one, being the absolute minimum, corresponds to the C_2 structure (figure 4), whereas the second, local minimum of C_s symmetry is above the first one by an amount of 5-8 kcal/mol. For this reason, we further analyse only the cross section of the APES of the systems along the Q_{a_2} coordinate leading to the equilibrium minimum energy of the C_2 configuration. The a_2 mode of distortions is illustrated in figure 5.

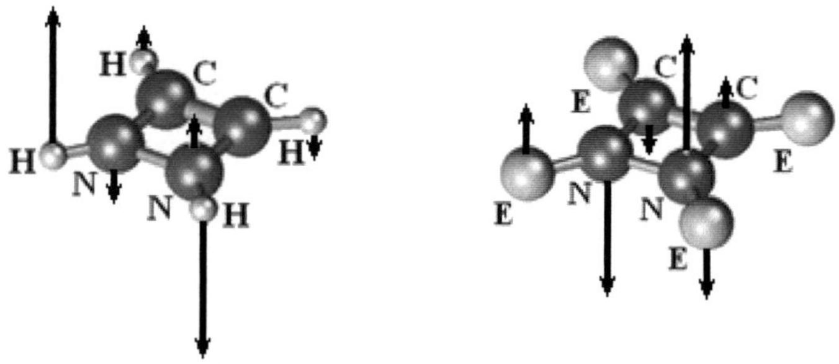

Figure. 5. Puckering a_2 normal modes in the $C_2N_2H_4$ (left) and $C_2N_2E_4$ (E=F, Cl, Br) systems; the arrow lengths show expected relative (not absolute) displacements.

The calculations [80] show that the geometry of the molecules under consideration varies significantly when bloating. First, it refers to the N-N-C=C dihedral angles that characterize the puckering of the planar heterocyclic rings. The dihedral angles E-N-N-E decrease in the series of halogen-substituted molecules from 45.1° in $C_2N_2F_4$ to 40.2° and 38.6° in $C_2N_2Cl_4$ and $C_2N_2Br_4$, respectively. The dihedral angles E-C=C-E, on the contrary, increase from F to Cl-, and Br-substituted compounds. It is also seen that there is a significant decrease in the values of the imaginary vibrational frequencies, which is due to an increase in the reduced mass of vibrations and the subsequent change of the vibrational a_2 mode.

Several lowest electronic terms with the same multiplicity of the $C_2N_2E_4$ molecules, and their electronic configurations in the planar (C_{2v}) geometry are presented in table 4. The numbers "1" and "-1" are used to denote electrons in two opposite spin orientations. It is seen that the ground state in all the considered molecules is the closed-shell spin-singlet, and the lowest excited $^1B_1$ and $^1A_2$ terms are produced by the one-electron excitations $b_1 \rightarrow a_1$ and $a_2 \rightarrow a_1$ from HOMO-1 b_1 and HOMO a_2 to the LUMO a_1, respectively. Calculations with a larger number of configurations show that the next excited $^1A_2$ states are much higher in energy (more than 10 eV), and their PJT mixing with the ground state is included in the p parameters introduced below.

Table 4. Main electronic configurations, wavefunctions weight coefficients, and relative energies ΔE (eV) of several electronic terms of the $C_2N_2E_4$ molecules in the optimized planar C_{2v} nuclear configuration calculated at the CAS(4,3) level of theory; 1 and -1 in the power of single-populated orbitals indicate two spin orientations.

State symmetry	**Main Electronic Configuration**	**Weight coefficients**			
		$C_2N_2H_4$	**$C_2N_2F_4$**	**$C_2N_2Cl_4$**	**$C_2N_2Br_4$**
$^1A_1$	$b_1^2 a_2^2$	0.9999	0.9995	0.9997	0.9997
ΔE		0.00	0.00	0.00	0.00
	$b_1^1 a_2^2 a_1^{-1}$	-0.7071	-0.7071	-0.7071	-0.7071
$^1B_1$	$b_1^{-1} a_2^2 a_1^1$	0.7071	0.7071	0.7071	0.7071
ΔE		6.55	4.72	4.48	3.51
	$b_1^2 a_2^{-1} a_1^1$	-0.7071	0.7071	-0.7071	0.7071
$^1A_2$	$b_1^2 a_2^1 a_1^{-1}$	0.7071	-0.7071	0.7071	-0.7071
ΔE		7.03	5.67	5.09	4.08

As mentioned above, the PJTE theory states that the only reason of spontaneous symmetry breaking followed by the low-symmetry distortion of any molecular system in the nondegenerate electronic state is the sufficiently strong PJT mixing between this state and one or more higher in energy electronic states [1, 3]. If $|\Gamma\rangle$ and $|\Gamma'\rangle$ are the wavefunctions of the nondegenerate ground and excited states and the energy gap between them is Δ, then the condition of instability of the ground state with respect to the nuclear displacements Q is [1] (cf. Eq. (4)):

$$\Delta < \frac{\left(F_{\Gamma\Gamma'}\right)^2}{2K_0} \tag{7}$$

where $F_{\Gamma\Gamma'}$ is the linear vibronic coupling constant between the ground and excited states, and K_0 is the primary force constant,

$$F_{\Gamma\Gamma'} = \left\langle \Gamma \middle| \left(\frac{\partial H}{\partial Q}\right)_0 \middle| \Gamma' \right\rangle \tag{8}$$

$$K_0 = \left\langle \Gamma \middle| \left(\frac{\partial^2 H}{\partial Q^2}\right)_0 \middle| \Gamma \right\rangle \tag{9}$$

and H is the adiabatic electronic Hamiltonian of the system which includes all Coulomb interactions between electrons and nuclei. The value of $F_{\Gamma\Gamma'}$ is nonzero only if two states$|\Gamma\rangle$ and $|\Gamma'\rangle$ obey the symmetry restriction: $\Gamma \otimes \Gamma'$ contains the symmetry of Q. At small Q the total force constant K which determines the curvature of the adiabatic potential energy surface (APES) in the direction of nuclear displacement Q, is equal to

$$K = K_0 - \frac{(F)^2}{2\Delta} \tag{10}$$

If the condition of instability (1) holds, then the value of K becomes negative.

In any polyatomic system, the number of excited states is practically infinite, and there are always excited states of relevant symmetry, albeit at large energy gaps Δ. Their contribution to the PJTE problem may be rather small, just lowering the curvature of the ground state from K_0 to $K_0 - p$, $p > 0$, and not producing instability [3, 84]. Denote by K_1 and K_2 the primary force constants in the ground and the active excited states, respectively, and by p_1 and p_2 the corrections to them produced by the PJT coupling to all the higher excited states of appropriate symmetry [3, 84]:

$$W_{\Gamma\Gamma} = -2\sum_n \frac{|F_{\Gamma n}|^2}{E_n - E_\Gamma} Q^2 = -p_1 Q^2 \tag{11}$$

$$W_{\Gamma'\Gamma'} = -2\sum_n \frac{|F_{\Gamma' n}|^2}{E_n - E_{\Gamma'}} Q^2 = -p_2 Q^2 \tag{12}$$

Then, the APES ε(Q) can be obtained from the secular equation of the perturbation theory:

$$\begin{vmatrix} \frac{1}{2}(K_1 - p_1)Q^2 - \varepsilon & FQ \\ FQ & \frac{1}{2}(K_2 - p_2)Q^2 + \Delta - \varepsilon \end{vmatrix} = 0 \tag{13}$$

The roots of this secular equation are:

$$\varepsilon_{gr,ex} = \frac{1}{4}(K_1 - p_1 + K_2 - p_2)Q^2 + \frac{\Delta}{2} \mp \frac{\Delta}{2}\sqrt{1 + \frac{1}{4\Delta^2}\left(K_1 - p_1 - (K_2 - p_2)\right)^2 Q^4 - \frac{1}{\Delta}\left(\left(K_1 - p_1 - (K_2 - p_2)\right) - \frac{4F^2}{\Delta}\right)Q^2} \tag{14}$$

At small values of Q we get:

$$\varepsilon_{gr} = \frac{1}{2}\left(K_1 - p_1 - \frac{2F^2}{\Delta}\right) Q^2 \tag{15}$$

IOP Conf. Series: Journal of Physics: Conf. Series **833** (2017) 012001 doi:10.1088/1742-6596/833/1/012001

Then the condition of instability at Q=0 of the ground state due to the PJT vibronic coupling is:

$$\mathrm{K}_1 - p_1 < \frac{2F^2}{\Delta} \tag{16}$$

The direct *ab initio* calculation of the constants F, K_1-p_1 and K_2-p_2 encounters difficulties. In the present paper these constants were estimated by fitting the general formula, obtained from the vibronic coupling theory (Eq. (14)), to the *ab initio* data for the adiabatic potential curves (energy profiles) - cross-sections of the APES along the Q_{a_2} coordinates. These calculated energy profiles of the $C_2N_2E_4$ compounds in the ground and lowest excited states are shown in figure 6. In the configuration of C_{2v} symmetry all the four systems are unstable becoming stabilized along the puckered geometry of C_2 symmetry, while the PJT active excited states A_2 become destabilized. By fitting Eq. (14) to these energy profiles we can extract the PJT constants. Estimated in such a way, the values of the PJT parameters are given in table 5.

Table 5. PJTE coupling constants F (in eV/Å), energy gaps Δ (in eV), primary force constants $K_1 - p_1$ and $K_2 - p_2$ and full force constants K_{gr}=K_1-p_1- ($2F^2$/ Δ) (in eV/Å^2) in the reference C_{2v} configuration of 1,2-diazetes $C_2N_2H_4$ obtained by fitting Eq. (14) to the ab initio calculated energy profiles.

Molecules	***K1-p1***	***K2-p2***	***F***	**Δ**	***Kgr***	***Kgr* (point-by-point)**
$C_2N_2H_4$	6.18	11.65	5.61	7.03	-2.77	-1.87
$_2N_2F_4$	5.11	5.25	6.77	5.67	-11.06	-10.27
$C_2N_2Cl_4$	2.41	3.29	4.05	5.09	-4.03	-4.10
$C_2N_2Br_4$	2.82	3.46	3.58	4.08	-3.46	-3.21

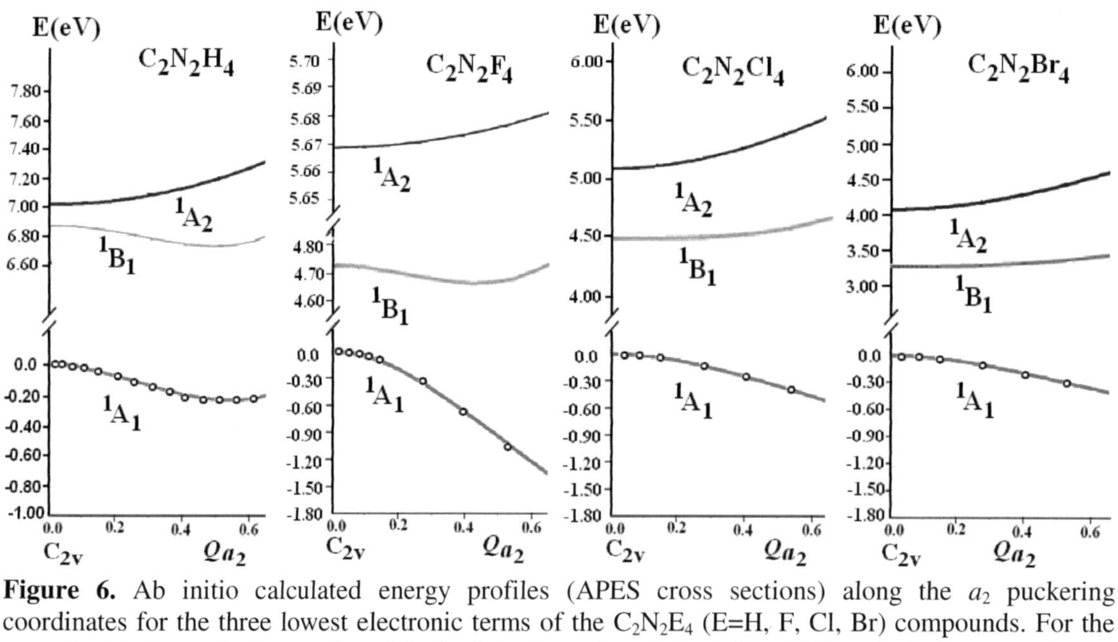

Figure 6. Ab initio calculated energy profiles (APES cross sections) along the a_2 puckering coordinates for the three lowest electronic terms of the $C_2N_2E_4$ (E=H, F, Cl, Br) compounds. For the ground state term A_1 the results of both ab initio and analytical (Eq. (14)) calculations are shown, respectively, by points and curves (the ground state minima for the stable puckered configurations are at larger distances not shown in the graphics).

The numerical data of tables 4 and 5 allow one to verify the condition of PJTE instability (10). For each of these systems we calculated the resulting value of the curvature K_{gr} according to Eq. (9). As we can see from table 5 (penultimate column), all these values are negative, hence the condition of puckering instability $K_1 - p_1 < \frac{2F^2}{\Delta}$ is fully satisfied. Furthermore, from figure 6 it is seen that in all the cases the ab initio calculated points of the energy profiles coincide with those estimated by the formulas of the general theory using the PJT parameters from table 6. Based on these results, and given the fact that K_1-p_1 >0, we may conclude that the instability of the high-symmetry C_{2v} nuclear configuration of the molecules under consideration with respect to the a_2 puckering modes and their distortions are due to the PJT mixing of the ground $^1A_1$ and the first excited $^1A_2$ states. It is important to note that these numerical estimates are not intended to prove that the puckering is of PJTE origin, this is done in the general theory [1, 2] where it is proved that any distortion of high-symmetry configurations is due to, and only to the PJTE; the proof here is to show that in the specific systems under consideration the revealed vibronic coupling between the specific electronic states describes the observed distortions with sufficient accuracy.

Table 6. Imaginary frequencies (in cm^{-1}) along a_2 and b_1 normal displacements in the cation, anion, dication, dianion, and excited triplet states of the $C_2N_2E_4$ series (dashes indicate no-imaginary frequencies).

Molecule	**Cations** $C_2N_2E_4^+$	**Anions** $C_2N_2E_4^-$	**Triplet state HOMO→ LUMO**	**Dications** $C_2N_2E_4^{2+}$	**Dianions** $C_2N_2E_4^{2-}$
Imag. Freq.	***a2/b1***	***a2/b1***	***a2/b1***	***a2/b1***	***a2/b1***
$C_2N_2H_4$	454.2/ --	999.8/939.3	1069.3/978.7	-- / --	942.1/971.6
$C_2N_2F_4$	600.1/489.4	1079.0/839.1	685.8/119.5	117.8/ --	Problem in calculation
$C_2N_2Cl_4$	327.9/213.8	419.3/351.2	873.4/689.7	-- / --	435.2/409.0
$C_2N_2Br_4$	293.8/189.7	665.6/53.75	-- / --	-- / --	45.0/121.3

In order to restore the high-symmetry planar configuration of the 1,2-diazetes under consideration we have to quench the PJTE that triggers their puckering. As mentioned in the Introduction, this can be done by either increasing the just energy gap Δ to the same active excited state (in our case $\Delta=\varepsilon(A_2) - \varepsilon(A_1)$ in the planar configuration) by weak coordination to external groups, or to change all the parameters of PJTE instability by adding or removing electrons. The point is to achieve planarization without destroying the basic cyclic structure and accompanying properties of the system (changing the ligands leads to *a priori* another chemical structure). In case of ionization, stable bulk materials with the tetra-heterocycles under consideration (or in similar problems with other PJT systems) can be obtained by compensating their charges with appropriate opposite-charged ions.

To try this possibility we performed ab initio calculations of the four tetra-heterocyclic 1,2-diazetes, $C_2N_2E_4$, E=H, F, Cl, Br, in their cation, anion, dication, dianion, and triplet (HOMO-LUMO) states of the planar configuration and revealed their a_2- and b_1-type instabilities, as well as the energy gaps Δ to their new lowest A_2 electronic states that could induce the PJTE puckering instability. The obtained imaginary frequencies are shown in table 6. We see that in the anion states the instability, in general, increases, more in the dianion than in the anion states, while in the cation states it decreases, and three of the four systems become stable in the dication state; one of them is stable also in the excited triplet state. The first (more obvious) reason for the weakening of the PJTE in these cases is the much higher position of the (changed) lowest excited $^1A_2$ state. The calculated new energy gaps $\Delta=\varepsilon(^1A_2) - \varepsilon(^1A_1)$ in the dications $C_2N_2E_4^{2+}$ are (in eV): 10.81, 9.65, 7.89, and 5.72 for E=H, F, Cl, and Br, respectively, compared to 7.03, 5.67, 5.09, 4.08 for the neutral molecules (table 5). The other parameters in the condition of instability (10) include the primary force constant K_1 which increases by removing electrons (excess inter-electron repulsions *ceteris paribus* soften the system), thus

IOP Conf. Series: Journal of Physics: Conf. Series **833** (2017) 012001 doi:10.1088/1742-6596/833/1/012001

weakening the PJTE too, while the vibronic coupling constant F formed by similar electronic states is expected to remain of the same order of magnitude.

On the basis of these calculations we can conclude that the planar configurations of the $C_2N_2E_4$ heterocycle compounds under consideration, distorted by the PJTE, can be restored in the dication state produced, for instance, by a "triple-decker sandwich" structure with two additional electron acceptors on both sides of the neutral cycle, or two negative ions in both sides of the dication, similar to some previous suggestions for other systems [75].

5. Conclusions

We accentuated several problem that demonstrate the specific features of polyatomic system with the JTE and PJTE that make them exceptional in view of possible applications in search for novel materials with sought for properties. One of them is related to computer simulation of expected properties of possible useful materials that in its most efficient approaches employs DFT methods. The latter, unfortunately, are not applicable to systems with dynamic JTE or PJTE. Computer simulation for such systems remains and open, unsolved problem.

Important effects emerge in the interaction of dynamic JTE and PJTE systems with external perturbations. It has been found recently that external perturbation *quench* the local JT tunnelling or over-the-barrier transitions between the minima of the APES, producing an additional (orientational) contribution to the observable properties which enhances them by several orders of magnitude. This property is demonstrated on permittivity, flexoelectricity, and electrostriction of paraelectric $BaTiO_3$ and similar systems with dynamic (disordered) JTE and PJTE.

In systems with electronic e^2 or t^3 orbital configurations a special arrangement of ground and lowest excited states produces two coexisting equilibrium configurations with different symmetry, dipole moment, and spin. This brings us to a whole class of materials with magnetic-dielectric bistabilities. Examples of the CuF_3 molecule and two crystals, $LiCuO_2$ and $NaCuO_2$, demonstrate this materials' property.

As examples of application of the JTE and PJTE theory to rationalize the origin of special structural features of matter and to design novel system with required properties, we mention a series of publications on two-dimensional (2D) and quasi 2D systems that use the PJTE in order to reveal the origin of their puckering out-of-plan distortions, and methods of restoration of the planar configuration by external perturbations. The latter were shown to include coordination to other species, oxidation, reduction, and substitution, as well as optical excitations. The detailed procedure on how this works is shown on the example of tetra-heterocycles of 1,2-diazetes, $C_2N_2E_4$, E=H, F, Cl, Br.

References

[1] Bersuker I B 2006 *The Jahn-Teller Effect* (Cambridge UK: Cambridge University Press).
[2] Bersuker I B and Polinger V Z 1989 *Vibronic interaction in molecules and crystals* (Heidelberg: Springer)
[3] Bersuker I B 2013 *Chem. Rev.* **113** 1351; Bersuker I B 2016 *Advances in chemical physics* vol 160, ed S A Rice and A R Dinner (New York: Wiley) pp 158-209
[4] Hohenberg P and Kohn W 1964 *Phys. Rev.* **136** B864
[5] Kohn W and Sham L J 1965 *Phys. Rev.* A **140** 1133
[6] Levy M 1979 *Proc. Nat. Acad. Sci. USA* **76** 6062; 1982 *Phys. Rev.* A **26** 1200; Lieb E 1983 *Int. J. Quantum Chem.* **4** 243
[7] Kryachko E S and Ludena E V 1990 *Energy Density Functional Theory of Many-Electron Systems* (Dordrecht: Kluwer)
[8] Bersuker I B 1997 *J. Comput. Chem.* **18** 260 (1997)
[9] Yu H S Li S L and Truhlar D G 2016 *J. Chem. Phys.* **145** 130901
[10] Garcia-Fernandez P, Wojdeł J C, Iniguez J and Junquera J 2016 *Phys. Rev.* B **93** 195137
[11] Erhard J, Bleiziffer P and Görling A 2016 *Phys. Rev. Lett.* **117** 143002
[12] van Leeuwen R and Gross E K U 2006 *Multicomponent Density-Functional Theory*, in *Time-*

Dependent Density Functional Theory, Lecture Notes in Physics vol 706 (Berlin, Springer) pp 93–106
[13] Resta R and Vanderbilt D 2007 Theory of Polarization: A Modern Approach, in *Physics of Ferroelectrics: a Modern Perspective*, ed. C. H. Ahn, K. M. Rabe, and J. M. Triscone (Heidelberg: Springer-Verlag); Landau L D and Lifshitz E M 1984 *Electrodynamics of Continuous Media* (Oxford: Pergamon Press); Kittel C 1996 *Introduction to Solid State Physics*, 7th edition (New York: Wiley).
[14] Bersuker I B, Gorinchoy N N and Polinger V Z 1984 *Theor. Chim. Acta*, **66** 161
[15] Bersuker I B 2015 *Appl. Phys. Lett.* **106** 022903
[16] Bersuker I B 2015 *Appl. Phys. Lett.* **107** 202904
[17] Bersuker I B 1966 *Phys. Lett.* **20** 589; Bersuker I B and Vekhter B G 1978 *Ferroelectrics* **19** 137; Bersuker I B 2013 *J. Phys. Conf. Series*, **428** 012028
[18] Polinger V Z, Garcia-Fernandez P and Bersuker I B 2015 *Physica B., Cond. Mater* **457** 296
[19] Polinger V Z 2013 *J. Phys. Conf. Series* **428** 012026
[20] Comes R, Lambert M and Guinner A 1968 *Solid State Commun.* **6** 715
[21] Hanske-Petitpierre O, Yacoby Y, Mustre de Leon J, Stern E A and Rehr J J 1992 *Phys. Rev.* B **44** 6700; Ravel B, Stern E A, Vedrinskii R I and Kraizman V 1998 *Ferroelectrics* **206-207** 407; Stern E 2004 *Phys. Rev. Lett.* **93** 037601
[22] Müller K A and Berlinger W 1986 *Phys. Rev.* B **34** 6130 (1986); Müller K A, Berlinger W, Blazey K W and Albers J 1987 *Solid State Commun.* **61** 21; Volkel G and Müller K A 2007 *Phys. Rev.* B **76** 094105
[23] Zalar B, Laguta V V and Blinc R 2003 *Phys. Rev. Lett.* **90** 037601; Zalar B, Labar A, Seliger J and Blinc R 2005 *Phys. Rev.* B **71**, 064107; Blinc R, Zalar B, Laguta V V and Itoh M 2005 *Phys. Rev. Lett.* **94** 147601
[24] Jeong I-K, Lee S, Jeong S-Y, Won C J, Hur N and Llobet A 2011 *Phys. Rev.* B **84** 064125
[25] Burns G and Dacol F H 1981 *Ferroelectrics* **37** 661; 1982 *Solid State Commun.* **42** 9
[26] Bersuker I B 2012 *Phys. Rev. Lett.* **108** 137202
[27] Domracheva H E, Pyataev A V, Vorobeva V E and Zueva E M 2013 *J. Phys. Chem.* B **117** 7833
[28] Raymond O, Ostos C, Font R, Curiel M, Bueno-Baques D, Machorro R, Mestres L, Portelles J and Siqueiros J M 2014 *Acta Mater.* **66** 184
[29] Wieczorek K, Ziebinska A, Ujma Z, Szot K, Gorny M, Franke I, Koperski J, Soszynski A and Roleder K 2006 *Ferroelectrics* **336** 61
[30] Wada S, Hoshina T, Takizawa K, Ohishi M, Yasuno H, Kakemoto H, Tsurumi T, Moriyoshi C and Kuroiwa Y 2007 *J. Korean Phys. Soc.* **51** 878
[31] Kleemann W, Schafer F J, and Fontana M D 1984 *Phys. Rev.* B **30** 1148
[32] Polinger V and Bersuker I B 2017 *J. Phys. Conf. Series*, this issue
[33] Garcia-Fernandez P, Bersuker I B and Boggs J E 2006 *J. Chem. Phys.* **125** 104102
[34] Zubko P, Catalan G and Tagantsev A K 2013 Flexoelectric effect in solids (*Annual Review of Materials Research* vol 43) ed D. R. Clarke (Palo Alto: Annual Reviews) pp. 387–421
[35] Maranganti R, Sharma N D and Sharma P 2006 *Phys. Rev.* B **74** 014110; Maranganti R and Sharma P 2009 *Phys. Rev.* B **80** 054109
[36] Yudin P V, Ahluwalia R and Tagantsev A K 2014 *Appl. Phys. Lett.* **104** 082913
[37] Ma W and Cross L E 2001 *Appl. Phys. Lett.* **79** 4420; 2002 **81** 3440; 2006 **88** 232902
[38] Nguyen T H, Mao S, Yeh Y-W, Purohit P K and McAlpine M C 2013 *Adv. Mater.* **25** 946
[39] Biancoli A, Fancher C M, Jones J L and Damjanovic D *Nature Materials* 2014 **14** 224
[40] Narvaez J and Catalan G, 2014 *Appl. Phys. Lett.* **104** 162903
[41] Li Y, Shu L, Zhou Y, Guo J, Xiang F, He L and Wang H 2013 *Appl. Phys. Lett.* **103** 142909
[42] Sakai H, Fujioka J, Fukuda T, Okuyama D, Hashizume D, Kagawa F, Nakao H, Murakami Y, Arima T, Baron A Q R, Taguchi Y and Tokura Y 2011 *Phys. Rev. Lett.* **107**

137601
[43] Bhattarcharjee S, Bousquet E and Ghosez Ph 2009 *Phys. Rev. Lett.* **102** 117602
[44] Hong J, Catalan G, Scott J F and Artacho E 2010 *J. Phys. Condens. Matter* **22** 112201
[45] Burcsu E, Ravichandran G and Bhattacharya K 2000 *Appl. Phys. Lett.* **77** 1698
[46] Surowiak Z, Fesenko E G and Skulski R 1999 *Arch. Acoust.* **24** 391
[47] Kahn O and Martinez C J 1998 *Science* **279** 44
[48] Sauvage, J.-P. (Editor) 2001 *Molecular machines and motors*, Series *Structure and Bonding*, vol 99 (Berlin, Springer)
[49] Feringa B L, van Delden R A, Koumura N and Geertsema E M 2000 *Chem. Rev.* **100** 1789-1816
[50] Rocha A R et al. 2005 *Nat. Mater.* **4** 335-339
[51] Hauser A, Enachescu C, Daku M L, Vargas A and Amstutz N 2006 *Coord. Chem. Rev.* **250** 1642
[52] Renz F et al. 2000 *Angew. Chem. Int. Ed.* **39** 3699
[53] Garcia-Fernandez P and Bersuker I B 2011 *Phys. Rev. Lett.* **106** 246406
[54] IharaY and Ryokichi T 1984 *Bull. Chem. Soc. Jpn.* **57** 2829
[55] Hoffmann M M, Darab J G, Palmer B J and Fulton J L 1999 *J. Phys. Chem. A* **103** 8471
[56] Bilc D I, Orlando R, Shaltaf R, Rignanese G-M, Iniguez J and Ghosez P 2008 *Phys. Rev.* B **77** 165107; Dovesi R, Orlando R, Civalleri B, Roetti C, Saunders V R and Zicovich-Wilson C M 2005 *Z. für Kristallographie* **220** 571
[57] Berger R and Tergenius L-E 1994 *J. Alloys and Comp.* **203** 203
[58] Pickardt J, Paulus W, Schmalz M and Schöllhorn C 1990 *J. Sol. Sta. Chem.* **89** 308
[59] Owens F J 1999 *Physica* C **313** 65
[60] Lee S et al. 2008 *Nature* **451**, 805
[61] Singh D J 1994 *Phys. Rev. B* **49** 1580
[62] Miyata R. et al. 2002 *Chem. Phys. Lett.* **365** 406
[63] Park H, Wang X, Nie S, Clinte R and Cao J 2005 *Solid State Comm.* **136** 559
[64] Novoselov K S *et al.* 2004 *Science* **306** 666
[65] Geim A K and Novoselov K S 2007 *Nat. Mater.* **6** 183Cahangirov S, Topsakal M, Akturk E, Sahin H and Ciraci S 2009 *Phys. Rev. Lett.* **102** 236804; Tao L, Cinquanta E, Chiappe D, Grazianetti C, Fanciulli M, Dubey M, Molle A and Akinwande D 2015 *Nature Nanotechnology Let.* **10** 227
[67] Pacile D *et al.* 2008 *Appl. Phys. Lett.* **92** 133107;Song X, Hu J and Zeng H 2013 *J. Mater. Chem.* C **1** 2952; Ersan F, Gokoglu G and Akturk E 2014 *Appl. Surf. Sci.* **303** 306
[68] Peng Q et al. 2013 Comput. Mater. Sci. **68** 320; Zhang C W et al.2012 Chem. Phys. Lett. **548** 60
[69] Reich E S 2014 Nature 506 19
[70] Al Balushi Z Y, Wang K, Ghosh R K, Vila R A, Eichfeld S M **et al.** 2016 Nature Materials 15 1166; Zhang C W et al 2012 Chem. Phys. Lett. 548, 60
[71] Li J, Medhekar N V and Shenay V B 2013 *J. Phys. Chem.* C **117** 15842; Guo, H et al 2014, J. Phys. Chem C118 7242; Cunningham G *et al.*2012 *ACS Nano* **6** 3468; Ataca C, Sahin H and Ciraci S 2012 *J. Phys. Chem.* C **116** 8983; Komsa H P and Krasheninnikov A V 2012 *J. Phys. Chem. Lett.* **3** 3652; Tongay S *et al.*2014 *Nat. Commun.* **5** 3252; Rasmussen F A and Thygesen K S 2015 *J. Phys. Chem.* C **119** 13169; Ersan F, Cahangirov S, Gökoğlu G, Rubio A and Aktürk E 2016 *Phys. Rev.* B **94** 155415
[72] Bersuker I B 2001 *Chem. Rev.* **101** 967
[73] Bersuker I B and Stavrov S S 1988 *Coord. Chem. Rev.* **8** 1
[74] Sergeeva A P and Boldyrev A I 2010 *Organometallics* **29** 3951; Pokhodnya K, Olson C, Dai X, Schulz D L, Boudjouk P, Sergeeva A P and Boldyrev A I 2011 *J. Chem. Phys.* **134** 1; Ivanov A S, Bozhenko K V and Boldyrev A I 2012 *Inorg. Chem.* **51** 8868
[75] Jose D and Datta A 2012 *J. Phys. Chem.* C **116** 24639
[76] Liu Y, Bersuker I B, Garcia-Fernandez P and Boggs J E 2012 *J. Phys. Chem.* A **116** 7564; Liu

Y, Bersuker I B and Boggs J E 2013 *Chem. Phys.* **417** 26
[77] Jose D and Datta A 2011 *Phys. Chem. Chem. Phys.* **13** 7304; Mandal T K, Jose D, Nijamudheen A and Datta A 2014 *J. Phys. Chem. C* **118** 12115 (2014); Jose D and Datta A 2014 *Acc. Chem. Res.* **47** 593; Nijamudheen A, Bhattacharjee R, Choudury S and Datta A 2015 *J. Phys. Chem. C* **119** 3802; Chowdhury C, Jahiruddin S and Datta A 2016 *J. Phys. Chem. Lett.* **7**, 1288
[78] Boltrushko V, Krasnenko V and Hizhnyakov V 2015 *Chem. Phys.* **460** 90; Krasnenko V, Boltrushko V and Hizhnyakov V 2016 *J. Chem. Phys.* **144** 134708
[79] Soto J R, Molina B and Castro J J 2015 *Phys. Chem. Chem. Phys.* **17** 7624
[80] Hermoso W, Ilkhani A R and Bersuker I B 2014 *Comp. Theor. Chem.* **1049** 109; Ilkhani A R, Hermoso W and Bersuker I B 2015 *Chem. Phys.* **460** 75; Ilkhani A R, Gorinchoy N N and Bersuker I B 2015 *Chem. Phys.* **460** 106; Gorinchoy N N and Bersuker I B 2017 *J. Phys. Conf. Series*, **833** 012010; doi :10.1088/1742-6596/833/1/012010
[81] Liu Y, Wang Y and Bersuker I B 2016 *Nature- Scientific Reports* **6** 23315
[82] Hatanaka M 2016 *J .Phys. Chem.* A **120** 1074
[83] Breton G W and Martin K L 2002 *J. Org. Chem.* **67** 6699; Pansuriya A M, Savant M M, Bhuva C V, Singh J, Kapuriya N and Naliapara Y T 2010 *J. Heterocyclic Chem.* **47** 513; Jursic B S and Zdravkovski Z 1995 *Int. J. Quantum Chem.* **56** 115
[84] Liu Y, Bersuker I B, Zou W and Boggs J E 2010 *Chem. Phys.* **376** 3

Jahn-Teller effect in molecular electronics: quantum cellular automata

B Tsukerblat,[1*] A Palii,[2,3] J M Clemente-Juan[4] and E Coronado[4]
[1]Ben-Gurion University of the Negev, Beer-Sheva, Israel
[2]Institute of Problems of Chemical Physics, Chernogolovka, Russia
[3]Institute of Applied Physics, Academy of Sciences of Moldova, Kishinev, Moldova
[4]Instituto de Ciencia Molecular (ICMol), Universidad de Valencia, Paterna, Spain

*E-mail: tsuker@bgu.ac.il

Abstract. The article summarizes the main results of application of the theory of the Jahn-Teller (JT) and pseudo JT effects to the description of molecular quantum dot cellular automata (QCA), a new paradigm of quantum computing. The following issues are discussed: 1) QCA as a new paradigm of quantum computing, principles and advantages; 2) molecular implementation of QCA; 3) role of the JT effect in charge trapping, encoding of binary information in the quantum cell and non-linear cell-cell response; 4) spin-switching in molecular QCA based on mixed-valence cell; 5) intervalence optical absorption in tetrameric molecular mixed-valence cell through the symmetry assisted approach to the multimode/multilevel JT and pseudo JT problems.

Keywords:
Jahn-Teller effect, quantum-dot cellular automata, vibronic interactions, quantum computing, molecular electronics, mixed-valency, magnetic polyoxometalates, electron transfer, adiabatic approximation

1. Introduction

Nowadays one can observe a fast growth in computing speed in conformity with the well known "Moore's law" according to which the computer chips double their density approximately every two years. The technical challenges in manufacturing microelectronic devices and natural limit of the size (a single molecule) indicates that the age of the "Moore's law" goes to the end. A central idea of quantum computing is that a quantum spin ½ (or more generally, any two-state quantum system) can serve as a carrier of information. The most serious obstacle in the realization of quantum computing is the decoherence which should be sufficiently long or, at least, not shorter than duration of the gate logical operation. The conclusion of [1] (see exhaustive discussion by Winpenny [2]) is that, in principle, one could reach a decoherence time in a molecular magnet that would permit quantum information processing. Nevertheless, practical realization of quantum computing has not been reached yet mainly due to the conceptual obstacle related to decoherence. That is why an alternative route to proceed in this fundamental problem draws close attention of researchers and engineers working in microelectronics and quantum computing. This route exploits the principles of quantum computing based on the paradigm of QCA employing quantum dots or/and especially on molecular implementation of QCA, a new revolutionary computing paradigm which was proposed by Lent et al [3] as a physical implementation of an automaton at nano-scale using QCA (see development of the field in Refs. [4-15]). This discovery gave a strong impact to the development of the new multidisciplinary field of nanotechnology that combines physics, chemistry, material science, computer science, and electrical engineering.

The electronic QCA devices consuming extremely small amounts of electrical power and very small heat release are capable of performing computation at very high switching speeds. The idea of a four-dot QCA cell is schematized in figure 1. The cell consists of the four quantum dots occupied by two extra electrons (or holes) which can tunnel between the sites or, alternatively, it can be a

IOP Conf. Series: Journal of Physics: Conf. Series **833** (2017) 012002 doi:10.1088/1742-6596/833/1/012002

tetranuclear mixed-valence (MV) molecule. The information is encoded in the two antipodal charge configurations of the cell and transferred via Coulomb interaction between the neighboring cells. Just these two quasi-stable charge

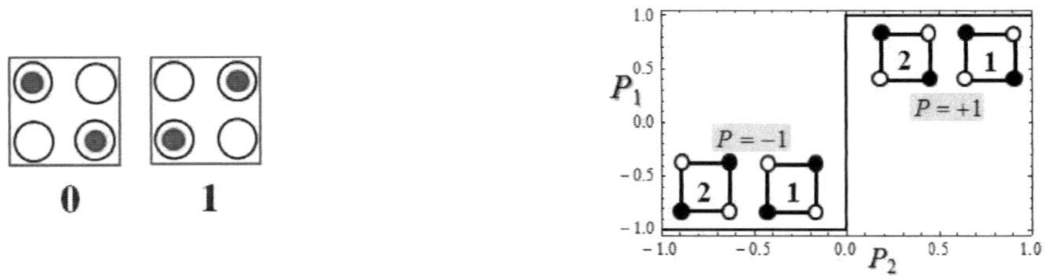

Figure 1. Left: charge distributions in a four-dot quantum cell or in a tetrameric MV molecule corresponding to the binary **0** and **1**. Right: strongly non-linear stepwise cell-cell response function (idealized).

distributions that conventionally are referred to as cell polarizations P = –1 and P = +1 encode the binary information **1** and **0** (figure 1). The arrays of the cells compose logical devices such as wires, majority gates, invertors, etc. from which more complex circuits can be designed. The architectures of these circuits had been proposed in underlying papers [3,4] and then discussed thoroughly [5-13]. Recently much attention has been paid to the role of the vibronic JT and pseudo JT interactions in functionality of such logical devices (see references in the text). Particularly it has been demonstrated that the vibronic coupling produces considerable self-trapping effect which significantly influences such key characteristic of the QCA as cell-cell response function. Hereunder we briefly discuss this and other related effects caused by the vibronic JT coupling in QCA.

2. Molecular quantum-dot cellular automata

Recently the idea of using molecular systems as QCA units has been proposed and discussed [9]. Molecular QCA offer promise of nanometer-scale devices with accompanying ultra-high device densities operating at room temperature and, providing at the same time options to control the key properties of the molecule by chemical means. A molecule acting as structured charge containers acceptable as a molecular QCA cell should have two or more positions in which the electrons occupying the nonbonding orbitals can be localized mimicking the dots and tunnel among these positions. As single molecule implementation of QCA, Lent and Tougaw [9, 16] proposed to use 1,4-dialyl butane radical cation consisting of butyl bridge linking two allyl groups accommodating electron. Then other attractive candidates for molecular cells have been proposed and discussed [9].

The proposal [13] to realize QCA cells by MV molecules is especially promising because they naturally have two or more sites with different oxidation degrees forming specific charge configurations while the electron-lattice interaction (vibronic coupling) can lead to the self-trapping and consequently to the barrier between localized configurations. This refers to a variety of MV systems studied in molecular magnetism, such as dimers, and mainly to large scale MV systems, like biologically important metal clusters and reduced polyoxometalates (e.g., Keggin and Wells- Dawson systems) accommodating two or several electrons delocalized over a metal network (see review articles [17- 19]).

Three examples of such molecules are shown in figure 2. These are the MV tetra–ruthenium complexes (figure 2, left), [2×2] Fe grid [20] and the tetra-ferrocene units coupled to the central cyclobutadiene which is bonded to Co fragment [21]. The last system contains two high-spin and two low-spin ferrous ions at opposite corners of the rhombic arrangement featuring thus matrix-like array that are most suitable for molecular implementation of the QCA concept.

A great potentiality of molecules as the leading candidates to be implemented in QCA is based on the following substantial advantages: (a) unlike cells composed of quantum dots, molecular cells of specified composition are identical and therefore have identical physical characteristics (electron

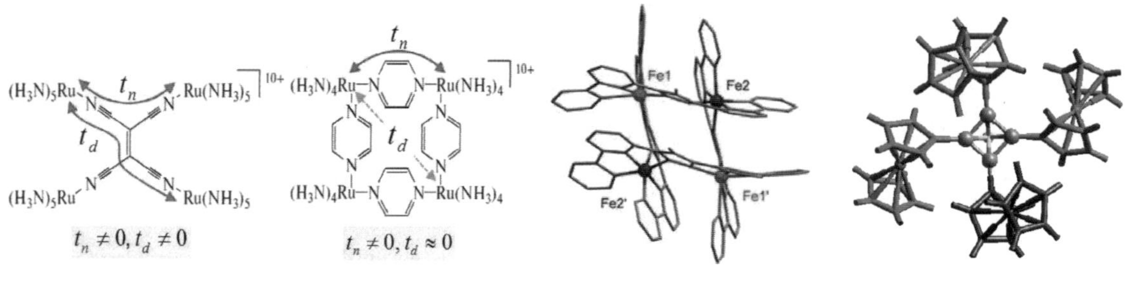

a b c

Figure 2. Molecular QCA cells: (a): two types of the tetra-ruthenium complexes exhibiting different electron networks of the transfer pathways for the electronic pair; (b) molecular structure of MV complex $[MeL_4Fe^{II}_4]^{4+}$ proposed in Ref. [20] as [2x2] Fe grid for QCA; (c) four ferrocene units coupled to the central cyclobutadiene which is bonded to Co fragment [21].

transfer integrals, vibronic parameters, etc.); (b) they can be engineered to have desired controlled characteristics with the use of chemical tailoring; (c) molecular cells are relatively large (but still nanoscale!) objects, much larger than ions and, thus, much easier for individual addressing while processing the information; (d) such molecules can be attached to different types of platforms by grafting clusters on solid surfaces, grafting monolayers, etc.

3. Jahn-Teller model of mixed-valence molecular cell

We will consider MV complexes assembled in the center-bridged and side-bridged square-planar tetramers as schematically shown in figure 3. The following consideration is applicable to the both cases of the geometric structure of the bridging ligands directly related to the electron transfer pathways.

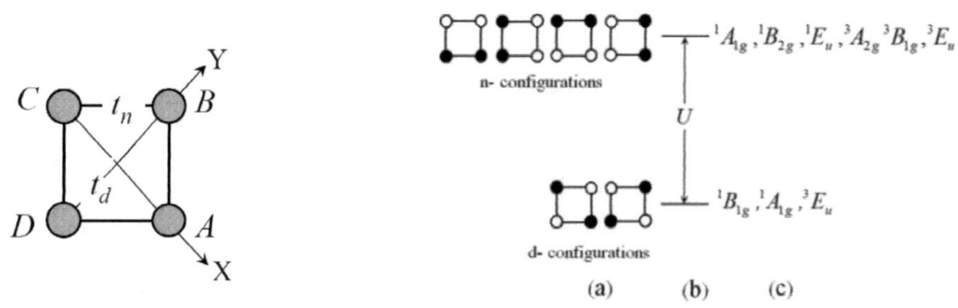

Figure 3. Left: schematic quantum cellular automata cell composed of four metal sites. Right: coulomb configurations (a), Coulomb energies (b) of a MV tetrameric system with two delocalized electrons and the group-theoretical assignment of the electronic terms (c).

The adopted molecular coordinate system is shown in figure 3, left. The Hamiltonian of the system is assumed to consist of three terms describing the main relevant interactions in MV system containing two electrons delocalized over four sites:

$$H = H_e + \sum_i \hbar\omega_i \left(Q_i^2 - \frac{\partial^2}{\partial Q_i^2} \right) + V . \quad (1)$$

Here H_e is the electronic (Hubbard type) Hamiltonian, including all interactions between electrons and ions provided that the reference configuration of the ions is fixed to the full-symmetric one ($\mathbf{D}_{4h}$). The next two terms are the operator of free vibrations and the electron-vibrational interaction Hamiltonian *V*. The vibronic interaction [22-24] is known to produce bistability in MV system which leads to a barrier between the two charge distributions and, therefore, it acts as one of the key factors in the QCA action. The main electronic interactions contributing to the Hamiltonian H_e are the following [25, 26]:

1. Coulomb repulsion between the two itinerant electrons which depends on whether these electrons are instantly localized at the adjacent sites (along the sides of the square) or they occupy the remote sites (antipodal positions) as shown in figure3. The Coulomb repelling forces tend to keep the electrons as far as possible from each other in order to minimize the energy, i.e. on the sites located at opposite vertices. These two localized configurations will be referred to as distant pairs belonging to the ground manifold. The difference in the energies of the Coulomb repulsion between the distant pairs and the neighboring ones (*d*- and *n*-pairs, the last have four possible configurations) will be denoted as *U* as illustrated in figure 3. The charge configurations in the two lowest states are assumed to encode the binary information in molecular QCA.

2. The transfer of the two electrons among the four sites. Only the one-electron transfer processes will be taken into consideration. The electron transfer parameters t_n and t_d correspond to the jumps between the distant and adjacent sites respectively (figure 3). The parameters of the isotropic exchange interaction acting within the localized configurations are normally smaller than the transfer integrals in MV compounds and therefore such magnetic exchange will be neglected.

Taking into account the interactions so far mentioned the electronic Hubbard-type Hamiltonian H_e of the cell can be represented as:

$$H_e = U_n \sum_{i,j} \sum_{\sigma,\sigma'} n_{i,\sigma} n_{i+1,\sigma'} + U_d \sum_{i,j} \sum_{\sigma,\sigma'} n_{i,\sigma} n_{i+2,\sigma'} + t_n \sum_{i,j} \sum_{\sigma} \left(a_{i,\sigma}^{+} a_{i+1,\sigma} + a_{i+1,\sigma}^{+} a_{i,\sigma} \right)$$

$$+ t_d \sum_{i,j} \sum_{\sigma} \left(a_{i,\sigma}^{+} a_{i+2,\sigma} + a_{i+2,\sigma}^{+} a_{i,\sigma} \right),$$

(2)

where $a_{i,\sigma}^{+}$ $(a_{i,\sigma})$ creates (annihilates) an electron occupying active orbital at the site *I* with spin projection σ, so that $n_{i,\sigma}$ are the occupation number operators. The summation in equation (2) is performed over the pairs of the neighboring sites (1st and 3rd terms) and over the pairs of the distant sites (2nd and 4th terms). The first term in equation (2) describes the interelectronic Coulomb repulsion (U_n) between the adjacent sites, while the second term (U_d) is that for the remote sites. The diagonal terms of the type of $n_{i,\uparrow} n_{i,\downarrow}$ responsible for the intra-site repulsion, are excluded. The last two terms describe the two types of the relevant one-electron transfer processes (figure 2a).

One can find the following spin-singlet and spin-triplet terms for the *d*-type configurations ${}^1B_{1g}(d), {}^1A_{1g}(d), {}^3E_u(d)$ (the low-lying group of levels) and the *n*-type configurations (excited Coulomb states) ${}^1B_{2g}(n), {}^1E_u(n), {}^1A_{1g}(n), {}^3A_{2g}(n), {}^3B_{1g}(n), {}^3E_u(n)$ as shown in figure 3. The explicit expressions for energy levels are found in Ref. [25]. The electron transfer processes in a bi-electron system separate spin-triplets and spin-singlets, thus resulting in an effective magnetic interaction between the electrons. This kind of the effective magnetic exchange caused by electron delocalization has been recognized in our earlier consideration of MV double reduced polyoxometales with Keggin and Wells-Dawson structures, and allowed us to explain the unusual diamagnetism of these compounds [27-29]. As to the MV square planar cell, it was shown that the energy pattern is symmetric with respect to the sign of t_n but it does depend on the sign of t_d.

It is reasonable to employ the vibronic model for MV systems formulated by Piepho, Krausz and Schatz [22] (conventionally referred to as PKS model),which deals with the independent "breathing" displacements of the ions around the sites of the electron localization. This model is simple but it properly reflects the main features of the vibronic coupling in MV systems, in particular, the presence of the potential barrier separating localized configurations. With the use of the group-theoretical classification [30], one can prove that in the case under consideration the PKS coordinates form the four-dimensional space which can be attributed to the bases of the following irreps of the point group $\mathbf{D}_{4h}$: A_{1g}, B_{1g}, E_u. One can obtain the following expressions for the PKS coordinates for a square-planar tetrameric system belonging to this point group:

$$Q_{A_{1g}} = \frac{1}{2}(Q_A + Q_B + Q_C + Q_D), \qquad Q_{B_{1g}} = \frac{1}{2}(Q_A - Q_B + Q_C - Q_D),$$
$$Q_{E_u x} = \frac{1}{\sqrt{2}}(Q_A - Q_C), \qquad Q_{E_u y} = \frac{1}{\sqrt{2}}(Q_B - Q_D). \tag{3}$$

In equations (3) Q_α $(\alpha = A_{1g}, B_{1g}, E_u x, E_u y)$ are the symmetry adapted PKS coordinates (actually, the normal coordinates of the system in the PKS model) composed of the four "breathing" vibrations Q_i associated with the sites (*i=A, B, C, D*) as shown in figure 3. In the case of the MV system under consideration, the linear vibronic interaction can be represented in the matrix form as follows:

$$V = \sum_\alpha \upsilon_\alpha \boldsymbol{O}_\alpha Q_\alpha, \tag{4}$$

where the parameters υ_α are the vibronic constants and the matrices $\boldsymbol{O}_\alpha$ are given in Ref. [25]. Taking into account the assignment of the electronic states, one can formulate the independent combined JT and pseudo JT vibronic problems, which can be symbolized as follows: $({}^1B_{1g} + 2{}^1A_{1g} + {}^1B_{2g} + {}^1E_u) \otimes (b_{1g} + e_u)$ for spin-singlets and $({}^3A_{2g} + {}^3B_{1g} + 2{}^3E_u) \otimes (b_{1g} + e_u)$ for spin-triplets terms, where the small letters are used for the vibrational irreps. One can see that the JT effect (see books and review articles [31-34]) is an inherent ingredient of the description of the quantum cell in QCA.

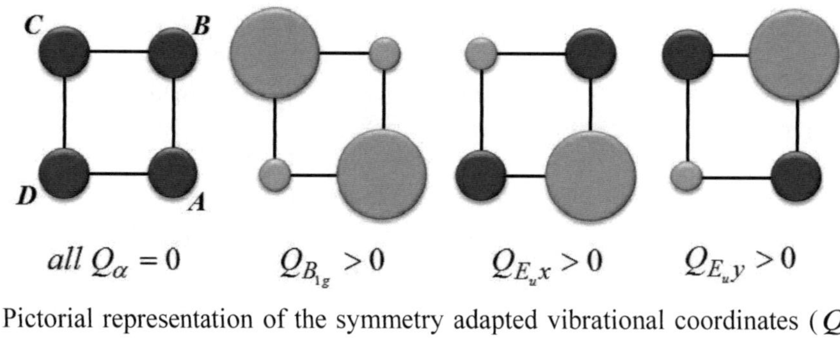

Figure 4. Pictorial representation of the symmetry adapted vibrational coordinates ($Q_\alpha > 0$) of a MV square-planar unit in PKS model. The balls mimic the sites, i.e. the metal ion and ligand surrounding: large green ball-expanded site, small blue ball-compressed site, medium grey balls-reference configuration (full-symmetric vibration $Q_{A_{1g}}$ is not shown).

4. Vibronic self-trapping and spin effects in mixed-valence molecular cell

Action of the cell as a unit keeping binary information assumes that the charges should be presumably localized at the antipodal sites. This is possible if the Coulomb repulsion suppresses considerably the transfer processes which are responsible for the delocalization of the electrons. Under the condition of strong Coulomb repulsion ($U >> |t_n|, |t_d|$), the full vibronic problem can be restricted to the ground manifold being thus reduced to one-mode $({}^1B_{1g} + {}^1A_{1g}) \otimes b_{1g}$ and ${}^3E_u \otimes b_{1g}$ subtasks [25]. This means that a pair of the levels with S=0 (${}^1B_{1g}$ and ${}^1A_{1g}$) are mixed through the only vibration b_{1g} and the same vibration is active in the S=1 term ${}^3E_u$. One thus arrives to the JT and pseudo JT problems for spin-triplets and spin-singlets correspondingly. One can see that in the framework of the PKS model the e- modes do not act within the ground manifold because they mix the ground Coulomb configurations of d-type with excited n-type configurations. The vibronic pseudo JT problem proves to be one-dimensional and therefore is essentially simplified. The adiabatic potentials are found to be:

$$\varepsilon_{\pm}(Q) = -\frac{\Delta}{2} + \frac{\hbar\omega}{2}Q^2 \pm \frac{1}{2}\sqrt{\Delta^2 + 4\upsilon^2 Q^2} ,$$
$$\varepsilon_{x,y}(Q) = -\Delta_1 + \frac{\hbar\omega}{2}Q^2 \pm \upsilon Q . \tag{5}$$

In equation (5) the functions $\varepsilon_+(Q)$ and $\varepsilon_-(Q)$ represent the two branches of the adiabatic potential for the $({}^1B_{1g} + {}^1A_{1g}) \otimes b_{1g}$ pseudo JT problem, while $\varepsilon_{x,y}(Q)$ is adiabatic surfaces for the static JT problem ${}^3E_u \otimes b_{1g}$. The following notations are used:

$$\frac{8t_n^2}{U} - \frac{16t_n^2 t_d}{U^2} \equiv -\Delta \quad , \quad \frac{4t_n^2}{U} \equiv \Delta_1 \quad . \tag{6}$$

Within the PKS model, the frequencies and the vibronic coupling parameters prove to be equal for all active vibrations (because they originate from the couplings of the same strength with the breathing modes at the equivalent centers). Depending on the interrelation between the gap Δ and the strength of the vibronic coupling υ, the lower sheet $\varepsilon_-(Q)$ can have a single minimum at $Q_{\pm} = 0$, providing weak coupling (or/and large gap Δ), $\upsilon^2 < \Delta\hbar\omega/2$ (figures 5a-c), or two minima at the points $Q_{\pm}$:

$$Q_{\pm} = \pm\sqrt{\frac{\upsilon^2}{\hbar^2\omega^2} - \frac{\Delta^2}{4\upsilon^2}} \tag{7}$$

under the condition of strong coupling (or/and small gap Δ), $\upsilon^2 > \Delta\hbar\omega/2$ (figures 5d-f). The energy of the single minimum of $\varepsilon_-(Q)$ in which the electronic pair is fully delocalized is just the electronic energy ($-\Delta$).The potential curves $\varepsilon_{x,y}(Q)$ represent two intersecting parabolas and exhibit two minima irrespectively of the value of the vibronic coupling (static JT effect) which are disposed at $Q_x = -Q_y = \upsilon/\hbar\omega$ and have the energies $-\Delta_1 - \upsilon^2/2\hbar\omega$. These two minima correspond to the localization of the electronic pair in the antipodal positions ("broken symmetry" states). In the case of static JT effect, the tunneling between the minima is forbidden (at least up to third order of perturbation theory with respect to electron transfer), and so the electronic pair is fully localized in each of the two minima irrespectively of the strength of the vibronic coupling.

IOP Conf. Series: Journal of Physics: Conf. Series **833** (2017) 012002 doi:10.1088/1742-6596/833/1/012002

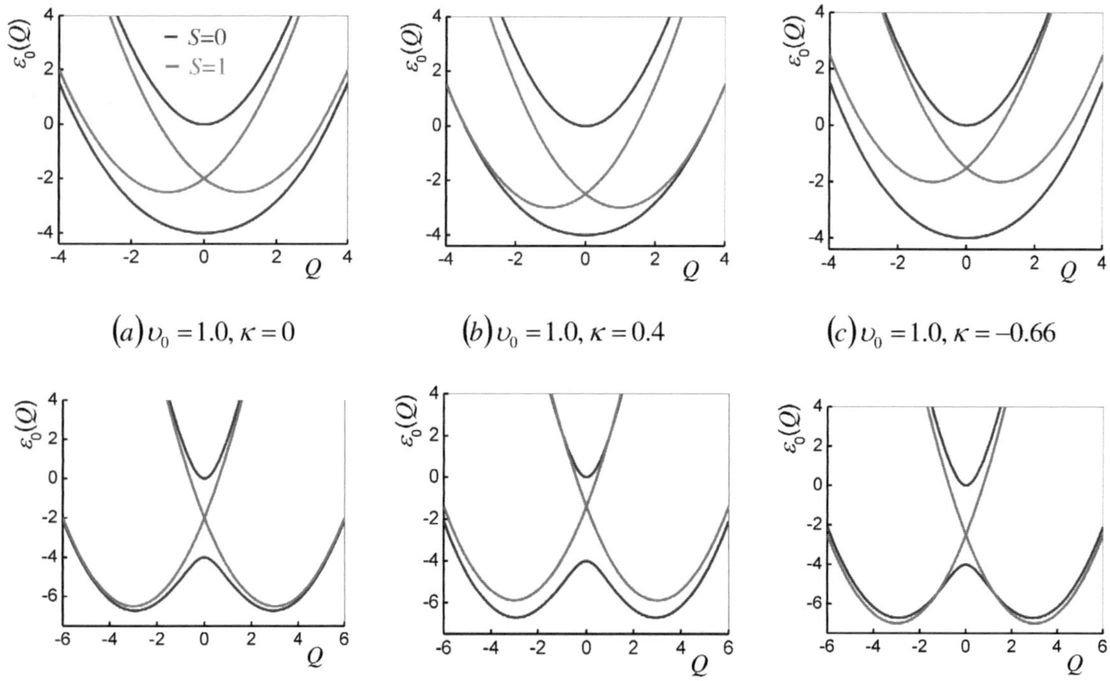

$(a)\,\upsilon_0 = 1.0, \kappa = 0$ $(b)\,\upsilon_0 = 1.0, \kappa = 0.4$ $(c)\,\upsilon_0 = 1.0, \kappa = -0.66$

$(d)\,\upsilon_0 = 3.0, \kappa = 0$ $(e)\,\upsilon_0 = 3.0, \kappa = -0.86$ $(f)\,\upsilon_0 = 3.0, \kappa = 0.4$

Figure 5. Adiabatic potentials of a square planar MV unit for the low lying levels belonging to the d configuration of the electronic pair: $\varepsilon_0(Q) = \varepsilon(Q)/\hbar\omega$ -dimensionless adiabatic energy, $\upsilon_0 = \upsilon/\hbar\omega$, $\Delta/\hbar\omega = 4$.

To distinguish different physical situations in figure 5 we use the dimensionless parameter $\kappa = (2\Delta_1 - \Delta)/\Delta_1$, which is expressed through t_d as $\kappa = 4t_d/U$ and serves as a measure of the position of the spin-triplet ${}^3E_e$ within the gap between two spin-singlets ${}^1B_{1g}$ and ${}^1A_{1g}$ in the electronic spectrum (providing $\kappa = 0$ the spin-triplet is located just in the middle of this gap). Figures 8a-c show that weak vibronic coupling ($\upsilon/\hbar\omega = 1.0$) effectively reduces the singlet-triplet gap (${}^1A_{1g}$, ${}^3E_u$). The most interesting case of strong vibronic coupling is illustrated in figures5 d-e. In the limit of strong coupling ($\upsilon/\hbar\omega = 3.0$) and providing $\kappa = 0$, the minima for the spin-singlet and spin-triplet states become practically degenerate. These minima correspond to the full localization of the charges along the diagonals of the square for both spin configurations. This is a result of the full quenching of the t_n processes by a strong vibronic interaction, which leads to the localization of the electronic pair and, consequently, to the paramagnetic ground state. Providing $t_d < 0$, the minima of spin-singlet states are deeper than those for the spin-triplet states although the singlet-triplet gap in the electronic spectrum (at $Q = 0$) is reduced. In the case of $t_d > 0$ the strong enough vibronic coupling leads to the spin-crossover of the levels, so that the spin-triplet state becomes the ground one. In the last case the system is locked at the minima.

5. "Intercommunication" between molecular cells: cell-cell response function within the Jahn-Teller model

In the previous section we have considered the electronic and vibronic properties of the isolated molecules which play role of the cells in QCA. Implementation of the logically significant OCA arrays or circuits requires consideration of the "intercommunication" between the molecular cells. It is essential that the charge configuration of the input cell is able to dictate the charge configuration of the two output

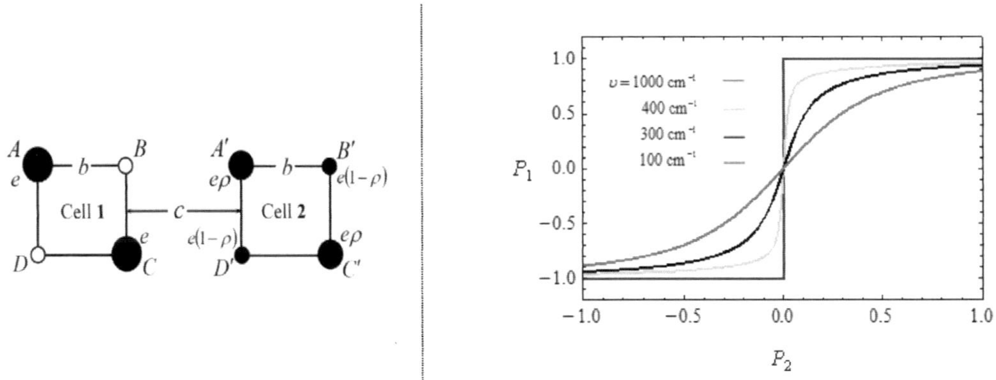

Figure 6. Left: geometry of interacting cells. Right: cell-cell response function calculated for different values of the vibronic coupling parameters and $\Delta = 800\ \mathrm{cm}^{-1}$, $\hbar\omega = 200\ \mathrm{cm}^{-1}$, $\kappa = -0.52$, $\upsilon = 310\ \mathrm{cm}^{-1}$, $b = 3.5$ Å, $c = 6$ Å.

cells. In the simplest case of a QCA wire one has to solve the non-stationary problem of the case of a QCA wire one has to solve the non-stationary problem of the signal propagation from the input cell in a definite state (binary "**1**" or "**0**") to the output. In a QCA fanout circuit an input signal is transformed into two or more output signals. Providing different cellular automata architectures one should assume that the state of a cell should be strongly influenced by the state of the neighboring cell (or cells), so that a slight polarization of a cell would induces nearly complete polarization of a neighboring cell. To characterize this kind of highly non-linear behavior of a bistable cell the use of the cell-cell response function has been proposed [35]. Since the vibronic coupling produces a trapping effect, it is expected to produce substantial influence on the cell-cell response function. Let us consider the polarization of the cell, a scalar quantity *P* introduced by Lent et al (see [3, 9]) as a normalized measure of the degree to which the electron densities are localized in antipodal positions along the diagonals *AC* or *BD* of a square planar four-site cell:

$$P = \frac{(\rho_A + \rho_C) - (\rho_B + \rho_D)}{\rho_A + \rho_C + \rho_B + \rho_D}, \tag{8}$$

where ρ_i are the electronic charges on the sites. If the sites *B* and *D* are empty $(\rho_B = \rho_D = 1,\ \rho_A = \rho_C = 0)$ the cell is fully polarized along *AC* and so $P = -1$. Alternatively, if the electrons occupy sites *B* and *D*, polarization $P = +1$. From the qualitative arguments one can conclude that in MV systems the effective electronic charges are the results of the competitive effects of the vibronic (JT and pseudo JT) coupling and transfer processes. The geometry of the interacting cells is shown in figure 6. One can introduce the values U_{AC} and U_{BD} which are the Coulomb interaction energies of cell **1** in the two antipodal localizations of the electronic pair with the cell **2** having a

IOP Conf. Series: Journal of Physics: Conf. Series **833** (2017) 012002 doi:10.1088/1742-6596/833/1/012002

certain polarization P_2 .The electronic densities are calculated in Ref. [25] within the quantum-mechanical (dynamical) approach to the vibronic problem. The vibronic coupling strongly modifies the cell-cell response function whose idealized (sharp step) view is shown in figure 1. From figure 6 (right) one can observe that the increase of the vibronic coupling parameter leads to the increase of the abruptness of the $P_1(P_2)$-curves, while at smaller values of υ the steepness of the curve is less pronounced, and even at $P_2 = \pm 1$ the polarization $|P_1|$ remains smaller than 1. This is evidently due to the fact that the vibronic coupling tends to localize the electrons (reducing their mobility) and, therefore, it decreases polarization of the cell. More detailed consideration, in particular, estimation of the parameters for some MV clusters is given in Ref. [25].

6. Spin-switching in molecular quantum cellular automata based on mixed-valence cell

The main efforts in the area of QCA have been focused on the analysis of electronic charge density distribution within the cell and its change due to polarization (electric quadrupole moment) induced by the Coulomb field of the neighboring cells. Much less attention has been paid on the role of spin effects in QCA. At the same time, there is a strong correlation between the molecular spin and the degree of localization in mixed valence clusters.

In Ref. [26] we focused on the study of spin-effects in a single square-planar MV cell comprising two electrons and in coupled molecular cells for QCA. Using the vibronic model it was demonstrated that the polarizabilities of the cell are different in spin-singlet and spin-triplet states of the electronic pair. Based on this inference the concept of spin-switching in molecular QCA has been proposed and the conditions under which this effect is feasible have been derived. In order to reveal these conditions we have performed a series of quantum-mechanical calculations of the vibronic energy levels of the isolated cell and of the cell subjected to the external polarizing field. To present the results in a descriptive way they are qualitatively discussed in terms of the adiabatic potential curves illustrating spin effects in switching cycle during operation in QCA. The spin-switching effect stemming from the theoretical model is shown to be dependent on the relative strength of the intra-cell electronic interactions (Coulomb repulsion energies and electron transfer parameters) and the vibronic coupling. This is expected to suggest a route for observation of spin-switching in which the non-magnetic vibronic ground state of the cell is changed to the magnetic one due to polarization induced in this cell by the neighboring one. The spin switching is shown to produce a significant effect on the cell-cell response function leading to the unusual stepwise shape of this function. Figure 7 shows the switching cycle of a cell subjected to the Coulomb field of the neighboring cell. The cycle is presented by plotting the adiabatic potentials for cells **1** and **2** calculated for different S values. Figure 7 shows only the lowest sheet $\varepsilon_-^{S=0}(Q)$of the adiabatic potential for S = 0. Depending on the spin-state of the polarized cell, one can distinguish two cases. Figure 7a shows the case in which the cell retains spin-singlet state when it becomes polarized. Figure 7b demonstrates a possibility of spin-switching. Thus, for a negative value of κ, $\kappa = -0.5$ (that is $t_d < 0$) the polarization of cell **2** is unable to change the ground spin state of cell **1** (figure7a), and so this state always remains nonmagnetic. It is worth to mention that the gap between the lowest levels with S = 0 and S = 1 is reduced due to partial vibronic localization of the system. On the contrary, in case of positive κ the polarization of cell **2** leads to the change of the ground spin state of cell **1** as shown in figure 7b for $\kappa = 0.5$. One can see that the system switches from a non-magnetic state (S_{gr} = 0 at $P_2 = 0$) to a magnetic one (S_{gr} = 1 at $P_2 = \pm 0.5$). This example demonstrates that for some ranges of parameters the spin-switching in the cell can be induced by the polarization of the neighboring cell.

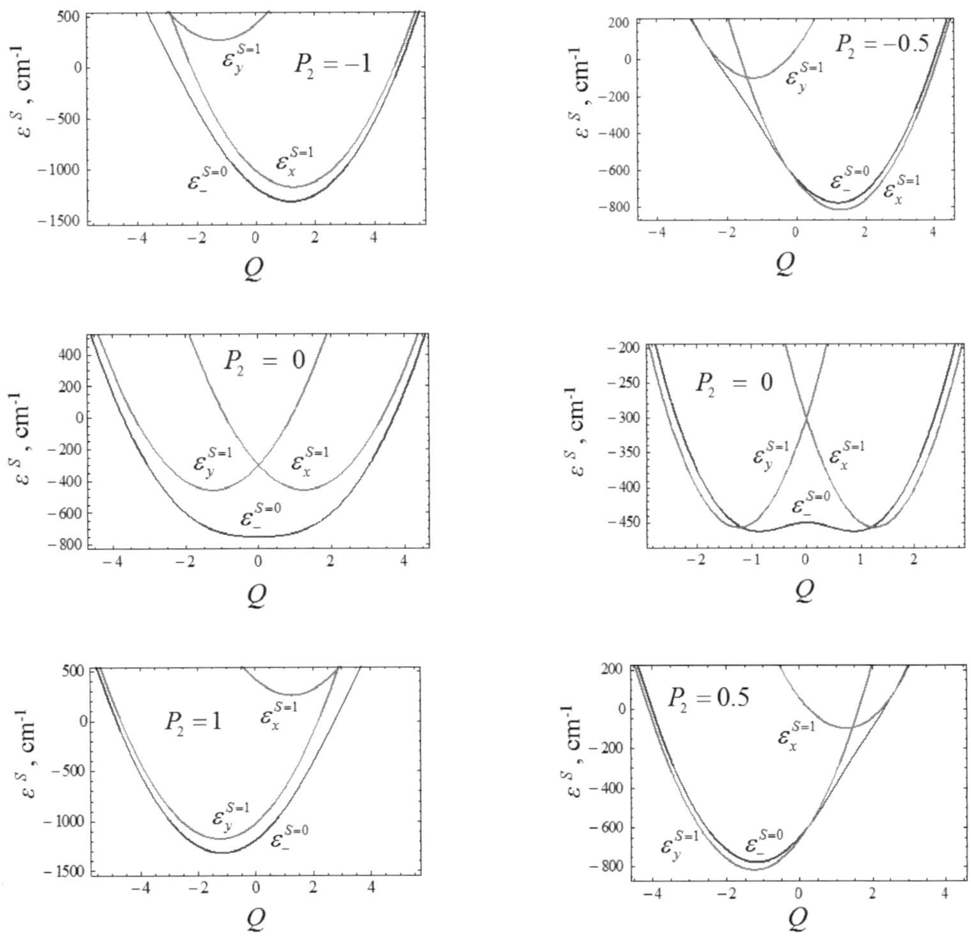

(a) (b)

Figure 7. Adiabatic potentials of the cell **1** calculated at different P_2: $8t_n^2/U = 600\,\mathrm{cm}^{-1}$, $\upsilon = 250\,\mathrm{cm}^{-1}$, and $\kappa = -0.5$ (a), $\kappa = 0.5$ (b); $S = 1$ – red lines, $S = 0$ – blue lines.

7. Beyond the Born-Oppenheimer approximation through the symmetry assisted approach

The adiabatic approximation (Born-Oppenheimer paradigm) is valid in the cases of strong vibronic coupling in the JT systems but can lose its accuracy providing moderate vibronic coupling. Moreover, by definition, this approximation (based on classical representation of the ionic movement) fails in the description of the discrete vibronic spectrum and, in particular, in the evaluation of the intervalence optical bands in MV systems. Understanding of limitations of the Born-Oppenheimer approximation gave rise to a new trend at the borderline between physics and chemistry that can be referred to as "Chemical theory beyond the Born- Oppenheimer paradigm" (see very recent book [36] with this indicative title). Alternatively, the quantum-mechanical approach usually leads to a complicated (often practically unsolvable) dynamic problem in multimode systems, especially in cases when several electronic states are coupled to the JT and pseudo-JT modes (see review in [32]). This occurs, for example, in the nanoscopic MV systems, which can be exemplified by polyoxometalates (like Keggin and Wells–Dawson structures [27-29]). Therefore, a precise solution of the quantum-mechanical vibronic problem is of primary importance for the evaluation of the shapes of the electron transfer optical absorption bands and quantitative analysis of the main parameters of tetrameric quantum cells.

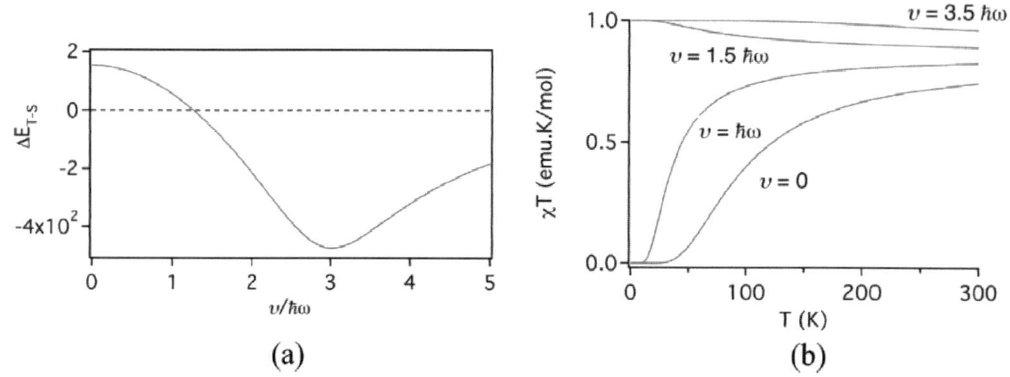

Figure 8. Illustration for the ground state of a bi-electronic square planar MV system in the case of $t_d > 0$: (a) singlet-triplet gap ΔE_{T-S} (in cm^{-1}) vs υ in the case of weak Coulomb repulsion, (b) temperature dependence of χT at different values of the vibronic coupling parameter shown in the plot and $|t_n|/U = 2,\ t_d/U = 2$.

To solve the multidimensional JT problem we have recently suggested [37-39] a theoretical tool which has been referred to as "symmetry assisted approach". The symmetry assisted approach allows to block the vibronic matrices to full extent in accordance with the point symmetry of the systems and thus to make tractable the complicated dynamic problems for these molecular cells. The mathematical details of the approach are described in Refs. [37-39]. These calculations are implemented in a specially designed software (which will be published elsewhere in a more general mathematical formulation). With this tool we also extended the theory of the molecular QCA cell beyond the limit of prevailing Coulomb repulsion in the electronic pair (as described in previous sections). Consequently the general pseudo JT problems for spin-singlet $({}^1B_{1g}, 2{}^1A_{1g}, {}^1B_{2g}, {}^1E_u) \otimes (b_{1g} + e_u)$ and spin-triplet states $({}^3A_{2g}, {}^3B_{1g}, 2{}^3E_u) \otimes (b_{1g} + e_u)$ in a square- planar bi-electronic system has been recently solved [39]. The obtained symmetry adapted electron-vibrational functions have been employed for the calculation of the profiles (shape- functions) of the charge transfer absorption bands in the tetrameric MV complexes and for the discussion of the magnetic properties. In Ref. [39] we presented also the characteristics of the performance of the evaluation which can be referred to as "index of efficiency" defined as the ratio (CPU time for blocked matrix) / (CPU time for full matrix). In particular, use of the symmetry adapted basis leads to a significant reduction of the CPU time, for example for N= 24 (number of the unperturbed vibrational levels) the CPU time with the use of symmetry is only 7.1% of the total time without application of the symmetry assisted approach.

To illustrate some results let us consider a particular case of a weak Coulomb repulsion or/and strong transfer, $|t_n|/U$, $|t_d|/U > 1$, positive t_d ($|t_n| = 2000\ \text{cm}^{-1}$, $t_d = 2000\ \text{cm}^{-1}$). The full sets of the vibronic levels as functions of the vibronic coupling for spin-triplets and spin-singlets in the case of $t_d > 0$ are given in Ref. [39]. Here we mention only that the low lying levels demonstrate spin-crossover (figure 8, left) which manifests itself in the temperature dependences of the magnetic susceptibility χT *vs* T (figure 8, right).

IOP Conf. Series: Journal of Physics: Conf. Series **833** (2017) 012002 doi:10.1088/1742-6596/833/1/012002

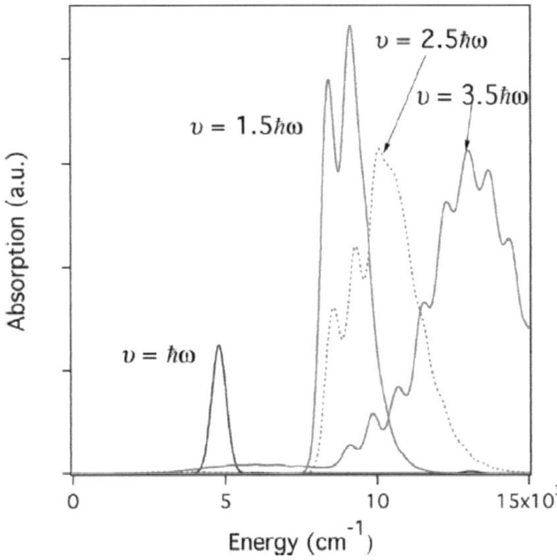

Figure 9. Shape-function of the charge transfer bands as functions of $\upsilon/\hbar\omega$ in the case of weak Coulomb repulsion and $t_d > 0$ (the same parameters as in figure 7) : blue-before spin-crossover (*S*=0 ground state); red-after spin-crossover (*S*=1 ground state).

The profiles of the intervalence absorption bands are given in figure 9. The shapes of the bands are given for the two sets of the vibronic parameters, the first one corresponds to the *S*=0 ground state (before spin-crossover point). Therefore, the dynamic pseudo JT vibronic problem in MV tetramers with the delocalized electronic pair is solved with the aid of the powerful symmetry-assisted approach which allows to eliminate the restrictions implied by the Born-Oppenheimer paradigm. The mathematical receipt for the quantum-mechanical treatment of the vibronic eigen-problem is based on the point symmetry consideration and, therefore, the total huge matrix of the JT Hamiltonian is blocked to maximum extent. More careful description of the shapes of the lines and optical bands entire requires consideration of the interaction with the thermal bath and relaxation processes [40].

8. Some new perspectives

Here we briefly indicate a new concept within the general trend of molecular implementation of QCA. It is propose to employ complex polyoxometalate anion $[V^{IV}_8V^{V}_4As_8O_{40}(H_2O)]^{4-}$ (briefly V_{12}) [41] as a quantum inverter. The metal skeleton of this polyoxometalate anion involves two mixed-valence squares with two delocalized electrons each as shown in figure 9. The two mixed-valence units can be associated with the two parallel cells coupled within the molecule via Coulomb repulsion. The Coulomb repulsion inside the squares is strong which ensures the antipodal electronic distributions, while the minimal Coulomb repulsion between the electrons in different cells is reached providing twisted configuration of charge distribution in different cells. This architecture of the cells in V_{12} is able to function of molecular quantum inverter as illustrated in figure 10 from which one can see that

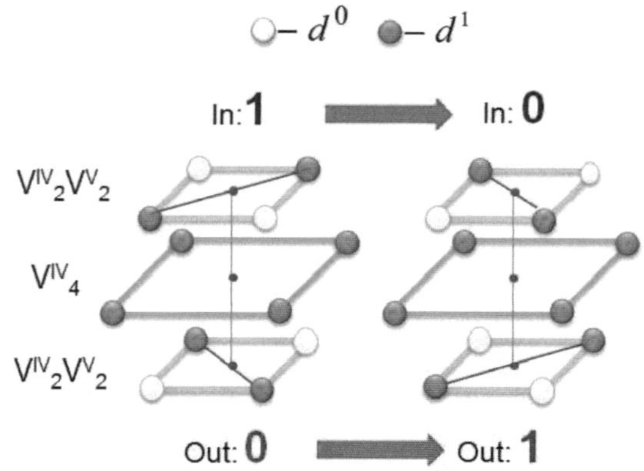

Figure 10. Scheme of quantum inverter in polyoxometalate V_{12} . Empty sites (V^V): yellow balls, sites occupied by electrons (V^{IV}): green balls.

in the favorable Coulomb configurations input "1" in one of the cell gives output "**0**" in the another cell. As distinguished from previous ideas on molecular QCA, in which a molecule was supposed to act as a quantum cell, the proposal reported here employs V_{12} as a logical gate. We estimated the Coulombic forces in different electronic distributions within the mixed-valence $V^{IV}_8V^V_4$ network, revealed the role of the electron transfer processes and vibronic coupling. Estimation of the energies of "twisted" and "eclipsed" charge configurations demonstrates feasibility of the quantum inverter function. More detailed results including ab initio evaluation of the main parameters will be published elsewhere.

9. Concluding remarks

In this short review we have summarized some recent results of the application of the JT vibronic model to the problem of QCA. Many of the systems so far mentioned attracted long lasting attention as molecular magnets for which the peak of studies occurred in the period 2000. The main attention was concentrated on the problem of single molecular magnets with promising applications as memory storage units. Then, these systems, especially MV clusters have been proposed as charge containers suitable for the molecular QCA, which revived interest in the study of these systems and gave rise to the new trends summarized in the present review article. Strong impact on new development has also been given by the new experimental techniques which has allowed to directly observe the key features of MV systems. For example, it is worth to mention the scanning tunneling microscopy (STM) [42] which gave direct access to the distributions of the electronic density in potential wells of the Fe(II)-Fe(III) system. From the theoretical point of view MV systems have been studied as objects in which relaxation processes and environmental decoherence [5,40] caused by the electron-thermal bath are important. This topic is closely related to the energy dissipations for the logic processes, a question of the conceptual significance and practical importance.

In the present review special attention has been paid to the role of the JT and pseudo-JT effects in charge trapping, encoding of binary information in the quantum cell and non-linear cell-cell response. We have also discussed the spin-switching in QCA, and intervalence optical absorption spectra of tetrameric MV square cells evaluated with the aid of the symmetry assisted approach to the multimode/multilevel JT problem. The results of recent studies in the area show that the JT and pseudo JT theory is very efficient at nanoscale and useful for the study of molecular units and corresponding processes in microelectronics. One can express also an optimistic expectation that the developed vibronic approach could be applicable in general problem of molecular information processing which is one of the most challenging problem in a wide field of knowledge ranging from molecular electronics down to supramolecular processes occurring in brains (see pioneering papers of Day [43, 44]).

Acknowledgments
B.T., A.P. and J.M.C-J were supported by Ministry of Education and Science of Russian Federation (agreement No. 14.W03.31.0001-Institute of Problems of Chemical Physics, Chernogolovka). A.P. thanks the Supreme Council for Science and Technological Development of the Republic of Moldova (CSSDT project 15.817.02.06F) for financial support. E.C. and J.M.C-J thank the EU (ERC Advanced Grant SPINMOL), the Spanish MINECO (grants MAT2014-56143-R, CTQ2014-52758-P and Excellence Unit Maria de Maeztu MDM-2015-0538), and the Generalidad Valenciana (Prometeo and ISIC Programmes of excellence). All authors are grateful to COST Action CA15128 "Molecular Spintronics (MOLSPIN)".

References
[1] Bertaina S, Gambarelli S, Mitra T, Tsukerblat B, Müller A and Barbara B 2008 Quantum oscillations in a molecular magnet *Nature* **453** 203–6
[2] Winpenny R E P 2008 Quantum information processing using molecular nanomagnets as qubits *Angew. Chem. Int. Ed.* **47** 2–5
[3] Lent C S, Tougaw P, Porod W and Bernstein G 1993 Quantum cellular automata *Nanotechnology* **4** 49–57
[4] Porod W, Lent C S , Bernstein G, Orlov A O, Amlani I, Snider G L and Merz J L 1999 Quantum-dot cellular automata: computing with coupled quantum dots, *Int. J. Electronics,* **86** 549–590
[5] Blair E P, Corcelli S A, Lent C S 2016 Electric-field-driven electron-transfer in mixed-valence molecules *J. Chem. Phys.* **145** 014307–12
[6] Tougaw P D and Lent C S 1994 Logical devices implemented using quantum cellular automata J. Appl. Phys. **75** 1818–25
[7] Tóth G and Lent C S 1999 Quasi-adiabatic switching for metal-island quantum-dot cellular automata *J. Appl. Phys.* **85** 2977–84
[8] Lent C S 2000 Bypassing the transistor paradigm *Science* **288** 1597–99
[9] Lent C S, Isaksen B and Lieberman M 2003 Molecular quantum-dot cellular automata *J. Am. Chem. Soc.* **125** 1056–63
[10] Lent C S, Tougaw P and Porod W 1993 Bistable saturation in coupled quantum dots for quantum cellular automata *Appl. Phys. Lett.* **62** 714–6
[11] Lent C S and Isaksen B 2003 Clocked molecular quantum-dot cellular automata *IEEE Trans. Electron Devices* **50** 1890–6
[12] Lieberman M, Chellamma S, Varughese B, Wang Y, Lent C, Bernstein G H, Snider G and Peiris F C 2002 Quantum-dot cellular automata at a molecular scale *Ann. N.Y. Acad. Sci.* **960** 225–239
[13] Braun-Sand S B and Wiest O 2003 Theoretical studies of mixed-valence transition metal complexes for molecular computing *J. Phys. Chem.* A **107** 285–91
[14] Orlov A O, Amlani I, Bernstein G H, Lent C S and Snider G L 1997 Realization of a functional cell for quantum-dot cellular automata *Science* **277** 928–30
[15] Timler J and Lent C S 2002 Power gain and dissipation in quantum-dot cellular automata *J. Appl. Phys.* **91** 823–31
[16] Lent C S and Tougaw P D 1993 Lines of interacting quantum-dot cells: a binary wire *J. Appl. Phys.* **74** 6227–33
[17] Clemente-Juan J M and Coronado E Magnetic clusters from polyoxometalate complexes *Coord. Chem. Rev.* **193–195** 361–394

[18] Kögerler P, Tsukerblat B and Müller A 2010 Structure-related frustrated magnetism of nanosized polyoxometalates: aesthetic beauty and properties in harmony *Dalton Trans.* **39** 21–36

[19] Clemente-Juan J M, Coronado E, Gaita-Ariño A 2012 Magnetic polyoxometalates: from molecular magnetism to molecular spintronics and quantum computing *Chem. Soc. Rev.* **41** 7464–78

[20] Schneider B, Demeshko S, Neudeck S, Dechert S and Meyer F 2013 Mixed-spin [2 × 2] Fe_4 grid complex optimized for quantum cellular automata *Inorg. Chem.,* 52: 13230–37

[21] Jiao J, Long G J, Rebbouh L, Grandjean F, Beatty A M and Fehlner T P 2005 Properties of a mixed-valence $(Fe^{II})_2(Fe^{III})_2$ square cell for utilization in the quantum cellular automata paradigm for molecular electronics *J. Am. Chem. Soc.* **127** 17819–31

[22] Piepho S B, Krausz E R and Shatz P N 1978 Vibronic coupling model for calculation of mixed valence absorption profiles, *J. Am. Chem. Soc.* **100** 2996–3005

[23] Brunschwig B S, Creutz C and Sutin N 2002 Optical transitions of symmetrical mixed-valence systems in the Class II-III transition regime *Chem. Soc. Rev.* **31** 168–184

[24] Piepho S B 1990 Vibronic coupling model for the calculation of mixed-valence line shapes: a new look at the Creutz-Taube ion *J. Am. Chem. Soc.* **112** 4197–206

[25] Tsukerblat B, Palii A ,Clemente-Juan J M and Coronado E 2015 Mixed-valence molecular four-dot unit for quantum cellular automata: vibronic self-trapping and cell-cell response, *J. Chem. Physics* **143** 134307–15

[26] Palii A, Tsukerblat B, Clemente-Juan J M and Coronado E 2016 Spin-switching in molecular quantum cellular automata based on mixed-valence tetrameric units *J. Phys. Chem.* C **120** 16994–05

[27] Borras-Almenar J J, Clemente-Juan J M, Coronado E and Tsukerblat B S. 1995 Mixed-valence polyoxometalate clusters. I. Delocalization of electronic pair in dodecanuclear heteropoly blues with Keggin structure, *Chem. Phys.* **195** 1–16

[28] Borras-Almenar J J, Clemente-Juan J M, Coronado E and Tsukerblat B S. 1995 Mixed-valence polyoxometalate clusters II. Delocalization of electronic pair in 18-site heteropoly blues with Wells-Dawson structure, *Chem. Phys.* **195** 17–28

[29] Borras-Almenar J J, Clemente-Juan J M, Coronado E and Tsukerblat B S. 1995 Mixed-valence polyoxometalate clusters III. Vibronic problema for the 2-electron reduced heteropoly blues with Keggin structure *Chem. Phys.* **195** 29–47

[30] Tsukerblat B 2006 *Group Theory in Chemistry and Spectroscopy* (NY: Dover Mineola)

[31] Englman R 1972 *The Jahn-Teller Effect in Molecules and Crystals* (London: Wiley)

[32] Bersuker I B and Polinger V Z 1989 *Vibronic Interactions in Molecules and Crystals* (Berlin: Springer-Verlag)

[33] Bersuker I B 2006 *The Jahn-Teller Effect* (Cambridge University Press)

[34] Bersuker I B 2013 Pseudo-Jahn-Teller effect-a two-state paradigm in formation, deformation and transformation of molecular systems and solids *Chem. Rev.* **113** 1351–90

[35] Lent C S and Tougaw P D 1993 Lines of interacting quantum-dot cells: a binary wire, *J. Appl. Phys.* **74** 6227–33

[36] Takatsuka K, Yonehara T and Hanasaki K. 2015 *Chemical Theory beyond the Born-Oppenheimer Paradigm: Nonadiabatic Electronic and Nuclear Dynamics in Chemical Reactions* (Singapore: World Scientific Publishing Co. Pte. Ltd.)

[37] Tsukerblat B, Palii A, Clemente-Juan J M and Coronado E 2011 A symmetry adapted approach to the dynamic Jahn-Teller problem, in Series: *Progress in Theoretical Chemistry and Physics*, **23** *Vibronic Interactions and the Jahn-Teller Effect* Eds: Atanasov A, Daul C and Tregenna-Piggott P L W (Berlin, Heidelberg: Springer) 39–58

[38] Tsukerblat B, Palii A, Clemente-Juan J M, Gaita-Ariño A and Coronado E 2012 Symmetry adapted approach to the dynamic Jahn-Teller problem: application to mixed-valence polyoxometalate clusters with Keggin structure *Int. J. Quantum Chem.* **112** 2957–64

IOP Conf. Series: Journal of Physics: Conf. Series **833** (2017) 012002 doi:10.1088/1742-6596/833/1/012002

[39] Clemente-Juan J M, Palii A, Coronado E and Tsukerblat B. 2016 Mixed-valence molecular unit for quantum cellular automata: beyond the Born-Oppenheimer paradigm through the symmetry assisted vibronic approach *J. Chem. Theory &Comp.* **12** 3545–60

[40] Blair E. P and Lent C S 2013 Environmental decoherence stabilizes quantum-dot cellular automata, *J. Appl. Phys.* **113** 124302–16

[41] Kögerler P Tsukerblat B and Müller A 2010 Structure-related frustrated magnetism of nanosized polyoxometalates: aesthetic beauty and properties in harmony *Dalton Transactions* **39** 21–36

[42] Lu Y, Quardokus R, Lent C S, Justaud F, Lapinte C and Kandel S A 2010 Charge localization in isolated mixed-valence complexes: an STM and theoretical study *J. Am. Chem. Soc.* **132** 13519–24

[43] Day P 1998 Molecular information processing: will it happen? *Proc. of the Royal Institution of Great Britain* Ed P Day (Oxford: Oxford University Press) **69** 85–106

[44] Day P 1990 Future molecular electronics: towards a supramolecular information processor *Chem. Br.* **26** 52–4

IOP Conf. Series: Journal of Physics: Conf. Series **833** (2017) 012003 doi:10.1088/1742-6596/833/1/012003

Manifestation of the Jahn-Teller effect in elastic moduli of strontium fluorite crystals doped with chromium ions

N S Averkiev[1], I B Bersuker[2], V V Gudkov[3,10], I V Zhevstovskikh[3,4], M N Sarychev[3], S Zherllitsyn[5], S Yasin[5,6], G S Shakurov[7], V A Ulanov[8] and V T Surikov[9]

[1] A.F. Ioffe Physical Technical Institute of the RAS, 194021, St.Petersburg, Russia
[2] Institute for Theoretical Chemistry, the University of Texas at Austin, TX 78712, Austin, USA
[3] Ural Federal University, 620002, Ekaterinburg, Russia
[4] M.N. Miheev Institute of Metal Physics, Ural Branch RAS, 620137, Ekaterinburg, Russia
[5] Hochfeld-Magnetlabor Dresden (HLD-EMFL), Helmholtz-Zentrum Dresden-Rossendorf, D-01314 Dresden, Germany
[6] American University of the Middle East (AUM), College of Engineering and Technology, Eqaila, Kuwait
[7] Kazan E.K. Zavoisky Physical-Technical Institute of the RAS, 420029, Kazan, Russia
[8] Kazan State Power Engineering University, 420066, Kazan, Russia
[9] Institute of Solid State Chemistry of Ural Branch of the RAS, 620990, Ekaterinburg, Russia

[10] E-mail: v.v.gudkov@urfu.ru

Abstract. Attenuation and phase velocity of ultrasound have been measured in strontium fluoride single crystal doped with chromium in the temperature range of 4 – 185 K at 56 -162 MHz. Anomalies have been found for all the normal modes corresponding to the non-vanishing elastic moduli of a cubic crystal. Interpretation of the observed anomalies has been done in the framework of relaxation in the system of Jahn-Teller (JT) complexes CrF_8 subject to full $T_{2g}\otimes\left(e_g+t_{2g}\right)$ JT problem. Relaxation time has been calculated from the experimental data on ultrasonic attenuation and adiabatic and isothermal contributions of the impurity subsystem to the total elastic moduli have been obtained.

1. Introduction

Recent study of II-VI and III-V crystals doped with 3*d* metals with degenerate orbital states proved the efficiency of ultrasonic experiment for investigating properties of the Jahn-Teller (JT) centers [1] (see, e.g., [2-5] and references therein). It provides independent and sometime unique information about electronic structure, dynamics and local symmetry of point defects in addition to obtained by traditional methods such as electron spin resonance, pulse techniques using electron spin echo, electron spin envelope modulation spectroscopy [6-10], and other methods [11-14]. In contrast with resonance methods, ultrasonic experiments deal with relaxation in the system of the JT impurities which occurs among equivalent JT distortions. The process of determination of relaxation contribution to total attenuation or elastic modulus should be done very accurately since theoretical description of JT effect (JTE) manifestation in an ultrasonic experiment relates only to the subsystem of the JT centers. However, the measured temperature dependence of attenuation $\alpha(T)$ is total attenuation which is represented as a sum of two components: the relaxation attenuation $\alpha_r(T)$ and the background one $\alpha_b(T)$. The last is assumed to be a smooth function (e.g., polynomial) in a wide temperature interval and coincides with $\alpha(T)$ at the ends of the interval. However, assumed form of $\alpha_b(T)$ is only approximation and nothing more which is desirable to be checked somehow.

One of the verification methods for the $\alpha_b(T)$ choice can be done with the help of the data on the temperature dependence of phase velocity v (or elastic modulus $c = \rho v^2$, where ρ is the density). It should be emphasized that the procedures of attenuation and phase velocity measurements are quite independent and are realized with different experimental methods. Moreover, the measurements of the phase velocity are almost independent of the bonding between the piezoelectric transducers and the studied specimen in contrast with the attenuation measurements and, therefore should be considered as more accurate ones. Thus, agreement of the results obtained with the help of these two independent techniques should serve as good confirmation of the $\alpha_b(T)$ choice.

The data on relaxation attenuation $\alpha_r(T) = \alpha(T) - \alpha_b(T)$ are used for determination of relaxation time τ. The dependence $\tau(T)$, in turn, provides information about the activation energy V_0 which is used for calculation the saddle points of the adiabatic potential energy surface (APES). Therefore, correctness of $\alpha_b(T)$ definition significantly influences on the final result.

At the same time, $\tau(T)$ obtained from data on attenuation can be used for calculation contribution of the impurity subsystem to the total dynamic elastic modulus c (dynamic means frequency dependent) and for construction the relaxed c^R and unrelaxed c^U versions of the modulus ($c^R = c_b + c_{JT}^T$ and $c^U = c_b + c_{JT}^S$, where c_b is the background modulus which represent the sum of all the subsystems except the JT one and assumed to be adiabatic modulus). Transformation from c^R to c^U takes place in the vicinity of the temperature T_1 which corresponds to the condition $\omega\tau(T_1) = 1$ (where ω is the cyclic frequency of the ultrasonic wave). So, the criteria of verification for the choice of $\alpha_b(T)$ and, consequently, criteria of correct $\tau(T)$ dependence should be: (i) shift of T_1 with frequency increase and (ii) asymptotic dependences of the dynamic moduli measured at difference frequencies at low temperature and high temperature (with respect to T_1) to c^R and c^U obtained at a certain frequency, since c^R and c^U are frequency independent versions of the modulus.

Temperature dependences of ultrasonic attenuation obtained for different normal modes in $SrF_2:Cr^{2+}$ were reported in [15]. The data on attenuation were used for calculation of relaxation time and evaluation of the extreme points of the APES for the full $T_{2g} \otimes (e_g + t_{2g})$ JT problem. Here we present new results on temperature dependences of the elastic moduli, prove the relaxation nature of the anomalies found at low temperatures and verify the choice of background attenuation using the method suggested above.

IOP Conf. Series: Journal of Physics: Conf. Series **833** (2017) 012003 doi:10.1088/1742-6596/833/1/012003

2. Experiment

2.1. Crystal properties

The single crystals of strontium difluorite were grown in Kazan E.K.Zavoisky Physical-Technical Institute of the RAS by Czochralski method in a helium atmosphere with addition of thermal decomposition products of Teflon. The chromium impurity was introduced into the melt as CrF_2 well-dried powder. A detailed study of the crystals [6] showed that the samples contain mostly centers of bivalent chromium.

Concentration of the impurities was determined in the Institute of Solid State Chemistry of the Ural Branch of the RAS by inductively coupled plasma mass spectrometer (ICP-MS) using quadrupole based instrument ELAN 9000 (Perkin-Elmer SCIEX) with standard operating and acquisition parameters. Chromium concentration proved to be $n_{Cr} = 1.6\times10^{19}$ cm^{-3} and significantly exceeded concentration of all the other measured impurities (Fe, Cu, Ni, Y).

The Cr^{2+} dopants in cubic crystals with fluorite structure replace the Sr^{2+} ions in the lattice site of O_h symmetry and are surrounded by eight fluorine ions in the corners of the cube. The ^5D term of a free Cr^{2+} ion is split by the cubic crystal field into the ground orbital triplet $^5T_{2g}$ and excited doublet ^5E_g. Due to the JTE, the cubic coordination of the Cr^{2+} ion in the degenerate state $^5T_{2g}$ is unstable with respect to twofold e_g and threefold t_{2g} type displacements of the environment. In linear approximation of the vibronic coupling, the dominant interaction with one of these two types of displacements results in the tetragonal or trigonal minima of the APES. Account of the quadratic coupling terms leads to orthorhombic minima involving both trigonal and tetragonal local modes. Investigation of ultrasonic attenuation [15] proved that this case takes place in $SrF_2:Cr^{2+}$.

2.2. Ultrasonic experiment description

Ultrasonic experiments were carried out at Dresden High Magnetic Field Laboratory and at Ural Federal University with the help of setups operating as a frequency variable bridge. The ultrasonic waves propagating along the $[110]$ axis were generated and registered by $LiNbO_3$ piezoelectric transducers at the frequencies of 55-162 MHz. There are no degenerate normal modes propagating along this crystallographic direction and all the non-vanishing elastic moduli of a cubic crystal c_{11}, c_{12} and c_{44} can be measured. Attenuation and phase velocity of the modes were determined by corresponding elastic modulus, taken as a complex variable c_β: $\alpha_\beta = (1/2)(\omega/v_\beta)\left[\mathrm{Im}\,(c_\beta)/\mathrm{Re}\,(c_\beta)\right]$, $v_\beta = \mathrm{Re}\sqrt{c_\beta/\rho}$, where index β denotes the type of the mode: one longitudinal $\beta = \ell$,and two shear waves: $\beta = T$ with polarization along $[111]$, $\beta = E$, $[1\bar{1}0]$. Accordingly, $c_\ell = (c_{11} + c_{12} + 2c_{44})$, $c_T = c_{44}$, $c_E = (c_{11} - c_{12})/2$. The $c_T = c_{44}$ modulus in the expression for c_β indicates that the wave produces trigonal distortions of the JT complex (i.e., distortions along the $\langle 111\rangle$ type axis), whereas c_E indicates trigonal distortions (along $\langle 001\rangle$). The expression for c_ℓ can be introduced in terms of symmetry moduli: $c_\ell = c_B + c_T + c_E/3$, where $c_B = (c_{11} + 2c_{12})/3$ is the bulk modulus describing the totally symmetric distortions. Thus, the local JTE e type mode manifests itself in attenuation and phase velocity of the shear wave polarized along the $\left[1\bar{1}0\right]$ axis and in the longitudinal wave, whereas the local t_2 mode indicates the anomalies in longitudinal and shear wave polarized along the $\left[001\right]$.

2.3. Ultrasonic attenuation and evaluation of relaxation time

Figure 1 shows temperature dependences of ultrasonic attenuation for the mode related to the c_{44} modulus. Similar dependences but with smaller peaks were obtained for other normal modes.

To calculate the temperature dependence of relaxation time using the data on attenuation, one should subtract background attenuation from the measured one and derive the relaxation attenuation: $\alpha_r(T)=\alpha(T)-\alpha_b(T)$. For the data measured at 55 MHz we used $\alpha_b(T)=1.73\times10^{-5}\times T^2+1.38\times10^{-8}\times T^4$. Relaxation time can be evaluated with the help of the following expression [2]:

$$\tau=\frac{1}{\omega}\left[\frac{\alpha_r(T_1)\cdot T_1}{\alpha_r(T)\cdot T}\pm\sqrt{\left(\frac{\alpha_r(T_1)\cdot T_1}{\alpha_r(T)\cdot T}\right)^2-1}\right]. \tag{1}$$

The signs before the square root should be chosen according to the statement that relaxation time is increased with temperature lowering. Relaxation time evaluated with the use of the data presented in figure 1 is given in figure 2.

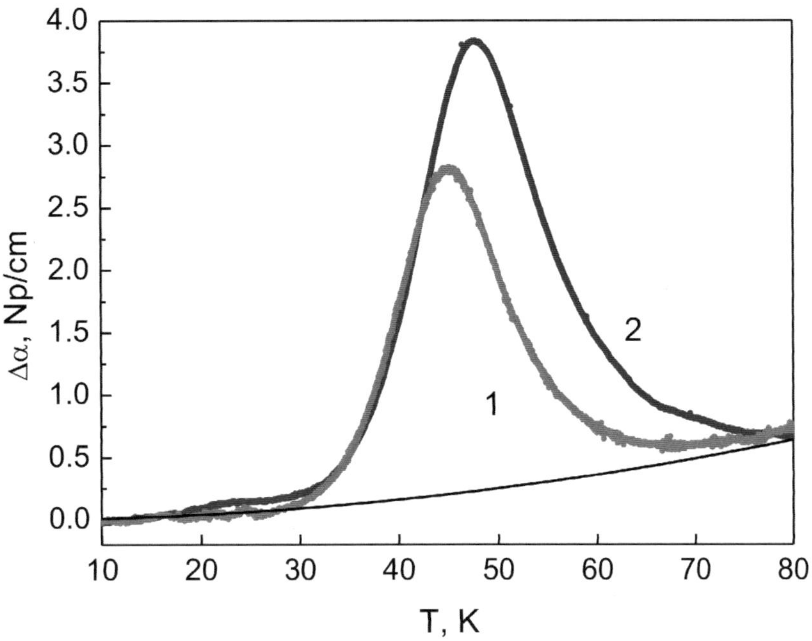

Figure 1. Temperature dependences of ultrasonic attenuation for normal mode related to the c_{44} modulus. Curve 1 was obtained at 56 MHz and curve 2 – at 105 MHz. $\Delta\alpha=\alpha(T)-\alpha(T_0)$, $T_0=4$ K is the reference temperature.

2.4. Dynamic, relaxed and unrelaxed moduli

Once we have relaxation time data, the relaxed and unrelaxed moduli can be evaluated [2]:

$$\frac{c^U - c_0}{c_0} = 2\frac{\Delta v}{v_0} + 2\frac{\alpha_r}{k_0}\frac{1}{\omega\tau}\,, \tag{2}$$

$$\frac{c^R - c_0}{c_0} = 2\frac{\Delta v}{v_0} - 2\frac{\alpha_r}{k_0}\omega\tau\,, \tag{3}$$

where $k_0 = \omega / v_0$ is the wave number, c_0 is dynamic modulus defined at $T_0 = 4\ \mathrm{K}$, $v_0 = v(T_0)$, $\Delta v = v(T) - v(T_0)$.

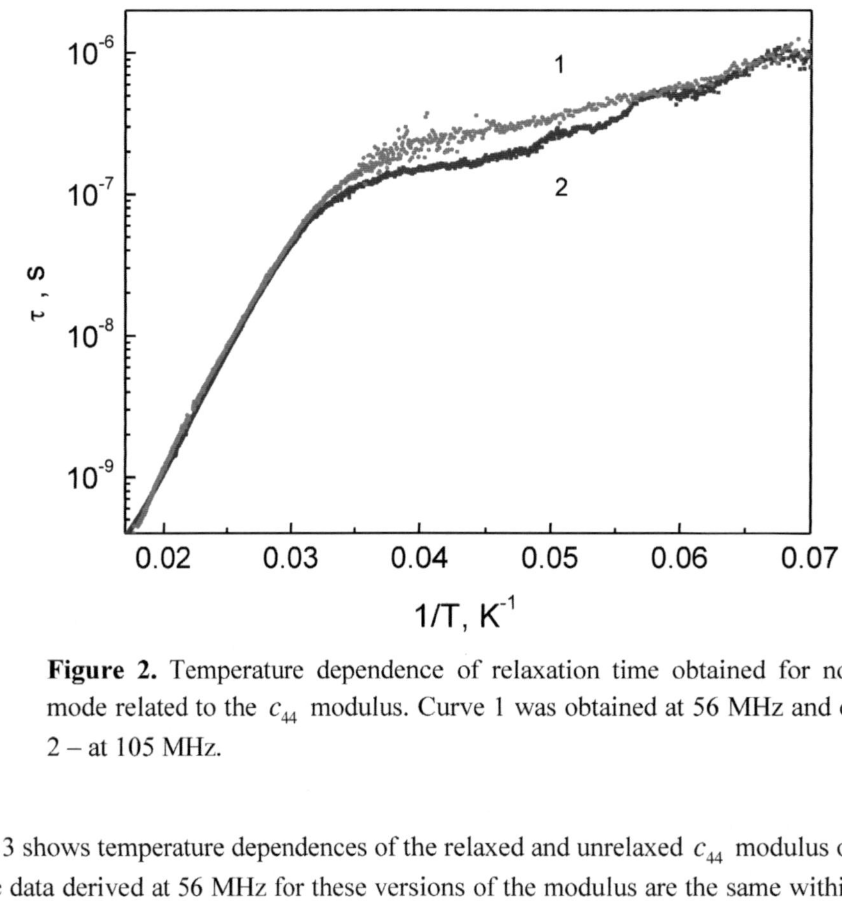

Figure 2. Temperature dependence of relaxation time obtained for normal mode related to the c_{44} modulus. Curve 1 was obtained at 56 MHz and curve 2 – at 105 MHz.

Figure 3 shows temperature dependences of the relaxed and unrelaxed c_{44} modulus obtained at 105 MHz. The data derived at 56 MHz for these versions of the modulus are the same within the accuracy of the measurements. Actually, it should be so, because c^R and c^U are frequency independent versions. In the contrast to them, dynamic modulus is frequency dependent and it manifests itself in transformation from one version to another at different temperatures for different frequencies of the wave.

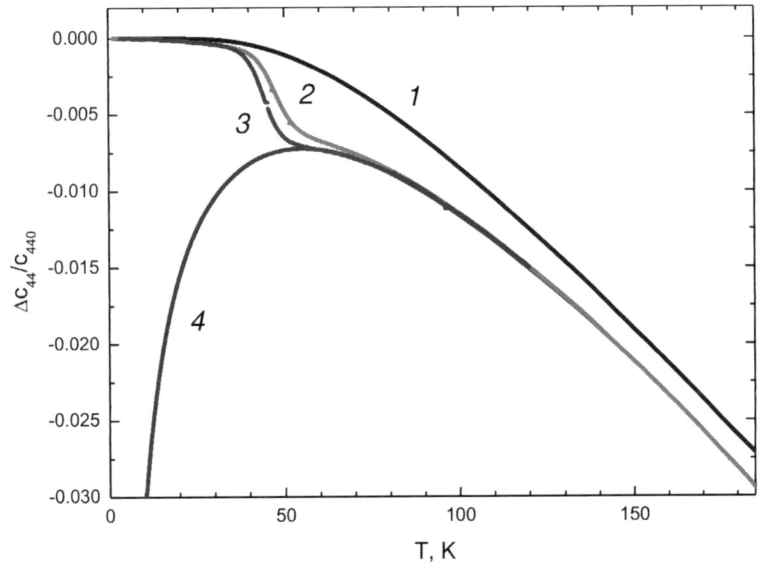

Figure 3. Temperature dependence of the dynamic modulus c_{44} (curve 2 – 105 MHz, 3 – 56 MHz), c_{44}^U (1 – 105 MHz), and c_{44}^R (4 – 105 MHz).

3. Conclusion

Attenuation and phase velocity in SrF_2:Cr^{2+} were measured at 56 – 162 MHz in the temperature range of 4 – 185 K. Anomalies were found for the normal modes related to the elastic moduli $c_\ell = (c_{11} + c_{12} + 2c_{44})$, $c_T = c_{44}$, $c_E = (c_{11} - c_{12})/2$. The anomalies were interpreted as relaxation between equivalent JT distortions of the CrF_8 complexes. Relaxation attenuation was derived by subtracting the background attenuation (assumed as smooth function which has small variation at low temperatures) from the measured attenuation. Relaxation time was evaluated using the data on relaxation attenuation which in good agreement proved to be the same for different frequencies. This fact justifies the correct choice of the background attenuation. Using the data on phase velocity and relaxation time, the relaxed and unrelaxed versions of the moduli were evaluated. Obtained for different frequencies, they proved to be the same within the accuracy of the measurements. Transformation of the dynamic moduli from relaxed to unrelaxed versions measured at different frequencies occurred at different temperatures. This fact indicates good accuracy of $\alpha(T)$ and $v(T)$ measurements and proves the relaxation origin of the observe anomalies in temperature dependences of both attenuation and elastic moduli.

Acknowledgments
The investigation was supported by the Russia Foundation for Basic Research (project No. 15-02-02750 a) and Ural Federal University Center of Excellence Radiation and Nuclear Technologies (Competitiveness Enhancement Program). We acknowledge the support from HLD at HZDR, member of the European Magnetic Field Laboratory (EMFL).

References

[1] Bersuker I B 2006 *The Jahn-Teller Effect* (Cambridge: Cambridge University Press)
[2] Gudkov V V 2009 Ultrasonic Consequences of the Jahn-Teller Effect *The Jahn-Teller Effect* eds H Koppel, D R Yarkony, H Barentzen (Heidelberg: Springer) pp 746-766
[3] Gudkov V V, Bersuker I B, Zhevstovskukh I B, Korostelin Yu V and Landman A I 2011 *J. Phys.: Condens. Matter* **23** 115401
[4] Gudkov V V and Bersuker I B (2012) Experimental Evaluation of the Jahn-Teller Effect Parameters by Means of Ultrasonic Measurements. Application to Impurity Centers in Crystals *Vibronic Interaction and the Jahn-Teller Effect* eds M Atanasov, C Daul, P L Treggenna-Piggot (Heidelberg: Springer) pp 143-161
[5] Averkiev N S, Bersuker I B, Gudkov V V, Baryshnikov K A, Zhevstovskikh I V, Mayakin V Yu, Monakhov A M, Sarychev M N, Sedov V E and Surikov V T (2014) *J. Appl. Phys.* **116** 103708
[6] Zaripov M M, Tarasov V F, Ulanov V A, Shakurov G S and Popov M L 1995 *Fiz.Tverd. Tela* **37** 806
[7] Oliete P B, Orera V M, and Alonso P J 1996 *Phys. Rev. B* **53** 3047
[8] Oliete P B, Orera V M, and Alonso P J 1996 *Phys. Rev. B* **54** 12099
[9] Hoffmann S K, Goslar J, and Lijewski S 2007 *J. Chem. Phys.* **127** 124705
[10] Azamat D V, Dejneka A, Lancok J, Trepakov V A, Jastrabik L, and Badalyan A G 2012 *J. Appl. Phys.* **111** 104119
[11] Oliveira G N P, Teixeira R, Mendonca T M, Silva M R, Correia J G, Lopes A M L and Araujo J P 2014 *J. Appl. Phys.* **116** 2239907
[12] Yadav R, Nair H S, Kumar A, Adiga S, Bhat H L, Yusuf S M and Elizabeth S 2015 *J. Appl. Phys.* **117** 093903
[13] Senanayake P S, Wells J-P R, Reid M F, Hughes-Currie R B, Berden G, Reeves R J and Meijerink A 2015 *J. Appl. Phys.* **117** 133109
[14] Krivobok V S, Nikolaev S N, Bagaev V S, Pruchkina A A, Onishchenko E E, Kolosov S A, Klevkov Yu V and Skorikov M L 2016 *J. Appl. Phys.* **119** 055704
[15] Zhevstovskikh I V, Bersuker I B, Gudkov V V, Averkiev N S, Sarychev M N, Zherlitsyn S, Yasin Sh,Shakurov G S, Ulanov V A and Surikov V T (2016) *J. Appl. Phys.* **119** 225108

Beyond Born-Oppenheimer treatment for the construction of triple-sheeted accurate diabatic Hamiltonian matrix of F+H_2 system

Bijit Mukherjee[1], **Saikat Mukherjee**[1], **K. R. Shamasundar**[2] **and Satrajit Adhikari**[1*]

[1]Department of Physical Chemistry
Indian Association for the Cultivation of Science
Jadavpur, Kolkata -700 032, INDIA
[2]Department of Chemistry
Indian Institute of Science Education and Research, Mohali
Sector 81, SAS Nagar 140306, India

E-mail: pcsa@iacs.res.in

Abstract. We intend to study the non-adiabatic interactions among the three lowest adiabatic states ($1^2A'$, $1^2A''$, and $2^2A'$) of F+H_2 triatomic reactive system in hyperspherical coordinates for a fixed hyperradius at $\rho = 7.5$ bohr as functions of hyperangles, θ ($0° \leq \theta \leq 90°$) and ϕ ($0° \leq \phi \leq 360°$). The adiabatic potential energy surfaces are calculated using MRCI level of methodology whereas the non-adiabatic coupling terms between those states are calculated from the analytic gradient methods implemented in MOLPRO quantum chemistry package. The ground ($1^2A'$) and the first excited ($1^2A''$) states exhibit conical intersection (CI) and seam of CI along C_{2v} geometries, whereas the first ($1^2A''$) and the second ($2^2A'$) excited states undergo Renner-Teller coupling at linear geometries. We carry out adiabatic-to-diabatic transformation (ADT) by solving ADT equations to obtain ADT angles for constructing single-valued, continuous and symmetric 3×3 diabatic potential energy matrix so that subsequent accurate scattering calculations can be performed.

1. Introduction

The elementary chemical reaction F+H_2 has been for decades of outstanding interest both from experimental[1–13] and theoretical[14–35] aspects. Though the system is limited to three atoms, yet it presents great challenges in obtaining a full quantum description of the reaction, but equally attracts rich experimental probes since this reaction is highly exoergic ($\sim$32 kcal mol^{-1}), the energy source for powerful chemical lasers and exhibits interesting quantum mechanical resonances. Early experiments date back to the end of sixties in the last century mainly initiated by studies using chemical lasers[1–3] and chemiluminescence techniques[4,5] followed by high resolution molecular beam experiments by Y. T. Lee and his group[6,7] which provided the first vibrationally resolved differential cross sections for its various isotopic variants. Vib-rotational resolved differential cross section experiments[8] and calculations[9] by Toennies *et al.* also confirmed previous observations. With the advancement of experimental sophistication, higher resolution molecular cross beam studies were carried out during the last decade to investigate energy and angular dependence of FH and FD products[10,11] and reaction resonances of this benchmark system.[12,13]

From the theoretical point of view, $F+H_2$ traces a long history of *ab initio* studies with attempts to develop accurate potential energy surfaces (PESs). For more than four decades until now, substantial number of PESs have appeared in the literature. H. F. Schaefer III group was one of the few who came up with the first *ab initio* PESs[14,15] (BOPS) in 1970s followed by empirical[16,17] and semiempirical[18,19] PESs of London-Eyring-Polanyi-Sato (LEPS) type given by Muckerman (M5)[16], Takayanagi and Sato (TS)[17] and Truhlar group (T5, T5A).[18,19] Subsequent renewed investigations by Truhlar and co-workers on the *ab initio* treatments of transition state for the title reaction produced modified PESs called SEC5[20] and SEC6.[21] The global *ab initio* PES constructed by Stark and Werner (SW)[22] could able to draw sophisticated theoretical analysis of the existing high resolution experimental works unlike the previous PESs that failed to account multiple experimental observations. In subsequent works, Werner *et al.* constructed a series of PESs[23–25] by taking successive corrections in terms of spin-orbit effect of the same system. On the basis of UCCSD(T) (spin unrestricted coupled cluster method with all single and double excitations and perturbative accounts of triple excitations) calculations, D. H. Zhang group gave another family of global PESs[26,27] that provide good estimates of exothermicity ($\sim$32 kcal mol^{-1}), resonance states in reactant channels[9,26] and reproduce experimental observations like reaction rate constant, branching ratios of the reaction $F+HD/H_2$, forward/backward scattering and peaks due to partial wave resonances of reaction F+HD.[27] For a complete overview and more detailed history of PESs constructed for this system, the authors request the readers to go through the reference section and references therein.

Owing to such wealth of theoretical investigation, it can be inferred that the reaction $F+H_2 \rightarrow HF+H$ possesses substantial difficulties for multiple reasons. Firstly, the strong correlation of the eleven electrons that makes the *ab initio* results highly sensitive to the choice on the level of electronic structure calculations and basis sets.[22] Secondly, the spin-orbit coupling effect due to the open-shell F atom, making this particular system one of the prototypes for its study.[23–28] Lastly, rich non-adiabatic interactions that are present in the system, the investigation of which is going on at present with great interest. Werner *et al.* calculated global PESs for the ground and excited states[24,25] based on which multisurface scattering calculations were performed.[29,30] In their work,[25] they concluded about the necessity of including three PESs for a complete description of the reaction. More recently, Das *et al.* have carried out thorough investigation in understanding the role of Jahn-Teller (JT) and Renner-Teller (RT) conical intersections (CIs) in this reaction.[31–35] While doing so, they have calculated a set of two diabatic PESs based on *ab initio* calculated non-adiabatic coupling terms (NACTs). It was shown that the JT CI is formed by a Σ-state, assigned as $1A'$ and one of the two Π-states (of the same symmetry) assigned as $2A'$, whereas the RT CIs are formed by the corresponding two degenerate Π-states, designated as $2A'$ and $1A''$ along the collinear HHF axis.[33] Thus, it is necessary to go beyond Born-Oppenheimer (BO) picture[36,37] to encapsulate the non-adiabatic interactions among the various electronic states that are present within the three state sub-Hilbert Space.

One has to take into account the NACTs for performing accurate multi-surface scattering calculations. NACTs along with adiabatic PESs are the two most important outcomes of the BO treatment. NACTs dictate the contributions from the excited electronic states on a molecular process even if it is occurring on the ground state. Since these coupling terms are often blown up in the neighborhood of a CI in the nuclear configuration space (CS), it is convenient to transform the kinetically coupled adiabatic Schrödinger equation (SE) to the *diabatic* representation where potential coupling elements appear as smooth functions of nuclear coordinates. This *adiabatic-to-diabatic* transformation requires two necessary conditions to be fulfilled. Firstly, the vector fields created by the NACTs have to satisfy the so called Curl Conditions.[38,39] Secondly, the non-removable components of the NACTs are approximately zero.[40] Baer[41] and his co-workers[42] have carried out detailed beyond BO treatment for two state sub-Hilbert space in which they have formulated diabatic potential energy matrix in terms of adiabatic PESs and adiabatic to diabatic transformation (ADT) angles.[43–46] Adhikari *et al.* have generalized the BO treatment for any three/four coupled electronic states, where the explicit form of NACTs and Curl- Divergence equation are derived in order to formulate diabatic Hamiltonian matrix.[47–53]

We carry out beyond BO treatment for the lowest three states ($1^2A'$, $1^2A''$, and $2^2A'$), which form a sub-Hilbert space, by calculating *ab initio* adiabatic PESs and NACTs, discuss about the nature of JT CI and RT interaction, obtain ADT angles by plugging those NACTs in the ADT equations and thereby, construct smooth, continuous and symmetric diabatic PESs at a fixed ρ as a function of θ and ϕ in hyperspherical coordinates.

2. Theory

A three-state electronic sub-Hilbert space has been considered ($1^2A'$, $1^2A''$, $2^2A'$), which forms a complete space for the present collision process, F+H_2. The CIs are present among these three states in the nuclear CS. The BO expansion[36,37] of the molecular wavefunction for this sub-Hilbert space and the total electron-nuclei Hamiltonian in the adiabatic representation are:

$$\begin{aligned}
|\Psi(s_e, s_n)\rangle &= \sum_{i=1}^{3} \psi_i(s_n)|\xi_i(s_e|s_n)\rangle, \\
\hat{\mathbf{H}} &= \hat{\mathbf{T}}_{s_n} + \hat{\mathbf{H}}_e(s_e|s_n), \\
\hat{\mathbf{T}}_{s_n} &= -\frac{\hbar}{2}\sum_i \left(\frac{\nabla^2_{s_n,i}}{m_i}\right), \\
\hat{\mathbf{H}}_e(s_e|s_n)|\xi_i(s_e|s_n)\rangle &= u_i(s_n)|\xi_i(s_e|s_n)\rangle,
\end{aligned} \tag{1}$$

where the eigenfunctions $[|\xi_i(s_e|s_n)\rangle]$ of the electronic Hamiltonian, $\hat{\mathbf{H}}_e(s_e|s_n)$, are defined by the sets of nuclear (s_n) and electronic (s_e) coordinates while the eigenvalues, $u_i(s_n)$ depend on nuclear coordinates. $\hat{\mathbf{T}}_{s_n}$ is the nuclear kinetic energy (KE) operator and the expansion coefficient, $\psi_i(s_n)$, are the nuclear wavefunctions. Using the BO expanded molecular wavefunction [Eq. (1)] for the given sub-Hilbert space, we can reduce the total molecular time independent SE into the following compact form of kinetically coupled nuclear equation:

$$-\frac{\hbar}{2m}\left(\vec{\nabla} + \vec{\tau}\right)^2 \psi + (\mathbf{u} - E)\,\psi = 0, \tag{2}$$

where the non-adiabatic coupling and adiabatic PES matrices have the following form:

$$\vec{\tau} = \begin{pmatrix} 0 & \vec{\tau}_{12} & \vec{\tau}_{13} \\ -\vec{\tau}_{12} & 0 & \vec{\tau}_{23} \\ -\vec{\tau}_{13} & -\vec{\tau}_{23} & 0 \end{pmatrix} \quad \text{and} \quad \mathbf{u} = \begin{pmatrix} u_1 & 0 & 0 \\ 0 & u_2 & 0 \\ 0 & 0 & u_3 \end{pmatrix}, \tag{3}$$

respectively. The non-adiabatic coupling matrix elements have the following definition:

$$\vec{\tau}_{ij} = \langle \xi_i(s_e|s_n)|\vec{\nabla}\xi_j(s_e|s_n)\rangle. \tag{4}$$

The validity of the three state sub-Hilbert space make it possible to transform the adiabatic nuclear SE [Eq. (2)] to the diabatic one through an orthogonal transformation, $\psi = \mathbf{A}\psi^d$. The required SE in diabatic representation becomes:

$$-\frac{\hbar}{2m}\boldsymbol{\nabla}^2\psi^d + (\mathbf{W} - E)\psi^d = 0, \qquad \mathbf{W} = \mathbf{A}^\dagger\mathbf{u}\mathbf{A} \tag{5}$$

IOP Conf. Series: Journal of Physics: Conf. Series **833** (2017) 012004 doi:10.1088/1742-6596/833/1/012004

under the condition:

$$\vec{\nabla}\mathbf{A} + \vec{\tau}\mathbf{A} = 0. \tag{6}$$

This equation is known as the ADT condition.[41] The model form of (3×3) **A** has to be an orthogonal matrix and the orthonormality conditions demand the fulfillment of six relations involving three independent variables, which are Euler like angles of rotation [$\Theta_{12}(s_n)$, $\Theta_{23}(s_n)$ and $\Theta_{13}(s_n)$], commonly called ADT angles. These are the natural requirement to construct the three-state **A** matrix for which product of three (3×3) rotation matrices[43], $\mathbf{A}_{12}(\Theta_{12})$, $\mathbf{A}_{23}(\Theta_{23})$, and $\mathbf{A}_{13}(\Theta_{13})$ are taken as shown below:

$$\mathbf{A}(\Theta_{12},\Theta_{23},\Theta_{13}) = \mathbf{A}_{12}(\Theta_{12})\cdot\mathbf{A}_{23}(\Theta_{23})\cdot\mathbf{A}_{13}(\Theta_{13}). \tag{7}$$

The above model form of **A** matrix [Eq. (7)] and the anti-symmetric form of $\vec{\tau}$ matrix [Eq. (3)] are substituted in Eq. (6), where simple manipulations as performed by Baer *et al.*[43] lead to the following differential equations for ADT angles:

$$\vec{\nabla}\Theta_{12} = -\vec{\tau}^{12} + \tan\Theta_{23}(\vec{\tau}^{13}\cos\Theta_{12} - \vec{\tau}^{23}\sin\Theta_{12}), \tag{8a}$$
$$\vec{\nabla}\Theta_{23} = -(\vec{\tau}^{13}\sin\Theta_{12} + \vec{\tau}^{23}\cos\Theta_{12}), \tag{8b}$$
$$\vec{\nabla}\Theta_{13} = -\frac{1}{\cos\Theta_{23}}(\vec{\tau}^{13}\cos\Theta_{12} - \vec{\tau}^{23}\sin\Theta_{12}), \tag{8c}$$

which in turn brings the explicit form[47] of $\vec{\tau}$ matrix elements in terms of the ADT angles:

$$\vec{\tau}_{12} = -\vec{\nabla}\Theta_{12} - \sin\Theta_{23}\vec{\nabla}\Theta_{13}, \tag{9a}$$
$$\vec{\tau}_{23} = \sin\Theta_{12}\cos\Theta_{23}\vec{\nabla}\Theta_{13} - \cos\Theta_{12}\vec{\nabla}\Theta_{23}, \tag{9b}$$
$$\vec{\tau}_{13} = -\cos\Theta_{12}\cos\Theta_{23}\vec{\nabla}\Theta_{13} - \sin\Theta_{12}\vec{\nabla}\Theta_{23}. \tag{9c}$$

The components of *ab initio* calculated NACTs are substituted in Eq. (8) to obtain the ADT angles which along with the adiabatic PESs, are used to calculate the diabatic potential matrix elements as given in Eq. (5). The explicit expressions of **W** matrix elements are:

$$\begin{aligned}
W_{11} &= u_1(\cos\Theta_{12}\cos\Theta_{13} - \sin\Theta_{12}\sin\Theta_{13}\sin\Theta_{23})^2 \\
&+ u_2(\sin\Theta_{12}\cos\Theta_{13} + \cos\Theta_{12}\sin\Theta_{13}\sin\Theta_{23})^2 + u_3\sin^2\Theta_{13}\cos^2\Theta_{23}, && (10a)\\
W_{22} &= u_1\sin^2\Theta_{12}\cos^2\Theta_{23} + u_2\cos^2\Theta_{12}\cos^2\Theta_{23} + u_3\sin^2\Theta_{23}, && (10b)\\
W_{33} &= u_1(\cos\Theta_{12}\sin\Theta_{13} + \cos\Theta_{13}\sin\Theta_{12}\sin\Theta_{23})^2 \\
&+ u_2(\sin\Theta_{12}\sin\Theta_{13} - \cos\Theta_{12}\cos\Theta_{13}\sin\Theta_{23})^2 + u_3\cos^2\Theta_{13}\cos^2\Theta_{23}, && (10c)\\
W_{12} &= u_1\sin\Theta_{12}\cos\Theta_{23}(\cos\Theta_{12}\cos\Theta_{13} - \sin\Theta_{12}\sin\Theta_{13}\sin\Theta_{23}) \\
&- u_2\cos\Theta_{12}\cos\Theta_{23}(\sin\Theta_{12}\cos\Theta_{13} + \cos\Theta_{12}\sin\Theta_{13}\sin\Theta_{23}) \\
&+ u_3\sin\Theta_{13}\sin\Theta_{23}\cos\Theta_{23}, && (10d)\\
W_{13} &= u_1(\cos\Theta_{12}\cos\Theta_{13} - \sin\Theta_{12}\sin\Theta_{13}\sin\Theta_{23})(\cos\Theta_{12}\sin\Theta_{13} + \sin\Theta_{12}\cos\Theta_{13}\sin\Theta_{23}) \\
&+ u_2(\sin\Theta_{12}\cos\Theta_{13} + \cos\Theta_{12}\sin\Theta_{13}\sin\Theta_{23})(\sin\Theta_{12}\sin\Theta_{13} - \cos\Theta_{12}\cos\Theta_{13}\sin\Theta_{23}) \\
&- u_3\sin\Theta_{13}\cos\Theta_{13}\cos^2\Theta_{23}, && (10e)\\
W_{23} &= u_1\sin\Theta_{12}\cos\Theta_{23}(\cos\Theta_{12}\sin\Theta_{13} + \sin\Theta_{12}\cos\Theta_{13}\sin\Theta_{23}) \\
&+ u_2\cos\Theta_{12}\cos\Theta_{23}(-\sin\Theta_{12}\sin\Theta_{13} + \cos\Theta_{12}\cos\Theta_{13}\sin\Theta_{23}) \\
&- u_3\cos\Theta_{13}\sin\Theta_{23}\cos\Theta_{23}. && (10f)
\end{aligned}$$

3. *Ab initio* calculation for PESs and NACTs

We have employed hyperspherical coordinates to map the triatomic potential energy surface of FH_2 in order to represent all the reaction arrangement channels evenhandedly. The hyperradius ρ solely determines the overall size of the traingle, whereas the two hyperangles θ ($0 \leq \theta \leq 90°$) and ϕ ($0 \leq \phi \leq 360°$) corresponding to spherical latitude and longitude, respectively, describe the shape of the triangle. All the *ab initio* calculations are carried out in terms of the interparticle distances of the three nuclei that are related to the hyperspherical coordinates[54] as:

$$R_1 = \frac{\rho}{\sqrt{2}} d_3 \left[1 + \sin\theta \cos(\phi + \epsilon_3)\right]^{1/2}, \quad (11a)$$

$$R_2 = \frac{\rho}{\sqrt{2}} d_1 \left(1 + \sin\theta \cos\phi\right)^{1/2}, \quad (11b)$$

$$R_3 = \frac{\rho}{\sqrt{2}} d_2 \left[1 + \sin\theta \cos(\phi - \epsilon_2)\right]^{1/2}, \quad (11c)$$

where $d_i = \sqrt{m_i(m_j + m_k)/\mu M}$, $M = \sum_i^3 m_i$, $\mu = \sqrt{(m_1 m_2 m_3/M)}$, $\epsilon_2 = 2\tan^{-1}(m_3/\mu)$ and $\epsilon_3 = 2\tan^{-1}(m_2/\mu)$. For F+$H_2$, $m_1 = m_F = 19$ amu, $m_2 = m_3 = m_H = 1.008$ amu, $M = 21.016$ amu, $\mu = 0.958$ amu, $d_1 = 1.379$, $d_2 = d_3 = 1$ and $\epsilon_2 = \epsilon_3 = 92.91°$. R_1 and R_3 are the bond distances between the F atom and the two H atoms whereas R_2 is between the two H atoms. At $\phi = 0°$ and 180°, the two F-H bonds are equivalent and the geometries attain C_{2v} symmetry for all $\theta's$. For, $\theta = 0°$, the triangle remains isosceles since it again goes to C_{2v} symmetry and the shape is constant for all $\phi's$, whereas $\theta = 90°$ corresponds to the collinear arrangement of the triatom. At the collinear geometries, the arrangement is HHF for $87.1° < \phi < 272.9°$, while for other values of ϕ, the arrangement is HFH.

We have calculated the three lowest adiabatic PESs ($1^2A'$, $1^2A''$, $2^2A'$) of the system and the NACTs between those states by MOLPRO quantum chemistry package[55] as functions of hyperangles, θ ($0 \leq \theta \leq 90°$) and ϕ ($0 \leq \phi \leq 360°$) for a fixed value of hyperradius $\rho = 7.5$ bohr. The *ab initio* data points for those three PESs are calculated in C_1 symmetry using Dunning's correlation consistent quintuple zeta ($cc - pV5Z$) basis set. The MCSCF calculations are performed involving the first three (3) electronic states firstly with nine (9) electrons being distributed over ten (10) active orbitals followed by same level of calculations using those orbitals with an active space of seven (7) electrons distributed over eight (8) active orbitals. Then, these orbitals are used as reference for internally contracted MRCI calculations with the same active space (7 elctrons in 8 active orbitals) for the three (3) electronic states. The three adiabatic states correlate to $^2\Sigma^+$ and $^2\Pi$ states in the collinear arrangements. The $^2\Sigma^+$ and one of the $^2\Pi$ states at collinear geometry form a symmetry allowed accidental CI at jacobi coordinates $R = 5.45$ bohr and $r = 1.4$ bohr, where r is the distance between the two H atoms and R being the distance between the center of mass of H_2 molecule and F atom. This is in agreement

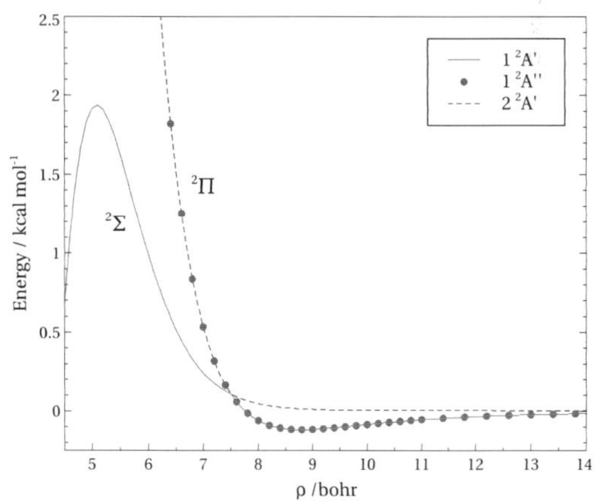

Figure 1. The three adiabatic potentials: $1^2A', 1^2A''$ and $2^2A'$ of F+H_2 system in the collinear arrangement (with $r = 1.4$ bohr) in hyperspherical coordinates. The CI between the $^2\Sigma$ and $^2\Pi$ states occurs at $\rho = 7.5$ bohr ($\theta = 90°$ and $\phi = 164.1°$).

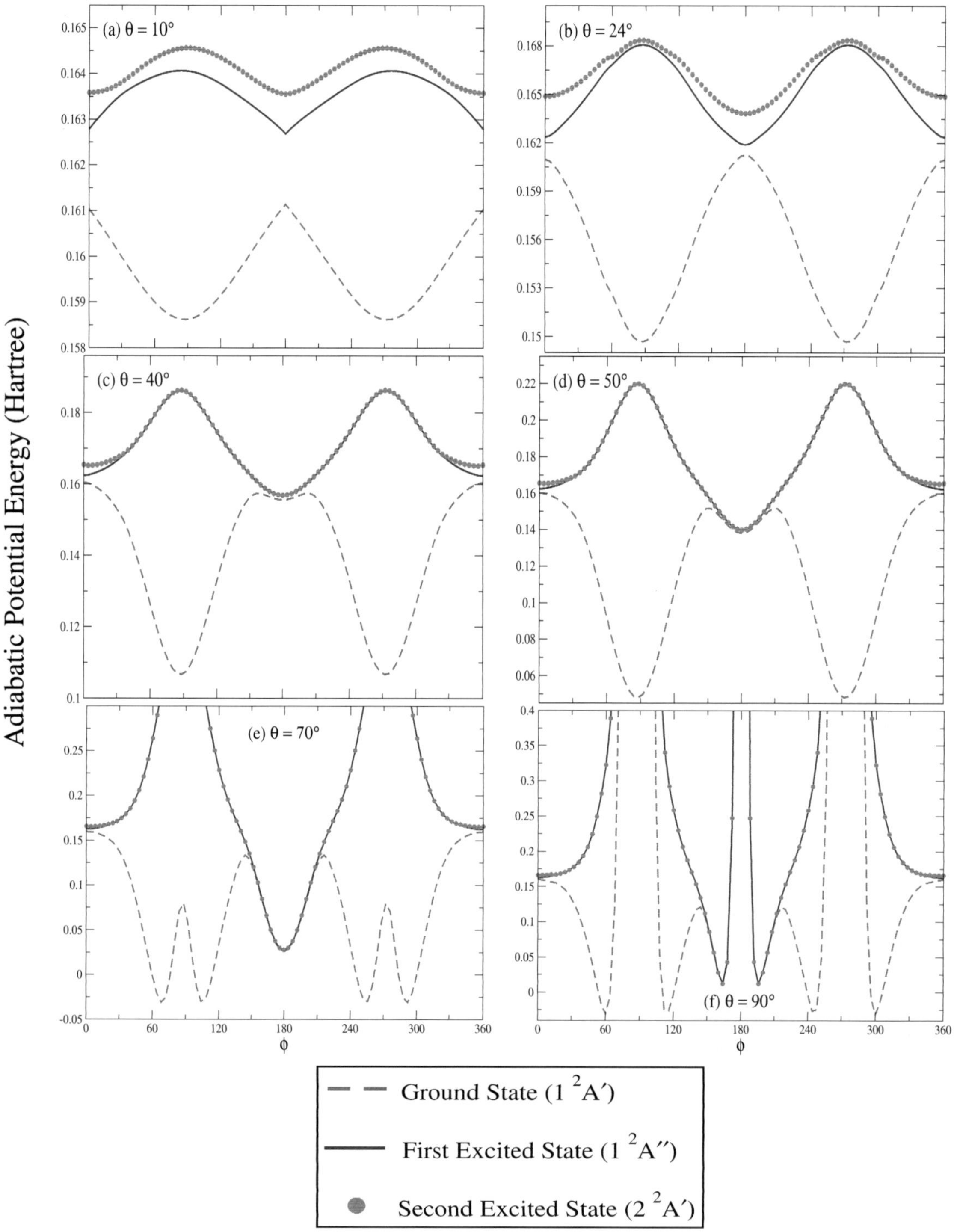

Figure 2. 1D cuts of three adiabatic singlet states as function of ϕ for different θs: (a) 10°, (b) 24°, (c) 40°, (d) 50°, (e) 70° and (f) 90°. The dashed red line represents the ground state ($1^2A'$), the blue solid line depicts the first excited state ($1^2A''$) and the megenda circles represent the second excited state ($2^2A'$). All the energies are in hartree scaled with respect to the asymptote. At around $\theta = 24°$ and $\phi = 180°$, the lowest two states interact strongly while at $\theta = 90°$, the upper two states exhibit RT interactions.

with the findings of Stark and Werner in their article on SW PES.[22] To map the result into the hyperspherical coordinates, we varied ρ and ϕ keeping θ at 90° such that the particles remain at collinear configurations. Figure 1 depicts collinear arrangements along the hyperradius ρ, where the location of CI is at $\rho = 7.5$ bohr, $\phi = 164.4°$ and $\theta = 90°$ (corresponding to $R = 5.45$ bohr and $r = 1.4$ bohr). At $\rho > 7.5$ bohr, the $^2\Pi$ state is lower than that of $^2\Sigma$ state, where $\rho < 7.5$ bohr, the two states cross each other. This is the main reason we perform calculations at $\rho = 7.5$ bohr in order to scan the nature of CIs in $\theta - \phi$ space at non- linear geometries through our study of NACTs. Calculations in $\theta - \phi$ space at other values of ρ are going on so that a set of global multisheeted diabatic PESs can be constructed.

Table 1. Comparison of the ground state energies of F+H_2 system in different geometries at $\rho = 7.5$. All the energies are in hartree units, scaled with respect to the asymptote.

θ	ϕ	Present work	SW
0°	0°	0.1723	0.1723
45°	60°	0.1064	0.1073
45°	120°	0.1182	0.1178
60°	60°	0.0320	0.0333
60°	120°	0.0599	0.0612
75°	60°	−0.0362	−0.0361
75°	120°	0.0014	0.0018
89°	60°	−0.0505	−0.0495
89°	120°	−0.0214	−0.0210

The ground state ($1^2A'$) energy of F+H_2 appears as 0.1722978 hartree at $\theta = 0°$ in which the two F-H bonds are equivalent having values 5.31 bohr and the H-H bond is 7.31 bohr leading to C_{2v} geometries for all values of ϕ. At the same geometries, the first and the second excited states have energies 0.1722928 and 0.1751432 hartree, respectively. The 1D potential energy curves of the three adiabatic states as function of ϕ for various θs are presented in Figure 2. We compare our ground state ($1^2A'$) PES of F+H_2 at $\rho = 7.5$ bohr, which has been calculated without the inclusion of spin-orbit coupling, with that of the SW PES for few geometries in Table 1. In Table 2 we compare the obtained collinear barrier with earlier calculations.

Table 2. Comparison of the calculated collinear barriers (in KCal mol^{-1}) with earlier calculations.

	Collinear barrier
This work	1.824
SW[22]	1.921
ASW[24]	1.971
FXZ[26]	2.222
SEC5[20]	1.570

We have employed State-Averaged MCSCF (SA-MCSCF) analytic gradient method implemented in MOLPRO to calculate the NACTs ($\tau_k^{ij}[k = \theta, \phi], i = j = 1, 2, 3, i < j, i \neq j$)

among the three states. The strong non-adiabatic interactions between the ground ($1^2A'$) and the first excited ($1^2A''$) states lead to one CI seam at $\phi = 0°$ for θ starting from close to 10° to all along 90°, and one CI at region $\phi = 180°$ and $\theta \approx 25°$. The former one at C_{2v} geometries is termed as C_{2v} seam. In this case, with increasing values of θ from 10° to 90°, the F-H_2 perpendicular distance shortens and finally reduces to zero at $\theta = 90°$, which is a collinear arrangement. The latter CI also arises at C_{2v} geometry but located around a specific point ($\phi = 180°$ and $\theta \approx 25°$), where the F-H distances become $\approx$ 2.8 bohr and H-H distance is $\approx$ 3.1 bohr.

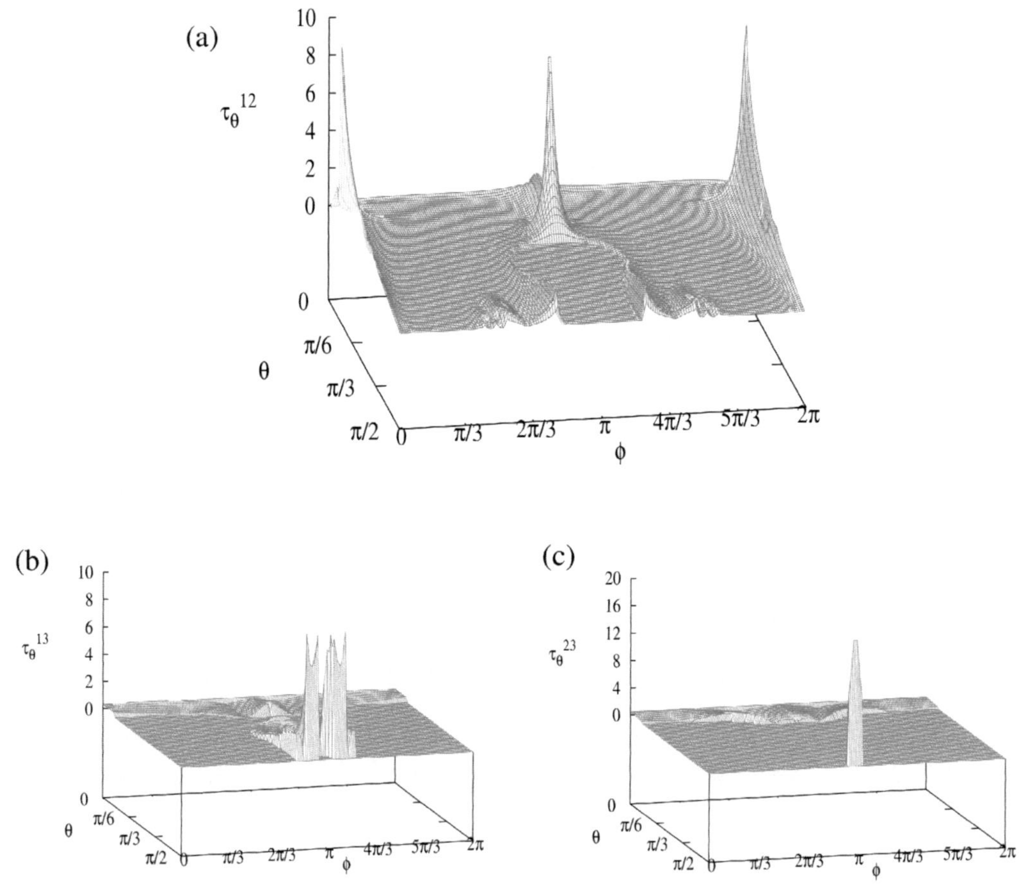

Figure 3. Functional forms of the τ_θ^{ij} as plotted in terms of θ and ϕ for $\rho = 7.5$ bohr. (a) τ_θ^{12}, representing the singularities at C_{2v} seams. (b) and (c) are τ_θ^{13} and τ_θ^{23}, depicting the PJT and RT interactions, respectively.

In addition, the first ($1^2A''$) and the second excited state ($2^2A'$) exhibit RT interactions which are evident at linear geometries for $\theta = 90°$ and ϕ close to 180°. Moreover, the ground state and the second excited state undergo pseudo Jahn-Teller (PJT) interaction at the same geometries ($\theta = 90°$ and ϕ close to 180°). Figures 3 and 4 display the θ and ϕ components of the NACTs, respectively, for each pair of states that enable us to figure out the JT-type and RT CIs as well as PJT interactions.

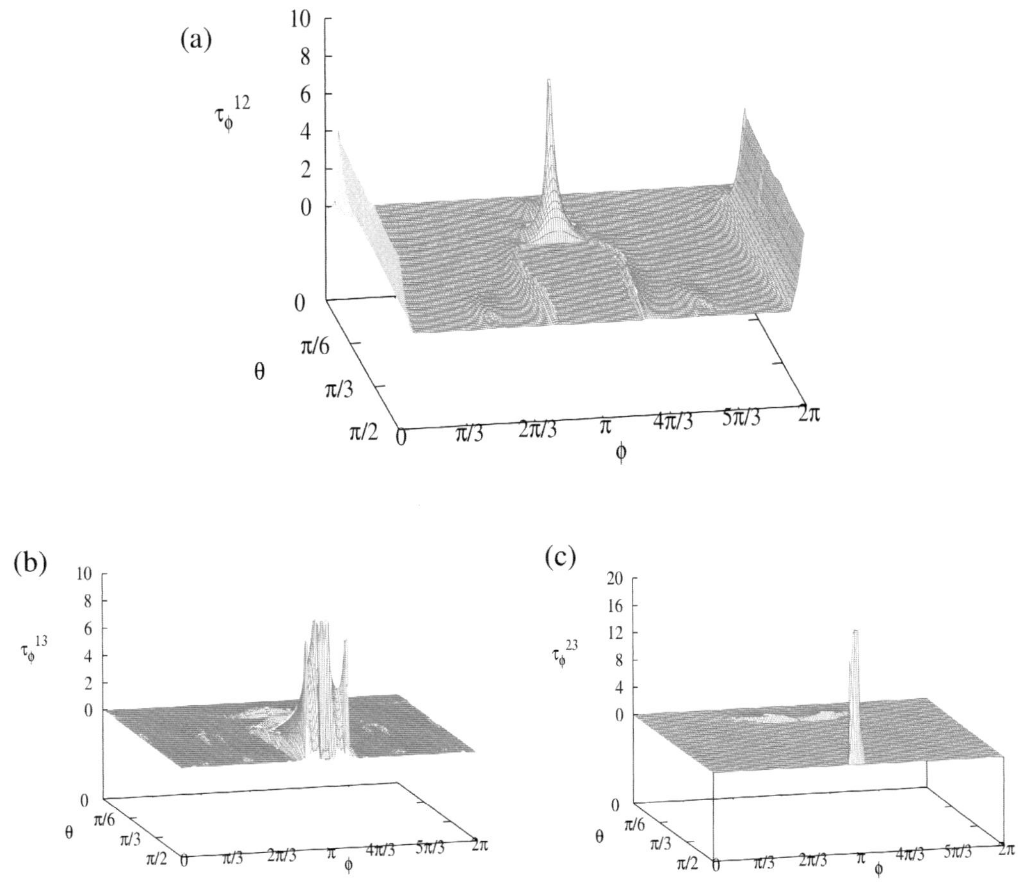

Figure 4. Functional forms of the τ_ϕ^{ij} as functions of θ and ϕ for ρ = 7.5 bohr. The ϕ components of NACT have similar features as that of θ component. (a) τ_ϕ^{12}, represents the C_{2v} singularities. (b) and (c), τ_ϕ^{13} and τ_ϕ^{23}, depict the PJT and RT interactions, respectively.

4. ADT and diabatic potential matrix

We substitute the *ab initio* calculated NACTs in Eq.(8) and integrate those differential equations by Backward Differentiation Formula over a 2D grid of geometries represented by 46 × 180 grid points in hyperspherical coordinates θ and ϕ. It is established that the solutions for the two sets of coupled differential equations (for θ and ϕ) are not unique and depend on the contour of integration.[51] Each contour will generate different sets of $\theta - \phi$ dependent ADT angles ($\Theta_{12}, \Theta_{13}, \Theta_{23}$). The various sets of ADT angles thus obtained due to different choices of paths show gauge invariance and the different ADT matrices appear related through orthogonal transformation.[51] In the present study, we consider a rectangular path, where the differential equations for the θ grid are integrated with positive increment from $\theta = 0°$ to $90°$ for each positive step integration of the differential equations for the ϕ grid from $0°$ to $360°$. Figure 5 displays the functional form of the calculated ADT angles for the ground-first excited state (Θ_{12}), ground-second excited states (Θ_{13}) and the first-second excited states(Θ_{23}). The ADT angle, Θ_{12} attains the numerical value of π at $\phi = 0°/360°$ establishing the JT-type CI seam at

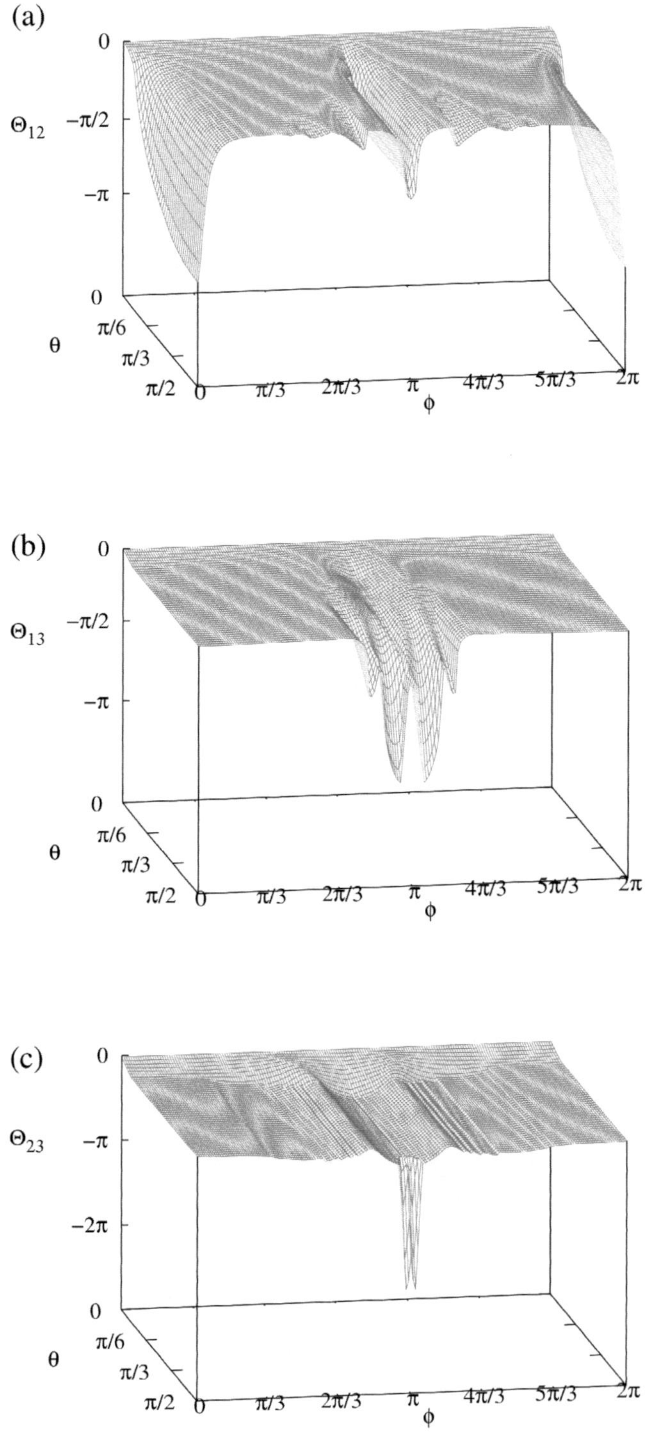

Figure 5. The functional forms of the ADT angles solved on a 2D contour integral as functions of θ and ϕ for $\rho = 7.5$ bohr. (a) Θ_{12} attains π in magnitude and in change in magnitude at the corresponding C_{2v} seams. (b) and (c) Θ_{13} and Θ_{23} attain π and 2π corresponding to the PJT and RT interaction respectively.

those geometries.[44,45] Moreover, at $\phi = 180°$, the overall change in numerical value of Θ_{12} is π at the asymptotic θ ($= 90$ °) depicting the other C_{2v} CI. On the other hand, Θ_{23} attains 2π in the region of RT coupling at $\theta = 90°$ and ϕ close to 180°.[46]

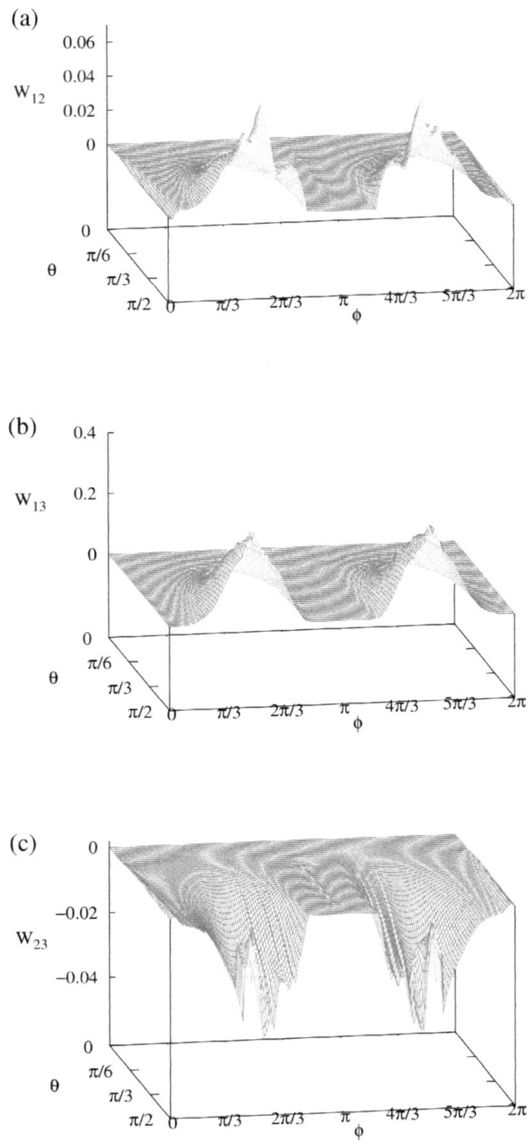

Figure 6. The diabatic coupling elements among the three states as functions of θ and ϕ.

Finally, in Figure 6, the elements of the diabatic potential matrix are displayed which appear as single-valued, continuous, smooth and symmetric. In the diabatic representation, the crossing of the lowest two PESs is evident from Figure 7 along the C_{2v} seams, while the coupling elements are now analytic functions of nuclear coordinates representing the inter-state mixing in the diabatic representation.

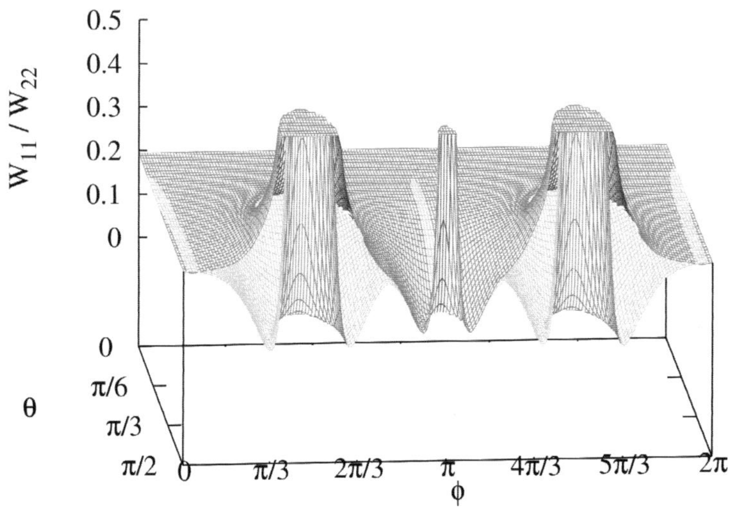

Figure 7. The crossing of diabatic PESs of ground and the first excited states as functions of θ and ϕ.

5. Summary

We have calculated the lowest three adiabatic states of F+H_2 namely, the $1^2A'$, $1^2A''$ and $2^2A'$ states without including the spin-orbit coupling and the NACTs between those states using MOLPRO quantum chemistry package in hyperspherical coordinates for a fixed hyperradius at $\rho = 7.5$ bohr. It has been found that the states interact at various regions of the nuclear CS resulting into CI seams, PJT interactions and RT couplings. The ground and the first excited states undergo JT-type interaction to form CI seam at ϕ close to 0° and 360° while θ from 10° all along to 90° and CI at the region $\{\phi = 180°, \theta \approx 25°\}$. The first and the second excited states show RT couplings at $\theta = 90°$ and ϕ around 180°, whereas the ground and the second excited states have PJT interactions found at regions close to $\{\theta = 90°, \phi = 180°\}$. These interactions are validated by the calculations of ADT angles which gave the theoretically predicted numerical value, i.e. π for JT-type CI[44,45] and 2π for RT coupling.[46] Since these three states constitute a sub-Hilbert space over the range of selected CS, the curl conditions are obeyed and the diabatic potential matrix elements are constructed using the obtained ADT angles. It turned out that the elements of the diabatic PES matrix are single-valued, smooth and symmetric and can be represented by analytic functions of the hyperangles θ and ϕ unlike the NACTs in adiabatic representation.

6. Acknowledgments

BM (File No: 09/080(0960)/2014-EMR-I) acknowledges CSIR India for research fellowship, SM thanks IACS for the same and SA acknowledges DST, India, through project no. File No. EMR/2015/001314 for the research funding. KRS acknowledges excellent hospitality by the other authors during his visit to IACS. BM, SM and SA also acknowledge IACS for CRAY super computing facility.

References

[1] Parker J and Pimental G C 1969 *J. Chem. Phys.* **51** 91–96
[2] Berry M J 1973 *J. Chem. Phys.* **59** 6229–6253
[3] Coombe R D and Pimental G C 1973 *J. Chem. Phys.* **59** 251–257
[4] Polanyi J C and Woodall K B 1972 *J. Chem. Phys.* **57** 1574–1586
[5] Perry D S and Polanyi J C 1976 *Chem. Phys.* **12** 419–431
[6] Neumark D M, Wodtke A M, Robinson G N, Hayden C C and Lee Y T 1985 *J. Chem. Phys.* **82** 3045–3066
[7] Neumark D M, Wodtke A M, Robinson G N, Hayden C C, Shobatake K, Sparks R K, Schafer T P and Lee Y T 1985 *J. Chem. Phys.* **82** 3067–3077
[8] Faubel M, Ruskin L, Schlemmer S, Sondermann F, Tappe U and Toennies J P 1994 *J. Chem. Phys.* **101** 2106–2125
[9] Baer M, Faubel M, Martinez-Haya B, Rusin L Y, Tappe U, Toennies J P, Stark K and Werner H J 1996 *J. Chem. Phys.* **104** 2743–2745
[10] Skodje R T, Skouteris D, Manolopoulos D E, Lee S, Dong F and Liu K 2000 *J. Chem. Phys.* **112** 4536–4552
[11] Qiu M H, Ren Z F, Che L, Dai D X, Harich S A, Wang X Y, Yang X M, Xu C X, Xie D Q, Gustafsson M, Skodje R T, Sun Z and Zhang D H 2006 *Science* **311** 1400–1443
[12] Wang X, Dong W, Qiu M, Ren Z, Che L, Dai D, Wang X, Yang X, Sun Z, Fu B, Lee S Y, Xu X and Zhang D H 2008 *Proc. Natl. Acad. Sci. U. S. A.* **105** 6227–6231
[13] Ren Z, Che L, Qiu M, Wang X, Dong W, Dai D, Wang X, Yang X, Sun Z, Fu B, Lee S Y, Xu X and Zhang D H 2008 *Proc. Natl. Acad. Sci. U. S. A.* **105** 12662–12666
[14] Bender C F, O'Neil S V, Pearson P K and SchaeferIII H F 1972 *Science* **176** 1412–1414
[15] Bender C F, Pearson P K, O'Neil S V and SchaeferIII H F 1972 *J. Chem. Phys.* **56** 4626–4630
[16] Muckerman J T 1981 *Theoretical Chemistry Advances and Perspectives* vol 6A (Academic Press, New York, 1981)
[17] Takayanagi T and Sato S 1988 *Chem. Phys. Lett.* **144** 191–193
[18] Brown F B, Steckler R, Schwenke D W, Truhlar D G and Garrett B C 1985 *J. Chem. Phys.* **82** 188–201
[19] Steckler R, Truhlar D G and Garrett B C 1985 *J. Chem. Phys.* **82** 5499–5506
[20] Lynch G C, Steckler R, Schwenke D W, Varandas A J C, Truhlar D G and Garrett B C 1991 *J. Chem. Phys.* **94** 7136–7149
[21] Mielke S L, Lynch G C and Truhlar D G 1993 *Chem. Phys. Lett.* **213** 10–16
[22] Stark K and Werner H J 1996 *J. Chem. Phys.* **104** 6515–6530
[23] Hartke B and Werner H J 1997 *Chem. Phys. Lett.* **280** 430–438
[24] Alexander M H, Manolopoulos D E and Werner H J 2000 *J. Chem. Phys.* **113** 11084–11100
[25] Li G, Werner H J, Lique F and Alexander M H 2007 *J. Chem. Phys.* **127** 174302
[26] Fu B, Xu X and Zhang D H 2008 *J. Chem. Phys.* **129** 011103
[27] Chen J, Sun Z and Zhang D H 2015 *J. Chem. Phys.* **142** 024303
[28] Che L, Ren Z, Wang X, Dong W, Dai D, Wang X, Zhang D H, Yang X, Sheng L, Li G, Werner H J, Lique F and Alexander M H 2007 *Science* **317** 1061–1064
[29] Zhang Y, Xie T X, Han K L and Zhang J Z H 2004 *J. Chem. Phys.* **120** 6000–6004
[30] Tzeng Y R and Alexander M H 2004 *J. Chem. Phys.* **121** 5183–5190

[31] Das A, Mukhopadhyay D, Adhikari S and Baer M 2011 *Chem. Phys. Lett.* **517** 92–97

[32] Das A, Mukhopadhyay D, Adhikari S and Baer M 2011 *Eur. Phys. J. D* **65** 373–381

[33] Das A, Mukhopadhyay D, Adhikari S and Baer M 2012 *Int. J. Quantum Chem.* **112** 2561–2570

[34] Das A, Sahoo T, Mukhopadhyay D, Adhikari S and Baer M 2012 *J. Chem. Phys.* **136** 054104

[35] Csehi A, Bende A, Halász G H, Vibók A, Das A, Mukhopadhyay D, Mukherjee S, Adhikari S and Baer M 2014 *J. Phys. Chem. A* **118** 6361–6366

[36] Born M and Oppenheimer J R 1927 *Ann. Phys. (Leipzig)* **84** 457–484

[37] Born M and Huang K 1954 *Dynamical Theory of Crystal Lattices* (Oxford University Press, Oxford)

[38] Baer M 2002 *Phys. Rep.* **358** 75–142

[39] Baer M 2006 *Beyond Born - Oppenheimer: Conical intersections and Electronic nonadiabatic coupling terms* (Wiley - Interscience)

[40] Matsunaga N and Yarkony D R 1998 *Mol. Phys.* **93** 79–84

[41] Baer M 1975 *Chem. Phys. Lett.* **35** 112–118

[42] Baer R, Charutz D, Kosloff R and Baer M 1996 *J. Chem. Phys.* **105** 9141–9152

[43] Top Z H and Baer M 1977 *J. Chem. Phys.* **66** 1363–1371

[44] Baer M and Englman R 1992 *Mol. Phys.* **75** 293–303

[45] Alijah A and Baer M 2000 *J. Phys. Chem. A* **104** 389–396

[46] Halasz G J, Vibók A, Boer R and Baer M 2006 *J. Chem. Phys.* **125** 094102

[47] Sarkar B and Adhikari S 2006 *J. Chem. Phys.* **124** 074101

[48] Sarkar B and Adhikari S 2008 *J. Phys. Chem. A* **112** 9868–9885

[49] Paul A K, Sardar S, Sarkar B and Adhikari S 2009 *J. Chem. Phys.* **131** 124312

[50] Paul A K, Ray S, Mukhopadhyay D and Adhikari S 2011 *J. Chem. Phys.* **135** 034107

[51] Mukherjee S, Bandyopadhyay S, Paul A K and Adhikari S 2013 *J. Phys. Chem. A* **117** 3475–3495

[52] Mukherjee S and Adhikari S 2014 *Chem. Phys.* **440** 106

[53] Mukherjee S, Mukhopadhyay D and Adhikari S 2014 *J. Chem. Phys.* **141** 204306

[54] Billing G D and Muckerman J T 1989 *J. Chem. Phys.* **91** 6830–6838

[55] Werner H J, Knowles P J, Knizia G, Manby F R, Schütz M *et al.* 2010 Molpro, version 2010.1, a package of ab initio programs see http://www.molpro.net

Manifestation of quantum rotor orbital excitations in Raman spectra of Jahn-Teller crystal LaMnO

o ale a O usmartse a u el usmartse

Lebe e nstitute of hysics Russian ca emy of ciences Mosco Lenins y prosp Russia

epartment of hysics Lou hborou h ni ersity Lou hborou h L T

nstitute for Theoretical an pplie lectro ynamics Russian ca emy of ciences Mosco hors aya street Russia

ational Research ni ersity i her chool of conomics Mosco Russia

-mail n o al e a sci l ebe e ru

bstract Materials consistin of Jahn-Teller JT ions such as cuprates an man anites isplay many outstan in properties inclu in hi h temperature supercon ucti ity an colossal ma netoresistance There the role of JT e ect althou h i ely reco ni e is still elusi e ere e sho that these materials ha e ibronic excitations relate to local eformations rotatin aroun JT ions in the ynamic limit arisin from linear electron- ibrational couplin in the Mexican hat potential pro le Their ener y epen s on total an ular momentum hich is quanti e as in quantum rotors e foun them in the representati e JT compoun of orthorhombic man anites LaMnO ince the Mexican hat potential ener y surface is ouble- alue they sho up near the roun an excite states of JT ions Recently by usin spectroscopic ellipsometry technique e sho e that they appear in the excite state in the form of si eban s accompanyin the electron transition bet een the JT split orbitals at nei hborin Mn ions ere by usin Raman scatterin technique e sho that they also exist near the roun state The foun quantum rotor excitations may play an important role in many unusual properties obser e in these materials

ntro uction

Orbital e eneracy for example the t ofol e eneracy of e symmetry electronic orbitals of u or Mn ions in cubic symmetry lea s to peculiar beha iour hen there is linear electron-lattice couplin The linear electron-lattice couplin can be trace to the Jahn-Teller theorem hich states that a system such as a complex in a soli or a molecule hich has a roun state possessin any e eneracy other than ramers e eneracy can achie e a lo er ener y by a istortion hich remo es the e eneracy The un istorte eometry of u or Mn ions in sixfol coor ination is octahe ral an the orbitals split into a oublet e an a triplet t n the un istorte cubic en ironment the roun state ith the unpaire electron in the oublet is orbitally e enerate The t o e enerate e orbitals an transform as x y an x y un er the octahe ral roup O_h respecti ely They are sai to ha e e- or -type symmetry The octahe ral complex may achie e lo er ener y by the three possible tetra onal istortions alon the three istinct cubic axes of hich only t o name as an are in epen ent They transform un er the octahe ral roup O_h li e the electronic a e functions an respecti ely The symmetri e an isplacements of

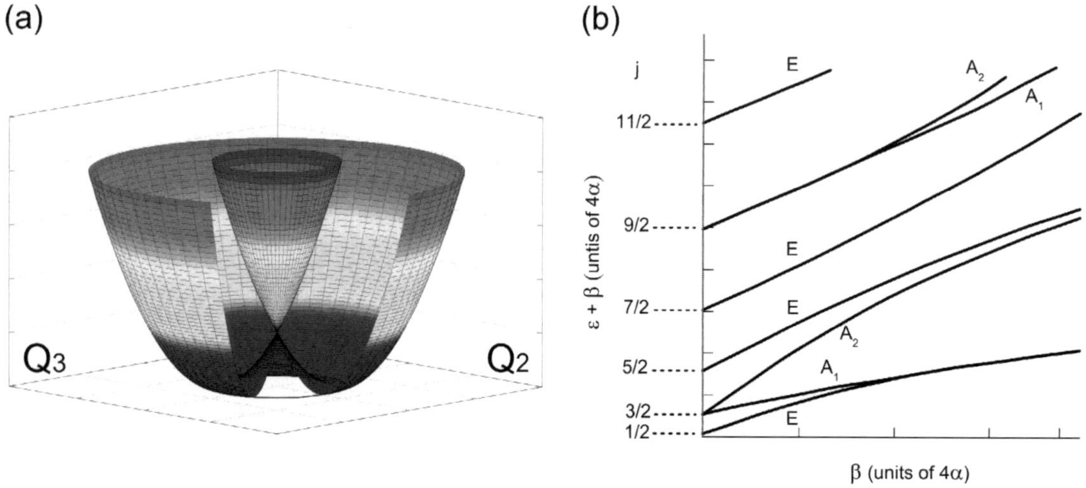

Figure 1. (a) Double-valued "Mexican hat" adiabatic potential energy surface for linear vibronic coupling (dynamic limit) in the normal coordinates Q_2 and Q_3. The "Mexican hat" perturbed by third-order (anharmonicity) terms has three discrete minima at the bottom of the through, separated by anharmonicity barriers (not shown). (b) Energies ε of the "Mexican hat" potential as functions of the barrier parameter β. The energy splitting parameter α is used in the quantization of discrete energy levels. Values of the angular momentum j=1/2, 3/2, 5/2, ... are given in the dynamic limit. Results after O'Brien [2].

the neighbours of transition-metal ions $Cu^{2+}(3d^9)$ or Mn^{3+} $(3d^4)$ in an octahedral environment correspond to the tetragonal normal modes of E symmetry.

The total (vibronic) energy of such JT complex is the sum of the JT energy and the elastic energy. It can be represented by the adiabatic potential-energy surface

$$V = \pm A\sqrt{Q_2^2 + Q_3^2} + \frac{1}{2}M\omega^2\left(Q_2^2 + Q_3^2\right), \tag{1}$$

where M is a mass of a corner atom, and ω is a circular vibrational frequency in the Q_2 or Q_3 normal modes (in the absence of JT effect). It is convenient to transform Eq. (1) to polar co-ordinates, $\rho \equiv \sqrt{Q_2^2 + Q_3^2}$ and $\phi \equiv \tan^{-1}\left(\frac{Q_2}{Q_3}\right)$

$$V = \pm A\rho + \frac{1}{2}M\omega^2\rho^2. \tag{2}$$

The adiabatic potential-energy surface represented by Eqs. (1) and (2) has some important features. First, it is double valued. The twofold degeneracy is lifted here, as for a given distortion we have two states, corresponding to the lower and upper sheets of the potential-energy surface. Second, the ground state energy is now lower by the JT energy, $E_{\rm JT} = -\frac{A^2}{2M\omega^2}$. And third, both potential-energy surfaces are independent of an angle ϕ, and the number of equivalent minimum-energy configurations is infinite. This means that the ground state energy is not uniquely defined, but occurs at any point of the circle with the radius $|\rho_0| = \frac{A}{M\omega^2}$ along the trough at the bottom of the lower potential-energy surface. This potential-energy surface for a doubly degenerate electronic state of e symmetry coupled to tetragonal E- symmetry vibrational modes has the so-called "Mexican hat" structure (see Fig. 1(a)). There are no barriers to prevent rotations in ϕ, which leads to the oscillations in the configuration of an octahedron without change in energy. This behaviour is known as the dynamic JT effect.

o e er the in nite e eneracy of the roun state achie e for all alues tan⁻ — in linear couplin may be remo e if one ta es into account hi her-or er couplin terms connectin the e enerate electronic states hich are of qua ratic or hi her or er in the isplacements an n particular it as sho n that the qua ratic ibronic couplin essentially perturbs the roun state of the e Jahn-Teller systems This ill sin le out iscrete alues of resultin in the appearance of three minima in the trou h of the Mexican hat The speci c alues of the an les o epen on the etails of the hi her or er terms The minima on the lo ener y surface of the eforme Mexican hat are separate by potential barriers urprisin ly by ab initio calculations it as sho n that the e ect of the thir -or er cubic terms to the barrier hei ht bet een the minima is hi her than or of the same or er of ma nitu e of the qua ratic contributions

The e ect of small cubic perturbations on the lo est ener y state of the system of the correct symmetry cos as consi ere by O rien This perturbation for the lo er potential-ener y surface −M cos ill lea to the three iscrete minima at ° if or ° ° if n at a su ciently lo temperature the system may become localise in one of them associate ith the tetra onal static lattice istortion f the minima on the lo ener y surface of the eforme Mexican hat are separate by relati ely lo potential barriers compare ith temperature tunnelin bet een the three potential ells an rapi reorientation in the con uration of the octahe ra can still occur n this ynamical re ime the thir -or er anharmonicity terms act as a perturbation on the freely-rotatin Mexican hat The in uence of the barriers on the stability of the roun state as consi ere by O rien ho rst in esti ate the quantum ener y le els in this system Their ei en alues associate ith the an irre ucible representations of the symmetry roup ere estimate The results are presente in the ia ram as a function of the potential barrier hei ht see i b There is also the ener y splittin parameter use in the quanti ation of iscrete ener y le els s follo s from the ia ram the ener y le el is associate ith the roun state ith the quantum number of the an ular momentum aroun the potential trou h n the ynamic limit ener ies of the rotor excitations enerate a sequence etermine by the qua ratic epen ence on alues of the an ular momentum These le els are associate ith the ynamic limit of rotor excitations Their ener y le els shift ith increasin the barrier hei ht in a static limit Thus for the t o extreme limits of the barrier hei ht e can ha e a ynamic fast-rotation limit for an a static slo -rotation limit for

Orthorhombic rare-earth man anites RMnO R La r m u Tb y are re ar e to be mo el JT compoun s exhibitin cooperati e JT e ect These systems are pero s ites ith the Mn ions ha in the t e electronic con uration mon them LaMnO crystal is famous as an insulatin parent compoun for ope man anites exhibitin colossal ma netoresistance e ect ts hi h-temperature phase is nearly cubic an at the orbital or erin temperature T_{OO} it un er oes a transition into the orthorhombic orbitally or ere bnm structure ith an antiferro istorsi e or erin of MnO octahe ra n our recent stu y of polarise Raman scatterin in the unt inne LaMnO crystal e isco ere that the multi-or er O feature obser e there up to the fourth or er is the si eban s on the lo -ener y mo e at cm⁻ This speci es that the lo -ener y mo e at cm⁻ stems from the tunnelin transition bet een the potential ener y minima arisin near the JT Mn ion ue to lattice anharmonicity rom the relation for the tunnelin frequency $\frac{}{M}$ usin for cm⁻ an ta in for the li an mass M the mass of an oxy en atom e estimate $_{JT}$ e n a ition in the ielectric function spectra of the unt inne LaMnO crystal e obser e excitations hose ener y spectrum is escribe in terms of the total an ular momentum ei enstates an is quanti e as in quantum rotors foun in JT centers They appear clearly as narro si eban s accompanyin the e electron transition bet een the JT split orbitals at nei hbourin Mn ions These results are presente to ether ith ne experimental e i ence of photoluminescence foun in LaMnO This supports our interpretation implyin ibronic ori in of the quantum rotor orbital excitations The obser e quantum rotor orbital excitations can be relate to the upper sheet of the Mexican hat a iabatic potential-ener y surface The obtaine experimental e i ence su ests that LaMnO exhibits beha iour peculiar of the ynamic JT re ime here tunnelin an rapi reorientation in the octahe ral con uration can still occur That is LaMnO can be consi ere as bein e ecti ely in the re ime of the O rien s ia ram see i b here the lattice anharmonicity acts as a perturbation only

n this paper e present the results of the polarise rst- an hi her-or er Raman scatterin in the unt inne LaMnO crystal measure un er i erent po er of laser excitation part from the anomalously stron perio ically repeate up to the fourth or er O feature e ere able to i entify a itional ea features un er hi h laser excitation po er hich can be associate ith the quantum rotor orbital excitations The foun beha ior can be relate to the lo er sheet of the Mexican hat a iabatic potential-ener y surface specifyin the roun state properties These results are in a reement ith our pre ious optical stu y in icatin the ynamic beha ior of the JT e ect in LaMnO in the re ime of the O rien s ia ram see i b

xperimental approach

LaMnO sin le crystals ere ro n usin the crucible-free oatin - one metho in a mirror furnace equippe ith an arc lamp The as- ro n LaMnO crystals are sin le phase o e er they exhibit hea ily t inne omain patterns belo the orbital or erin temperature T_{OO} in the orthorhombic *Pbnm* structure The particular pattern an the si e of the omains epen on local temperature ra ients an mechanical stresses experience urin the ro th sta e e ere able to et ri of the t ins from an essential part of the sample olume follo in the proce ure escribe in etail in our earlier stu y t as etecte by sin le-crystal -ray i raction that the percenta e of et innin at some areas as as hi h as an o er the entire surface it as at least The orthorhombic irection as i enti e as perpen icular to the crystal surface The sample as also characterise by ma netometry measurements e etermine the antiferroma netic transition temperature T at hich is typical for nearly oxy en-stoichiometric LaMnO crystals

The sample surface as polishe to optical ra e Raman scatterin spectra ere acquire usin a Raman-microscope spectrometer LabR M R oriba Jobin on equippe ith a ratin monochromator an a liqui -nitro en coole etector The LabR M R spectrometer has hi h spectral resolution cm⁻ pixel at nm lar e spectral ran e of Raman shift from to cm⁻ an unique e e lter technolo y e e laser operatin at the a elen th nm e ith appropriate lters as use for excitation un er i erent irra iation po er The LaMnO crystal sample as mounte on a col n er of the micro-cryostat coole ith liqui nitro en e use the near-normal bac -scatterin eometry ith an a ustable confocal pinhole The propa ation irection of the inci ent an scattere li ht as parallel to the orthorhombic axis in bnm symmetry notations of the unt inne LaMnO crystal The Raman scatterin spectra ere recor e in the istincti e principle polarisations usin eneric linear polarisation of the laser ra iation hile rotatin the sample in the x plane

xperimental results

factor- roup analysis of the orthorhombic crystal structure of LaMnO bnm h ith f u unit cell i es Raman-acti e phonon mo es namely an symmetry

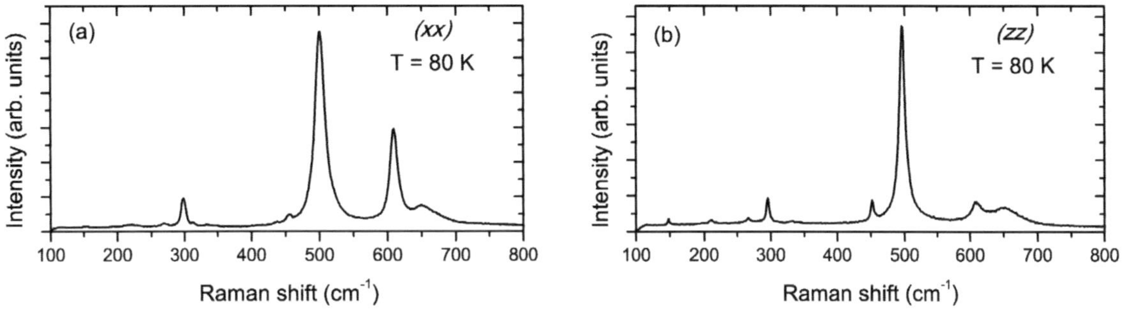

i ure The lo -temperature micro-Raman spectra T measure on the oriente x surface of the unt inne LaMnO crystal ith the inci ent laser li ht polarise alon a the x axis an b the axis in bnm symmetry notations The e e laser operatin at the a elen th nm e as use for excitation at lo -le el laser irra iation po er

mo es The Raman scatterin tensor etermines the selection rule for Raman scatterin The normal mo es of istinct symmetry coul be obser e un er irra iation ith a polarise li ht at a i en scatterin eometry ccor in ly on the oriente crystal surface the symmetry phonons can be re istere in the parallel xx yy an polarisations of the inci ent an scattere li ht hereas the an phonons contribute in the crosse x x y y an xy yx polarisations respecti ely ll etails on the i erent symmetry normal mo es probe by the Raman scatterin on the oriente x surface of the unt inne LaMnO crystal are i en in our pre ious stu y s a result e ere able to i entify the symmetry mo es at an cm^{-} an the symmetry mo es at an cm^{-} i ures a an b illustrate the lo -temperature micro-Raman spectra measure on the unt inne LaMnO crystal for the t o selecte polarisations of the inci ent e e laser irra iation at laser irra iation po er alon the main x an crystallo raphic irections respecti ely ere the pronounce hi h-frequency mo es cm^{-} an cm^{-} are assi ne to the in-plane ibrations ue to the JT in-phase anti-symmetric an symmetric stretchin of the corner-share oxy en octahe ra in LaMnO respecti ely One may notice that the hi hest-frequency cm^{-} mo e is quite notably pronounce in the xx polarisation hereas it is almost entirely suppresse in the polarisation e oul li e to note that our Raman phonon spectra are in a reement ith those of an earlier Raman stu y of an unt inne LaMnO sin le crystal by aitoh *et al.* si nifyin the similar crystal quality n our pre ious etaile temperature Raman stu y of the unt inne LaMnO crystal e n that the temperature coe cient of the full i th at half maximum M of the anti-symmetric JT mo e is about cm^{-} hereas for the symmetric JT mo e it is about cm^{-}

n the one-phonon ran e of Raman scatterin at hi her frequencies an a itional broa pea ispose at aroun cm $^{-}$ is a ailable see i s a an b t cannot be assi ne to a normal symmetry mo e in LaMnO crystal structure The excitation is also ea ly infrare - acti e ue to the parity-brea in e ect Moreo er the excitation emonstrates its eneral nature in all iso alent Mn rare-earth RMnO R La r m u Tb y o oxi es here it is present in the Raman spectra aroun the same frequency s follo s from our pre ious Raman scatterin stu y the excitation at cm^{-} is not a sin le mo e it is compose of symmetry mo es lie *et al.* associates the Raman feature at cm^{-} ith the maximum in the phonon ensity-of-states O of oxy en ibrations accor in ly to polarisation properties an pre ictions base on lattice- ynamics calculations n

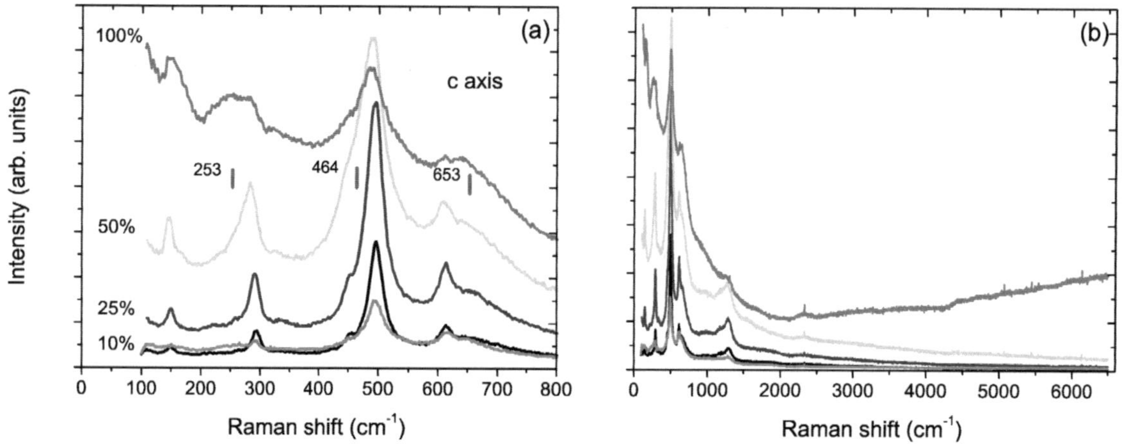

i ure epen ence of the Raman scatterin spectra measure at in parallel polarisations on the oriente x surface of the unt inne LaMnO crystal on the laser excitation po er a in the one-phonon spectral ran e an b in the exten e spectral ran e up to cm⁻ The local actual temperatures un er the i erent laser excitation po er estimate from the i th of the phonon at cm⁻ blac cur e The Raman scatterin measurements repeate at laser po er ma enta cur e after nishin the cycle

exten e structure represente by the three broa features is e elope in the secon -or er Raman scatterin see for example i b t emonstrates stron polarisation epen ence so that only one broa feature of symmetry is isibly pronounce in the -polarise spectra see for example i b t as sho n in our pre ious etaile stu y of the Raman scatterin in LaMnO that in the multi-phonon spectral ran e the O multiple excitations appear ith a perio of about cm⁻ up to the fourth or er shifte by about cm⁻ from the central laser line

i ures a an b sho epen ence of the Raman scatterin spectra on the laser excitation po er measure at in parallel polarisations on the oriente x surface of the unt inne LaMnO crystal in the one-phonon spectral ran e an in the exten e spectral ran e up to cm⁻ sin the linear temperature coe cient of the M of the mo e at cm⁻ cm⁻ obtaine from our pre ious stu y the local actual temperatures ue to the sample heatin un er the laser irra iation ith i erent po er ere estimate e estimate that the sample heatin is ne li ible un er the laser irra iation ith the laser po er The obtaine Raman scatterin is consistent ith the ata sho n in i b Thou h the o erheatin e ect is appreciable un er the laser irra iation ith the laser po er here the local actual temperature in the laser spot is estimate to be o e er the Raman scatterin measurements repeate at laser po er after nishin the cycle in icate that there ere no irre ersible chan es urin the Raman measurements of the LaMnO crystal un er the use hi h-le el po er of the laser irra iation One can follo the e ect of broa enin of the mo e at cm⁻ ith increasin temperature from about un er the laser irra iation ith the laser po er an see i a ere in the exten e spectral ran e the O feature sho s up to the thir or er on the ecreasin scatterin bac roun see i b o e er one can notice that a itional broa features appear here un er the laser po er e tte the obtaine Raman spectra ith a set of Lorent ians an etermine that the a itional features appear at an cm⁻ t the same time in the exten e spectral ran e the O feature becomes almost suppresse

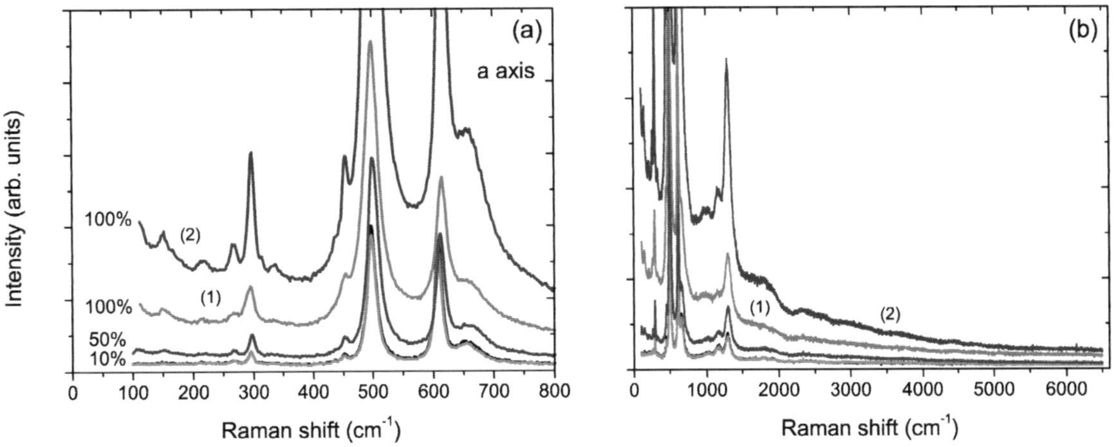

i ure epen ence of the Raman scatterin spectra measure at in parallel xx polarisations on the oriente x surface of the unt inne LaMnO crystal on the laser excitation po er a in the one-phonon spectral ran e an b in the exten e spectral ran e up to cm⁻ The local actual temperatures un er the i erent laser excitation po er estimate from the i th of the phonon at cm⁻ blac cur e an The Raman scatterin measurements repeate at laser po er ma enta cur e after nishin the cycle

by temperature o e er interestin ly a broa Raman scatterin sho s up on the increasin bac roun abo e cm⁻ ith some iscernible superstructure in it see i b The obser e beha ior is close to the orbital or erin temperature T$_{OO}$ into the pseu o-cubic hi h-temperature phase Therefore it can be associate ith the n erprints of the ynamic JT e ect in LaMnO here the ener etic barriers in the Mexican hat potential pro le become thermally accessible

i ures a an b sho epen ence of the Raman scatterin spectra on the laser excitation po er measure at in parallel xx polarisations on the oriente x surface of the unt inne LaMnO crystal in the one-phonon spectral ran e an in the exten e spectral ran e up to cm⁻ The local actual temperatures un er the i erent laser excitation po er ere estimate from the M temperature epen ence of the phonon at cm⁻ imilarly e estimate that the sample heatin is ne li ible un er the laser irra iation ith the laser po er here the obtaine Raman scatterin results are consistent ith the results sho n in i a o e er the o erheatin e ect is appreciable un er the laser irra iation ith the laser po er here the local actual temperature in the laser spot is estimate to be The Raman scatterin measurements repeate at laser po er after nishin the cycle in icate no irre ersible chan es urin the Raman measurements of the LaMnO crystal un er the use po er of the laser irra iation ere one can follo the e ect of broa enin of the mo e at cm⁻ ith increasin temperature from about to about un er the laser irra iation ith the laser po er an see i a n the exten e spectral ran e the O feature is relati ely more pronounce than in the correspon in -polarise Raman spectra an sho s up to the thir or er on the ecreasin scatterin bac roun see i b

i ures a an b emonstrate the Raman scatterin spectra measure un er the hi h laser irra iation po er at i erent temperatures in parallel xx polarisations on the oriente x surface of the unt inne LaMnO crystal in the one-phonon spectral ran e an in the exten e spectral ran e up to cm⁻ respecti ely ue to stron o erheatin e ect

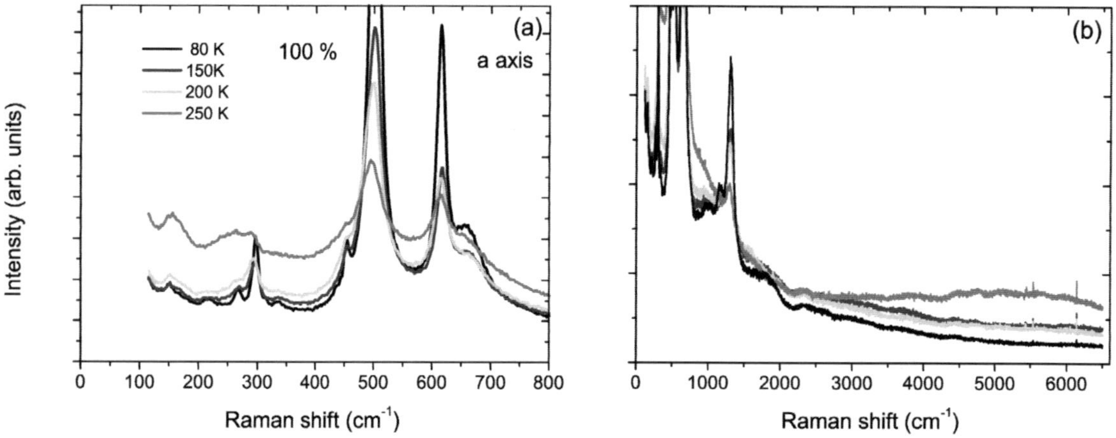

i ure The Raman scatterin spectra measure un er the hi h laser excitation po er at i erent temperatures in parallel xx polarisations on the oriente x surface of the unt inne LaMnO crystal a in the one-phonon spectral ran e an b in the exten e spectral ran e up to cm⁻ The local temperatures in the laser spot estimate from the M of the phonon mo e at cm⁻ for the measurements at for the measurements at for the measurements at an for the measurements at

un er the hi h laser irra iation po er the local temperatures in the laser spot are i erent from the referre temperatures These temperatures estimate from the M of the phonon mo e at cm⁻ are for the measurements at for the measurements at for the measurements at an for the measurements at One can notice chan es in the one-phonon ran e un er the laser po er for the measurements at similar to those obser e in the -polari e spectra un er the laser po er in i a t the same time a broa Raman scatterin sho s up on the increasin bac roun abo e cm⁻ ith some iscernible superstructure in it see i b e uess that the obser e beha ior close to orbital or erin temperature T_{OO} can be associate ith the n erprints of the ynamic JT e ect in LaMnO here the ener etic barriers in the Mexican hat potential pro le become thermally accessible

n i s a an b the Raman scatterin spectra measure un er the hi h laser irra iation po er in parallel xx an polarisations on the oriente x surface of the unt inne LaMnO crystal in the spectral ran e up to cm⁻ an in the exten e spectral ran e from cm⁻ to cm⁻ are isplaye These Raman spectra correspon to the ata sho n in i s a an b an i s a an b ere the local actual temperatures estimate from the i th of the phonon at cm⁻ are relati ely close to orbital or erin temperature T_{OO} for the measurements at for the measurements at

The obser e beha iour in the one-phonon ran e an in the exten e spectral ran e can be associate ith the n erprints of the ynamic JT e ect in LaMnO crystal One can notice that similar chan es are obser e in parallel xx an polarisations on the oriente x surface of the unt inne LaMnO crystal in the one-phonon spectral ran e an in the exten e spectral ran e The ne structure in i b is represente by a set of the narro lines hich are repro ucible in both polarisations

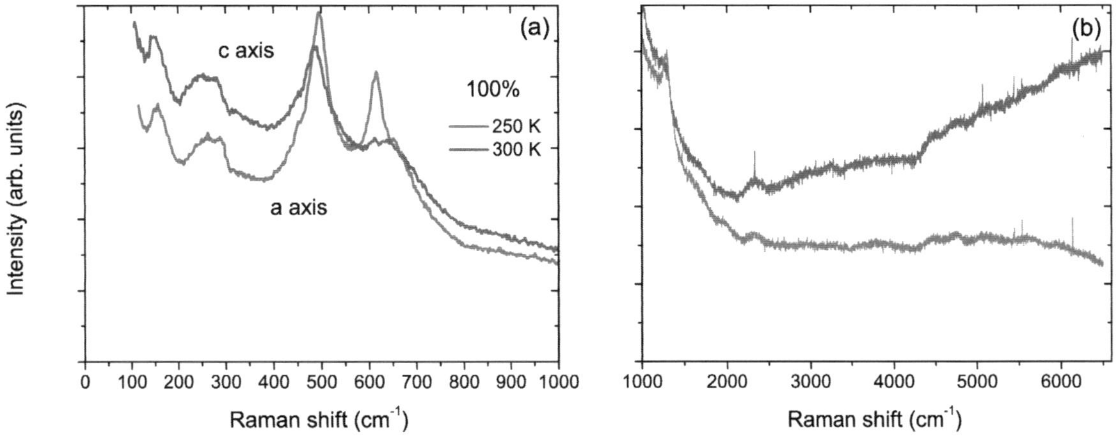

i ure The Raman scatterin spectra measure un er hi h laser excitation po er in parallel xx an polarisations on the oriente x surface of the unt inne LaMnO crystal in a the spectral ran e up to cm⁻ an b the exten e spectral ran e from cm⁻ to cm⁻ The local actual temperatures estimate from the i th of the phonon at cm⁻ for the measurements at for the measurements at

iscussion an conclusions

e consi er that the narro lines appearin in the hi h-ener y Raman scatterin spectra close to the orbital or erin temperature T_{OO} are relate to quantum rotor orbital excitations in a Mexican hat potential pro le see i a The quantum rotor orbital excitations are enuine electronic- ibrational ibronic excitations Their ei en alues associate ith the an irre ucible representations of the symmetry roup ere earlier estimate by O rien an represente in the ia ram as a function of the potential barrier hei ht see i b s follo s from the ia ram the ener y le el is associate ith the roun state ith the quantum number of the an ular momentum aroun the potential trou h There is also the ener y splittin parameter use in the quanti ation of iscrete ener y le els n the ynamic limit ener ies of the rotor excitations enerate a sequence etermine by the qua ratic epen ence on alues of the an ular momentum The symmetry states are characterise by the an ular momentum quantum numbers an n the an symmetry states are characterise by the an ular momentum quantum numbers The tunnelin splittin is estimate at cm⁻ from our recent stu y of the anomalous multi-or er Raman scatterin in LaMnO ere usin cm⁻ e e aluate the quantum rotor ener ies obeyin the square la The estimate ener ies of the an symmetry states correspon in to mo ulus of the an ular momentum quantum numbers are pointe on the top scale of i s a c i ures a c sho the Raman scatterin spectra measure at in parallel polarisations on the oriente x surface of the unt inne LaMnO crystal on the laser excitation po er also sho n in i s a an b The local actual temperature un er laser irra iation po er estimate from the i th of the phonon at cm⁻ is of rather close to the orbital or erin temperature T_{OO} ere the a itional features pop up in the one-phonon spectral ran e at an cm⁻ an in the exten e spectral ran e abo e cm⁻ The obser e beha iour can be associate ith the n erprints of the ynamic JT e ect in LaMnO crystal e pre ict that the features e i ent at hi h ener ies at cm⁻ e cm⁻ e cm⁻ e an cm⁻ e coul be

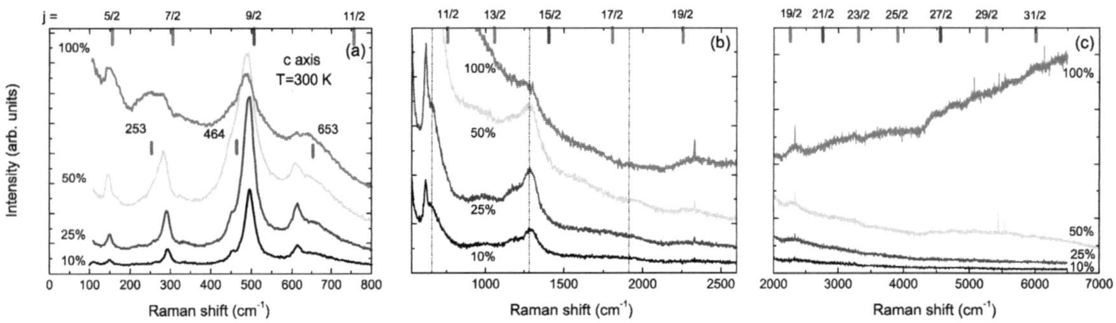

i ure The Raman scatterin spectra measure at in parallel polarisations on the oriente x surface of the unt inne LaMnO crystal on the laser excitation po er a in the one-phonon spectral ran e b in the spectral ran e from to cm^{-} an c in the spectral ran e from to cm^{-} The local actual temperatures un er the i erent laser excitation po er estimate from the i th of the phonon at cm^{-} The ash- otte lines in icate the O main pea position perio ically repeate in the multi-phonon ran e The re an blue bars on the top scale enote the estimate ener ies of the an symmetry quantum rotor orbital excitations in the ynamic limit respecti ely

associate ith the quantum rotor orbital excitations in the ynamic limit for the alues an respecti ely lso the a itional features obser e at lo ener ies at an cm^{-} coul be associate ith the quantum rotor orbital excitations in the ynamic limit for the alues an correspon in ly e oul li e to note that the qualitati e a reement ith the quantum rotor pea ener ies an the e i ent features in the Raman scatterin spectra is rather oo

e oul li e to note that the results of the present Raman stu y are in a reement ith our recent spectroscopic ellipsometry stu y of the unt inne LaMnO crystal There e sho that the quantum rotor orbital excitations arise abo e the pronounce lo -ener y optical ban locate at e n accor ance ith our pre ictions the orbital excitations for the alues an appear as a oublet at e the orbital excitations for the alues an appear as a oublet at e an the orbital excitations for the alues an ill appear as a oublet at e in the a-axis room-temperature spectra ere the qualitati e a rement ith the pea ener ies obser e in the a-axis spectra at an e is surprisin ly oo The quantum rotor orbital excitations obser e abo e the hi h-spin electronic transition at e can be relate to the upper sheet of the Mexican hat a iabatic potential-ener y surface

rom the results of the present Raman stu y an our earlier optical stu ies e can conclu e that LaMnO system can ha e lo - an hi h- temperature ynamic limit The hi h-temperature limit of the ynamic JT e ect in LaMnO occurs close to the orbital or erin temperature T_{OO} here the ener etic barriers in the Mexican hat potential pro le become thermally accessible This is illustrate by the present Raman scatterin stu y s a result apart rom the anomalously stron perio ically repeate O feature e ere able to i entify a itional ea features un er hi h laser excitation po er hich can be associate ith the quantum rotor orbital excitations The foun beha ior can be referre to the lo er sheet of the Mexican hat a iabatic potential-ener y surface specifyin the roun state properties in the hi h-temperature ynamic limit o e er the lo -temperature ynamic limit of the JT e ect in LaMnO can also occur The ynamic beha iour at temperatures ell belo the orbital or erin temperature T_{OO} as reporte recently for another mo el material for a cooperati e JT e ect u e ar ue that superexchan e interactions bet een the Mn ions may ri e the system to the ynamic-li e beha iour The ma netic interactions in ma netic oxi es an in particular in the LaMnO system are successfully escribe in the frame or of u el- homs ii mo el n our comprehensi e spectroscopic ellipsometry stu y of the unt inne LaMnO crystal e sho e that the a-axis inetic ener y of the e electrons associate ith the hi h-spin electronic transition bet een nei hbourin Mn ions at e increases by about me belo the eel temperature T This ener y is comparable ith the orbital or erin temperature of LaMnO T_{OO} me Therefore the superexchan e interaction may be crucial in ri in the e electrons into the re ime of the ynamic JT e ect ith a or able barriers for the tunnelin bet een the potential ener y minima ere one shoul ta e into account the anisotropy of the superexchan e interactions in LaMnO hich results in the ferroma netic or erin in the ab plane an antiferroma netic or erin alon the c axis The lo -temperature ynamic beha iour can be ri en by ferroma netic correlations in the ab plane y contrast the increase antiferroma netic correlations alon the c axis can lea to the ynamic-to-static transition at lo temperature

inally e note that the lo - an hi h-temperature beha iour of the LaMnO system in icates the ynamic re ime of the JT e ect hich can be escribe in the frame or of the O rien s ia ram see i b urther experiments are nee e to in esti ate quantum rotor orbital excitations in the JT compoun s ith JT ions such as Mn or u ions The isco ere orbital excitations of quantum rotors may play an important role in many unusual properties obser e in these compoun s upon opin such as hi h-temperature supercon ucti ity an colossal ma netoresistance

c no le ments
The authors ac no le e fruitful iscussions ith i hnya o e than albasho for ro in the crystals e also than ula o for et innin the crystal J trempfer e ino lou an M chul for characteri ation of the LaMnO sample This or as supporte by the Russian oun ation for asic Research pro ects - - an - -

References

Jahn an Teller *Proc. R. Soc. London*
O rien M M *Proc. Roy. Soc.*
ersu er *The Jahn-Teller Effect* e or Melbourne ape To n in apore ao aulo ambri e ni ress
arc a- ernan e ersu er ramburu J arriuso M T an Moreno M *Phys. Rev.*
aplan M an e hter *Cooperative Phenomena in Jahn-Teller Crystals* e or lenum ress
ersu er e hter an O urtso *Sov. Phys. Usp.*
u el an homs ii *Sov. Phys. JETP*
Ro r ue - ar a al J ennion M Moussa Mou en insar L an Re cole schi *Phys. Rev.* R
o ale a usmartse a O u el Ma simo u hnyy albasho M emi ho e ne a Trepa o usmartse an toneham M *J. Phys.: Condens. Matter*
o ale a u el otuce usmartse a O oryache ry nar emi ho Trepa o e ne a usmartse an toneham M *JETP*
albasho M arabashe Mu o s y a M an er o *J. Cryst. Growth*
o ale a Oles n r e M albasho M Mal u r yriou haliullin an eimer *Phys. Rev.*
mirno a *Physica*
lie M brashe M Lee - opo un Thomsen Men R L an hu *Phys. Rev.*

o ale a oris apo na L a artin J L opo ich or ano Mal u toneham M an eimer *Phys. Rev.*
aitoh O amoto Ta ahashi T Tobe amamoto imura T shihara Mae a a an To ura *Nature*
La er riere J Jan l Mu hin an ano u *Eur. Phys. J.*
lie M a ie Lit inchu en an - un Jan l La er iere J opo an ospo ino M M *Phys. Rev.*
lie M brashe M opo an a ie *Phys. Rev.*
eisenhofer J Leono remin M ant h hi na Mayr lamo nisimo an an er Marel *Phys. Rev. Lett.*
u el an homs ii *Sov. Phys. Uspekhi* *Usp. Fiz. Nauk*
Oles n r e M haliullin orsch an einer L *Phys. Rev.*
o ale a oris or ano Mal u rucher trempfer J onuma M e ino lou ernhar toneham M an eimer *Phys. Rev.*
o ale a oris ernhar ula o imeno albasho M haliullin an eimer *Phys. Rev. Lett.*

IOP Conf. Series: Journal of Physics: Conf. Series **833** (2017) 012006 doi:10.1088/1742-6596/833/1/012006

Structure and lattice dynamics of Jahn-Teller crystal $BiMnO_3$: ab initio calculation

D Nazipov[1*], A Nikiforov[1] and L Gonchar[1,2]

[1]Ural Federal University, 19 Mira St., Ekaterinburg, Russia
[2]Ural State University of Railway Transport, 66 Kolmogorov St., Ekaterinburg, Russia

*E-mail: qwear0@gmail.com

Abstract. In this work, ab initio study of crystal structure and phonon spectra of $BiMnO_3$ has been performed using density functional theory (DFT) with hybrid DFT/Hartree-Fock functionals in CRYSTAL14 code. Crystal structure parameters, infrared and Raman spectra of $BiMnO_3$ have been calculated for magnetic ordered monoclinic C2/c phase. Results are compared with existing experimental data. Our results are in agreement with last x-ray diffraction and optical experiments. Using isotopic substitution method ions involved in vibrations are determined. Calculated charge density maps show the antiferroelectric ordering of dipole moments of Bi^{3+}.

1. Introduction

Crystal compounds of $BiMO_3$ (M – transition metal ion or Al, Ga, In) range are under investigation because of various physical properties which could be used in electronics [1]. The Jahn-Teller crystal $BiMnO_3$ (BMO) is a bismuth-based distorted perovskite oxide like $BiFeO_3$. The trivalent bismuth ion possesses of $6s^2$ lone-pair electrons [2,3,4,5], which make this ion polarized. The transition-metal ions (Mn^{3+} or Fe^{3+}) are magnetic. The dipole moments of Bi^{3+} sublattice and the magnetic moment of transition metal sublattice are ordered at low temperatures [1-5,6,7]. This coincidence leads to widely discussed multiferroic properties of these compounds [1-7]. The sublattice of orbitally degenerated Mn^{3+} ions adds new properties to the correlation between polarization and magnetic subsystem. Thus, the $BiFeO_3$ is an antiferromagnetic crystal with ferroelectric (FE) ordering [5]. The $BiMnO_3$ is ferromagnetic (FM) and has no spontaneous polarization in the bulk samples [3, 4]. The FM ordering of $BiMnO_3$ is caused by orbital ordering of manganese sublattice due to Jahn-Teller effect [8,9]. Under Curie temperature, the centrosymmetric group C2/c of crystal structure leads to absence of ferroelectricity due to symmetry considerations [2, 3, 9]. An inversion symmetry operation causes an impossibility of FM+FE ordering in this crystal. The temperature and external pressure change the crystal structure to monoclinic and orthorhombic symmetry [3, 4, 10], but they cause no changes of spontaneous polarization in the bulk crystal. The experiments using Raman and infrared spectroscopy confirm these considerations [11]. But the phonon spectra are still not investigated in detail due to problems related to very low symmetry of the crystal in magnetic ordered phase. The Bi^{3+} dipole moments ordering is studied only in cases FM+FE ordering. The authors have studied the thin films [7] or considered magnetic symmetry breaking [12,13]. The presence of dipole moments is not yet studied neither theoretically nor experimentally.

The current investigation is devoted to *ab initio* calculations of $BiMnO_3$ crystal structure and optical spectra. We perform a detailed investigation of phonon spectra for the first time and describe the experimental data [11].

2. Computational Details

Calculations have been performed in program package CRYSTAL14 [14], within the framework of density functional theory (DFT) using linear combination of atomic orbitals (LCAO) approximation. Hybrid DFT/HF functional WC1LYP has been used with 16% of Hartree-Fock exchange energy. In this functional, the DFT-exchange part is calculated using Wu-Cohen method in generalized gradient approximation (GGA). It should be noted that this functional developed without the fitting parameters. The correlations are taken into account by the functional LYP. All-electron Gauss-type basis sets are used for description of Mn^{3+} and O^{2-} ions, 86-411d41G and 8-411d1, correspondingly [15]. Heavy Bi^{3+} ion is described by Hay-Wadt effective small-core pseudopotential that is taken from a source [16]. External $6s^2$ orbital of Bi^{3+} ion plays an important role in the formation of the distortion of the crystal lattice, so it is taken into account directly and was not included in the core. All calculations performed in spin-polarized mode in high-spin state S=2 for Mn^{3+}. The accuracy of calculations is determined by a set of tolerances for the overlap of wave functions, above which the Coulomb and exchange integrals are calculated. In this paper, the tolerance is selected to be 10^{-10} for Coulomb and 10^{-20} for the exchange integrals. The integration in reciprocal space is made with 4 x 4 x 4 grid using Monkhrost-Pack scheme. The energy per unit cell is calculated with an accuracy of 10^{-8} Hartree.

Phonon frequencies are calculated in the center of first Brillouin zone in harmonic approximation.

3. Results and Discussion

3.1. *Crystal structure*

The calculation of crystal structure parameters of $BiMnO_3$ has been made for monoclinic phase, C2/c space group, in ferromagnetic arrangement of spins. According to x-ray diffraction experiment, that performed for $BiMnO_3$ single crystal in magnetic ordered state, at $T < T_c$ there is a center of symmetry and monoclinic C2/c structure is realized, instead of the C2, as previously assumed [17]. The results of our calculation (table 1) are in a good agreement with the experiment [18] (the difference is less than 1%).

Table 1. Calculated and experimental [18] (in brackets) crystal structure parameters of $BiMnO_3$

Lattice constant	a	b	c	β
Value, Å	9.484(9.569)	5.589(5.602)	9.799(9.879)	110.96(111.02)

Atom coordinates in cell

Atom	Wyckoff symbol	x/a	y/b	z/c
Bi	8f	0.1351(0.1365)	-0.230(-0.217)	0.1181(0.1269)
Mn1	4e	0	0.2014(0.2113)	0.25
Mn2	4c	0.25	0.25	0
O1	8f	0.0993(0.0991)	0.1666(0.1776)	0.0794(0.0815)
O2	8f	0.3510(0.3524)	0.4496(0.4567)	0.1662(0.1646)
O3	8f	0.1447(0.1443)	0.4161(0.4342)	0.3782(0.3639)

Bond lengths, Å

Mn1-O1	2.20(2.20)	Mn2-O1	1.91(1.93)
Mn1-O2	1.91(1.90)	Mn2-O2	2.24(2.23)
Mn1-O3	1.95(1.97)	Mn2-O3	1.92(1.95)

IOP Conf. Series: Journal of Physics: Conf. Series **833** (2017) 012006 doi:10.1088/1742-6596/833/1/012006

3.2. *Phonon spectra*

The set of fundamental vibrations in gamma point of $BiMnO_3$, in monoclinic C2/c phase, looks as follows:

$$\Gamma = (13A_g + 14B_g)^{Ram} + (14A_u + 16B_u)^{IR} + (A_u + 2B_u)^{Acoust} \quad (1)$$

Table 2. Irreducible representations, frequencies of IR and Raman modes in cm^{-1} and calculated intensities in IR spectrum in a.u.

Infrared					Raman			
№	**Repr.**	**Calc.**	**Exp. [11]**	**Calc. Intens.**	**№**	**Repr.**	**Calc.**	**Exp. [11]**
5	A_u	74	68	16	4	A_g	53	50
6	B_u	76	75	59	7	B_g	81	75
8	B_u	85	92	255	10	B_g	93	84
9	A_u	88	101	0.5	11	B_g	102	93
12	A_u	107	115	200	13	A_g	108	99
16	B_u	145	143	282	14	A_g	113	
17	A_u	189	159	64	15	B_g	137	133
19	B_u	201	196	14	18	A_g	196	180
20	B_u	218	219	200	21	B_g	223	
22	A_u	229		11	23	B_g	232	
24	B_u	241	237	112	26	A_g	262	242
25	A_u	244		2	28	A_g	287	278
27	B_u	273	261	213	29	B_g	287	
30	A_u	291		1	34	B_g	346	
31	B_u	316	284	66	36	A_g	358	347
32	B_u	321	308	1000	37	A_g	360	
33	A_u	331	321	374	39	B_g	373	
35	A_u	351	345	312	42	B_g	395	
38	B_u	368	369	249	44	A_g	422	400
40	B_u	379		335	45	A_g	450	424
41	A_u	381	407	34	46	B_g	456	440
43	A_u	421	426	7	47	A_g	462	
48	B_u	468	463	341	49	B_g	482	
50	A_u	485	486	261	53	A_g	523	
51	A_u	490		141	54	B_g	533	530
52	B_u	495	508	159	58	B_g	646	635
55	B_u	542	541	128	60	A_g	652	
56	B_u	562	561	401				
57	A_u	567	604	87				
59	A_u	649	628	54				

Among 60 modes, 27 even modes are active in Raman spectrum, 30 odd are active in infrared (IR) spectrum and 3 odd modes are acoustic. Our calculation predicts all the modes, their irreducible representations, a comparison of results with the experiment [11] is presented in table 2. The comparison with experimental data is complicated because not all the lines are identified in the experimental spectra. There are 25 lines in infrared spectrum, which are taken at T=10K, and 15 Raman lines, which are taken at room temperature in backscattering geometry. Both of the experimental spectra are obtained in the unpolarized configuration, so we compare with the results calculated for the powder sample. Also it should be noted that missing lines in experimental IR spectrum are low intensity lines or narrow doublets.

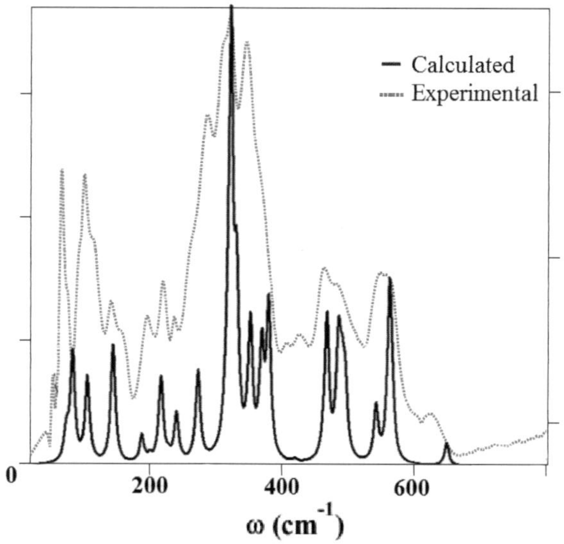

Figure 1. Calculated and experimental (T=10K) infrared spectrum

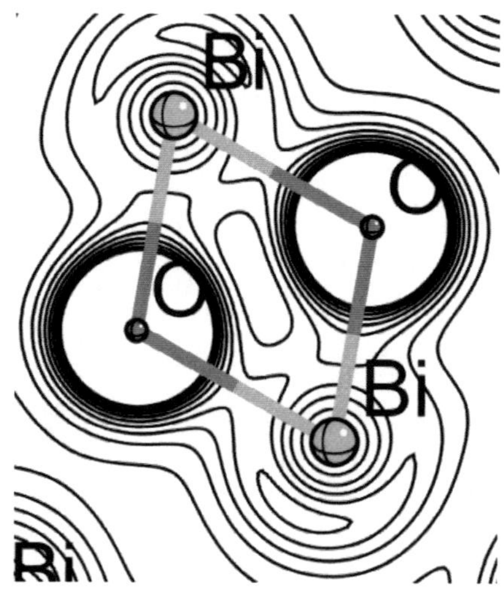

Figure 2. Charge density map in plane of two Bi^{3+} ions

In CRYSTAL package, the intensity of the modes I_p in infrared spectrum is computed according to

$$I_p = \frac{\pi}{3}\frac{N_A}{c^2} d_p |\vec{Z}_p|^2, \tag{2}$$

where N_A is the Avogadro's number, c is the speed of light, d_p is the degeneracy of the mode, $\vec{Z}_p$ is the mass-weighted effective mode Born charge vector. Calculated I_p for the IR-active modes for $BiMnO_3$ are presented in table 2.

The infrared absorbance spectrum $A(\omega)$, which is shown in figure 1, is calculated using following model:

$$A(\omega) = \sum_p \frac{I_p}{\pi} \frac{\gamma_p/2}{(\omega-\omega_p)^2 + \gamma_p{}^2/4}, \tag{3}$$

where ω_p is TO frequency, γ_p is the damping factor (γ_p=8 cm^{-1}), and I_p is integrated intensity computed as described above.

The phonon frequencies and a IR spectrum which have been obtained in this work are in a good agreement with the existing experimental data [11]. Thereby our results confirm that $BiMnO_3$ single crystal has centrosymmetric C2/c space group at low temperature.

3.2.1. Dipole moments. Using *ab initio* calculations, it is possible to draw a map of charge density in specific plane. Figure 2 shows a charge density around bismuth ions. There is a localization of negative charge near Bi^{3+} ion and formation of dipole moment.

3.2.2. Isotopic substitution. To determine the ions that take part in vibrations, the isotopic substitution method has been used. The phonon spectrum was recalculated with a changed mass of each symmetrically nonequivalent ion. Obtained frequencies were compared with the initial spectrum and frequency shifts have been determined. In figure 3, results of isotopic substitution are presented. Heavy Bi^{3+} ions are involved in vibrations with frequencies below 200 cm^{-1} approximately. Both positions of manganese ions take part in vibrations in a region of frequencies 200 to 400 cm^{-1}, but Mn2 sublattice is not involved in Raman-active even modes. Oxygen ions take part in all vibrations, predominantly above 400 cm^{-1}.

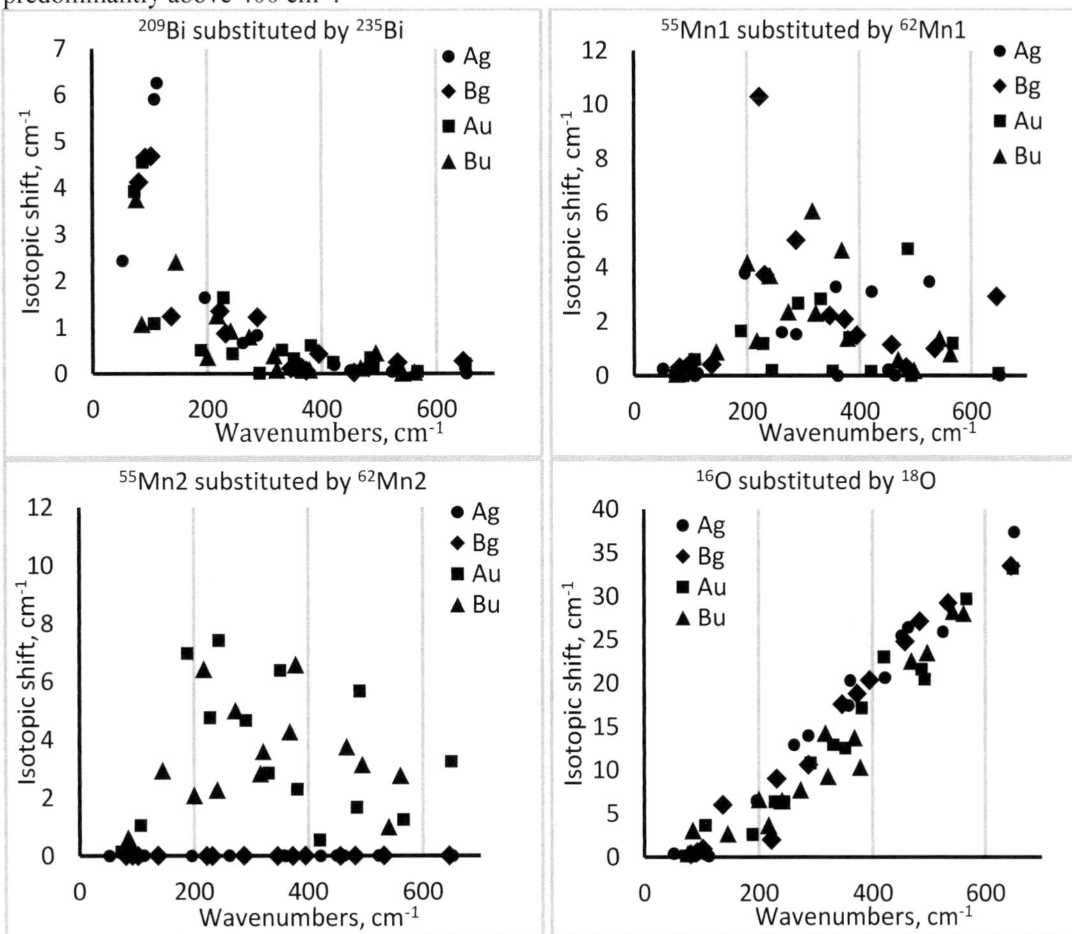

Figure 3. Isotopic substitution data for $BiMnO_3$

4. Conclusion

In this paper, crystal structure parameters of monoclinic centrosymmetric phase of $BiMnO_3$ were calculated from first principles. The phonon spectra were calculated and full set of fundamental vibrations, their frequencies and irreducible representations were obtained. Intensities in infrared spectrum were calculated. Despite that the symmetry of the investigated crystal is very low the results obtained are in a good agreement with existing experimental data. The phonon spectra are investigated using isotopic substitution method. Presence of dipole moments in bismuth sublattice is reproduced in calculations.

References

[1] Khomskii D 2006 *J. Magn. Magn. Mat.* **306** 1
[2] Pyatakov A and Zvezdin A 2012 *Phys. Usp.* **55** 557 (Russian text published in 2012 *Usp. Phys. Nauk* **182** 593)
[3] Kozlenko D, Belik A, Kichanov S, Mirebeau I, Sheptyakov D, Strässle Th, Makarova O, Belushkin A, Savenko B and Takayama-Muromachi E 2010 *Phys. Rev B* **82** 014401
[4] Kozlenko D, Dang N, Jabarov S, Belik A, Kichanov S, Lukin E, Lathe C, Dubrovinsky L, Kazimirov V, Smirnov M, Savenko B, Mammadov A, Takayama-Muromachi E and Khiem L 2014 *J. of Alloys and Compounds* **585** 741
[5] Kozlenko D, Belik A, Belushkin A, Lukin E, Marshall W, Savenko B and Takayama-Muromachi E 2011 *Phys. Rev B* **84** 094108
[6] Grizalez M, Martinez E, Caicedo J, Heiras J and Prieto P 2008 *Microelectronics Journ.* **39** 1308
[7] Moreira dos Santos A, Parashar S, Raju A, Zhao Y, Cheetham A and Rao C 2002 *Solid State Commun.* **122** 49
[8] Gonchar L, Nikitina T and Nikiforov A 2013 *JETP* **116** 812 (Russian Text published in 2013 *Zhurnal Eksperimental'noi i Teoreticheskoi Fiziki* **143** 935)
[9] Gonchar L and Nikiforov A, 2013 *Phys. Rev. B* **88** 094401
[10] Calestani G, Orlandi F, Mezzadri F, Righi L, Merlini M and Gilioli E 2014 *Inorg. Chem.* **53** 8749
[11] Mohamed W, Nucara A, Calestani G, Mezzadri F, Gilioli E, Capitani F, Postorino P and Calvani P 2015 Phys Rev. B **92** 054306
[12] Solovyev I and Pchelkina Z 2010 *Phys. Rev B* **82** 094425
[13] Solovyev I and Nikolaev S 2014 *Phys. Rev. B* **90** 184425
[14] URL: http://www.crystal.unito.it/
[15] URL: http://www.crystal.unito.it/basis-sets.php
[16] URL: http://www.tc.uni-koeln.de/PP/clickpse.en.html
[17] Moreira dos Santos A, Cheetham A, Atou T, Syono Y, Yamaguchi Y, Ohoyama K, Chiba H and Rao C 2002 *Phys. Rev. B* **66** 064425
[18] Toulemonde P, Bordet P, Bouvier P and Kreisel J 2014 *Phys. Rev.* B **89** 224107

IOP Conf. Series: Journal of Physics: Conf. Series **833** (2017) 012007 doi:10.1088/1742-6596/833/1/012007

Jahn-Teller crystals – new class of smart materials

M D Kaplan[1] and G O Zimmerman[2]

[1]Simmons College, Department of Chemistry and Physics, 300 The Fenway, Boston, MA 02115
[2]Boston University, Department of Physics, 590 Commonwealth Ave., Boston, MA 02215

E-mail: mkaplan@simmons.edu

Abstract. Jahn-Teller crystals represent a promising class in the search for new smart materials. Jahn- Teller multiferroics are of a special interest. We show that the properties of these crystals are not only "of interest for future applications", but are already used and protected by various patents. Special attention is paid to some new results on magnetic shape memory effects in dielectrics because the physics of the corresponding materials is not yet completely clarified.

1. Introduction

This paper is written to call attention to a set of materials, which have heretofore been neglected in the search for 'smart' materials. Those materials are Jahn-Teller materials which undergo drastic changes as a function of externally applied forces. The development of any technologically advanced society constantly requires new materials with specific properties that could be used for various applications – smart materials. The term "smart" means that the materials are characterized by properties that can be significantly and quickly changed by means of an external and controllable medium/force (temperature, pressure (hydrostatic or uniaxial), light intensity, magnetic or electric external fields, pH, etc.).

In order to further illustrate the concept, we list in figure 1 different properties of possible smart materials. This figure and the definition of "smartness" of the materials may facilitate the formulation of the criteria in the search for smart materials. Since most of the material properties are defined by their electronic energy spectra it should be clear that those materials with the electronic orbital degeneracies should be of primary interest. In those materials it is easy to significantly change the material properties with the application of external forces. One should also note that the most dramatic (anomalous) property changes take place in materials at the phase transitions. The above considerations point to the Jahn-Teller crystals–multiferroics as a likely class of compounds for the search for smart materials.

The term "smart material" was coined more than 40 years ago in connection with the discovery of the smart memory compound Nitinol (Ni-Ti alloy discovered at the Naval Research Laboratory (Maryland, USA) [1]). This material undergoes a first order structural phase transition from the cubic (austenite) to the tetragonal (martensite) phase as a function of temperature. Correspondingly Nitinol can be changed from a state with one shape to another shape and back, with the change of temperature

IOP Conf. Series: Journal of Physics: Conf. Series **833** (2017) 012007 doi:10.1088/1742-6596/833/1/012007

or external pressure. Therefore that material "remembers" its shape. The strain change at these transformations is huge – up to 7-10%.

Here we point out that in Jahn-Teller ferroelastics the spontaneous strain is also large [2] and can reach magnitudes similar to that of Nitinol. Similarly, materials with the first order Jahn-Teller structural transition exhibit hysteresis and memory effect.

It was found by [3] that big strains in smart memory materials could be related to domain reorientation in the martensite crystal phase. Soon after that magnetic shape memory materials were discovered [4]. In these materials the initial sample shape was recovered by an external magnetic field. It was suggested that the mechanism of the crystal memory is based on the domain magnetic reorientation accompanied by huge martensite reorientation magnetostriction. The reorientation of the domains, according to reference [5], depends significantly upon the magnetic anisotropy of the compounds.

Type of smart material	Input	Output
Piezoelectric	Deformation	Potential difference
Electrostrictive	Potential difference	Deformation
Magnetostrictive	Magnetic field	Deformation
Thermoelectric	Temperature	Potential difference
Shape memory crystals	Temperature	Deformation
Magnetic shape memory crystals	Magnetic field	Deformation
Photochromic	Radiation	Color change
Thermochromic	Temperature	Color change

Figure 1. Examples of types of smart materials and their properties.

2. Jahn-Teller magnetic shape memory materials

Jahn-Teller crystals are characterized by huge static and dynamic magnetostriction [6]. Figure 2 shows some experimental and theoretical results for static and dynamic magnetostriction of Jahn-Teller elastics [7, 8]. Due to the huge magnetostriction of Jahn-Teller crystals, domain reorientation by an external magnetic fields in these materials is also possible. That indicates that magnetic shape memory materials exist among the Jahn-Teller elastics. Moreover the mechanism of the domain reorientation in Jahn-Teller systems is quite well understood. Therefore the elastic anisotropy can play a much more significant role in the reorientation than the magnetic anisitropy. The magnetic shape memory effect in Jahn-Teller crystals can take place even in completely non-magnetic crystals. The magnetostriction of the paramagnetic systems is big enough for the reorientation processes.

We will here focus our discussion on the mechanism of the magnetic shape memory of the crystals. Analysis of the magnetoelastic properties of the Jahn-Teller crystals has shown [9] that the multisublattice crystals exhibit a property similar to metamagnetism in Ising antiferromagnets. That property was named metamagnetoelasticity – the overturn of a sublattice of the local distortions in sufficiently strong external magnetic fields. This effect was first experimentally observed in some rare earth compounds [10], then explained theoretically [11], and after that observed in different Jahn-Teller compounds (figures 3 and 4). It is important to mention that the experimental data corresponding to domain reorientation look very much like metamagnetoelasticity curves: the crystal strain jumps up in magnitude as a function of the external magnetic field when it reaches some critical value. However, the mechanisms of these two phenomena are different. In case of the domain

IOP Conf. Series: Journal of Physics: Conf. Series **833** (2017) 012007 doi:10.1088/1742-6596/833/1/012007

reorientation the magnetic field (Zeeman interaction) has to overcome the structural sublattice anisotropy, but at the metamagnetoelasticity the overturn of the sublattice distortions takes place when the Zeeman interaction becomes stronger than the intersublattice virtual phonon exchange interaction responsible for the cooperative Jahn-Teller effect. Naturally the magnitudes of the magnetic fields for these two phenomena are different. New magnetostriction experiments with the single domain crystals will elucidate the correct mechanism of the shape memory effect.

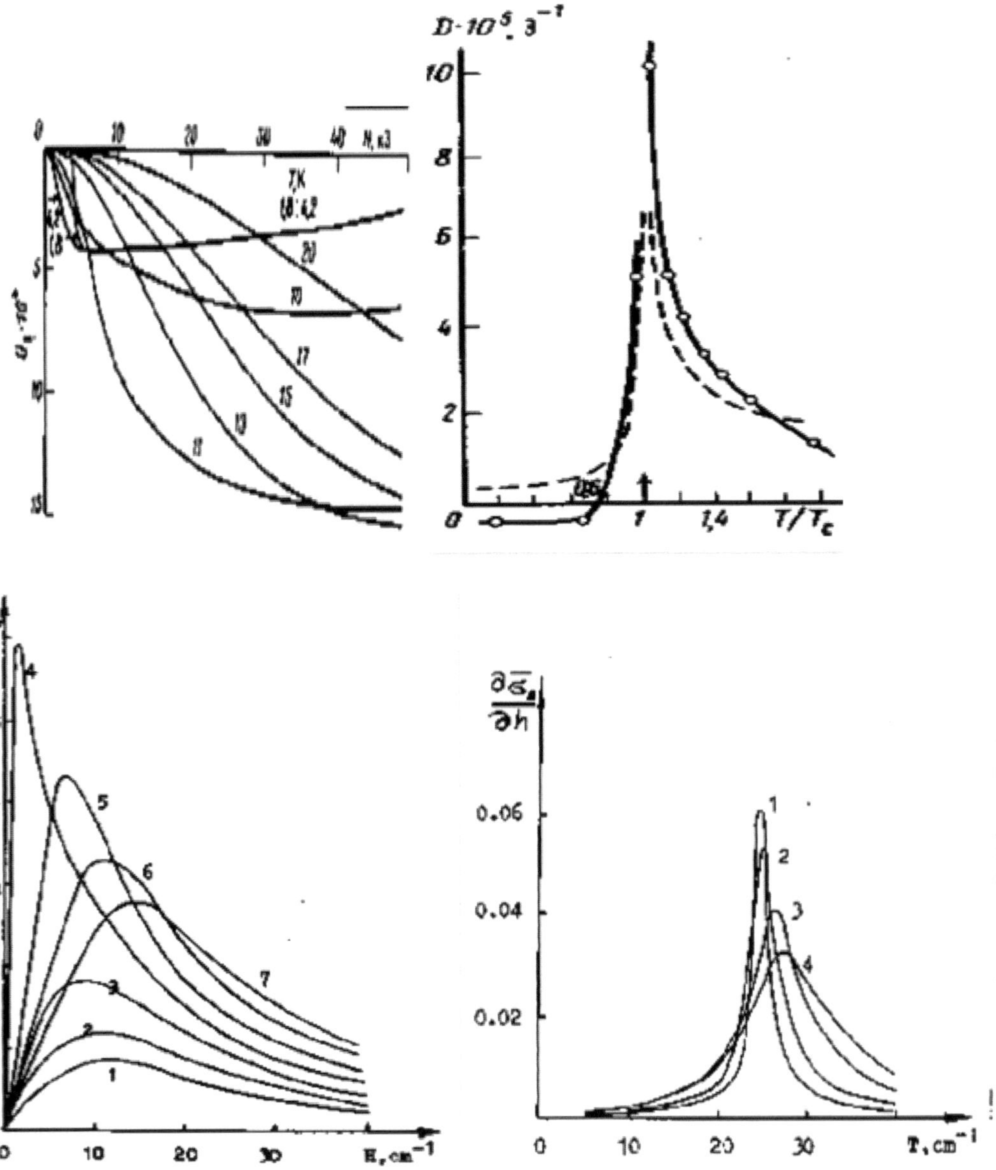

Figure 2. The top two graphs are experimental data by V.I Sokolov and Z.A. Kazey [7], the two graphs below are the theoretical data for the dynamic magnetostriction in $TbVO_4$ and $Tb_{1-x}Gd_xVO_4$ crystals by M.D. Kaplan and B.G. Vekhter [8].

While initially the magnetic smart memory was found among the semiconducting and metallic compounds, lately the Jahn-Teller dielectrics with XY-ordering attracted attention for that smart property [12, 13]. On one hand these types of magnetic crystals are undergoing a transition (figure 5) responsible for the jump of the crystal strain with the field. On another hand, it was shown in [13] that the strain change with the magnetic field is much faster in dielectrics than in conductive materials. The explanation of that is related to the absence of eddy currents in dielectrics. This property makes the Jahn-Teller dielectric crystals more attractive for different applications. On figures 6-8 different domain magnetostriction and hysteresis are shown for the XY-ordered MnV_2O_4 crystal. Similar effects were observed [14] and explained [15] in the colossal magnetoresistance $LaMnO_3$ compounds.

The above argument supports the statement that memory (temperature and/or magnetic field induced) is a smart property typical of the Jahn-Teller crystals. However in case of the magnetic shape memory effects the future magnetostriction measurements in single and multidomain crystal samples will elucidate the real mechanism responsible for these properties.

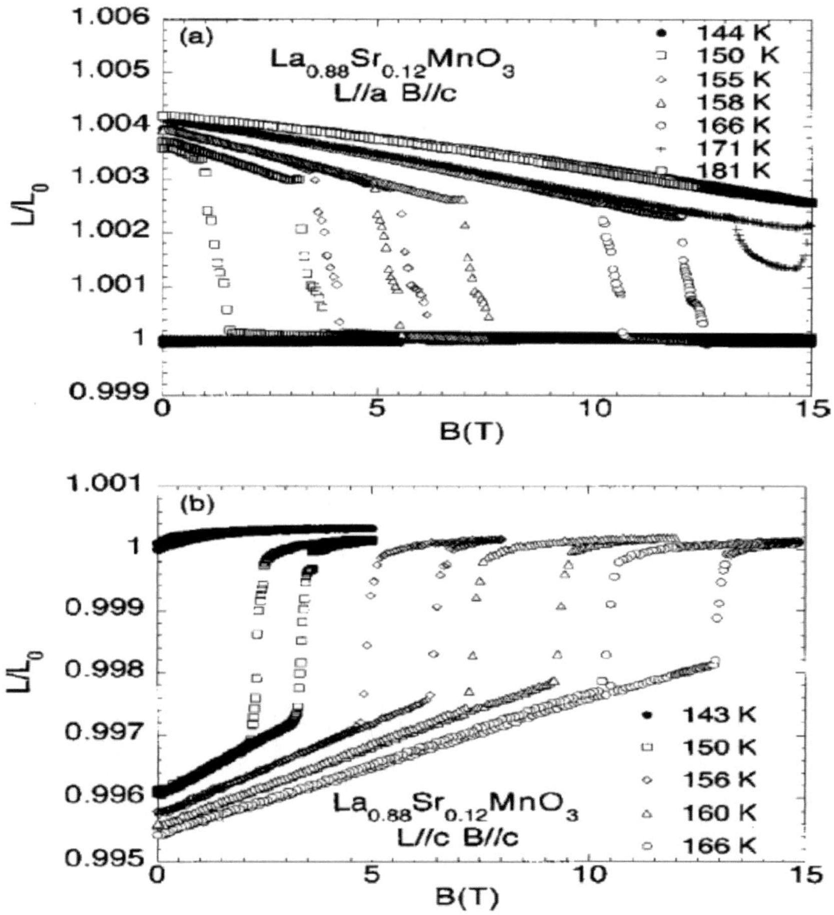

Figure 3. Magnetic field dependence of magnetostriction in colossal resistance compound [14].

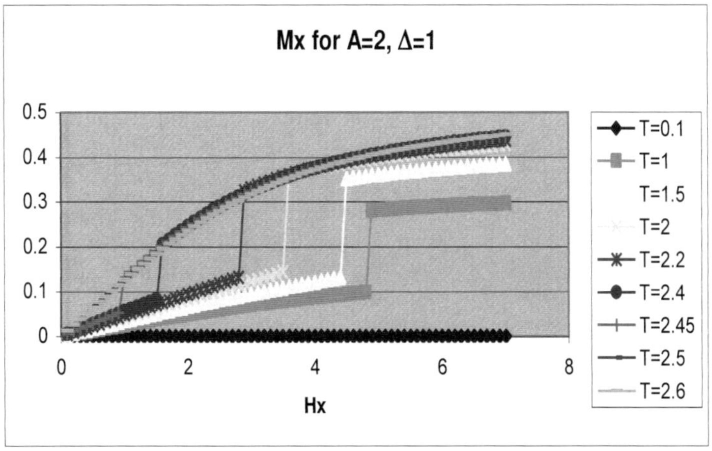

Figure 4. Metagnetoelasticity manifestation in XY-ordered materials [15] (A, H_x, T, and Δ are given in units of energy Δ).

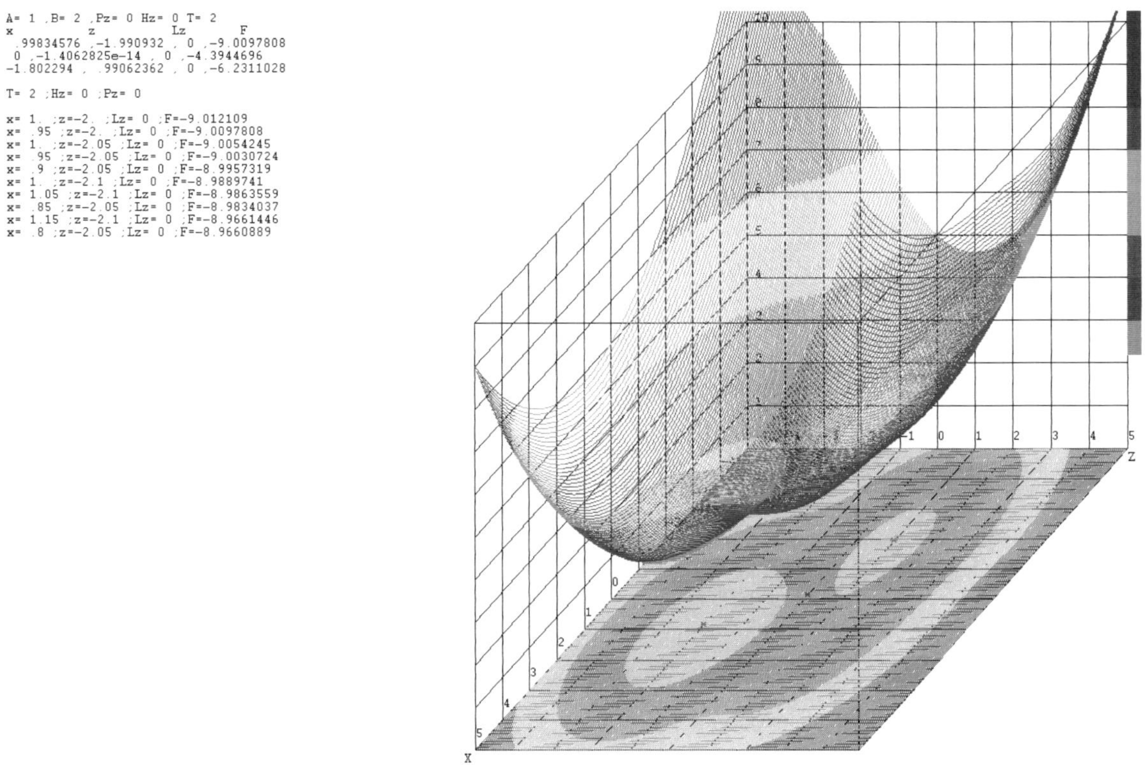

Figure 5. Example of a 3-dimensional surface of the crystal free energy as a function of the structural order parameters X and Z for different molecular field constants A and B at competing XY and ZZ structural orderings (all parameters are dimensionless, the minima of the free energy correspond to the different crystal spontaneous strains/shapes).

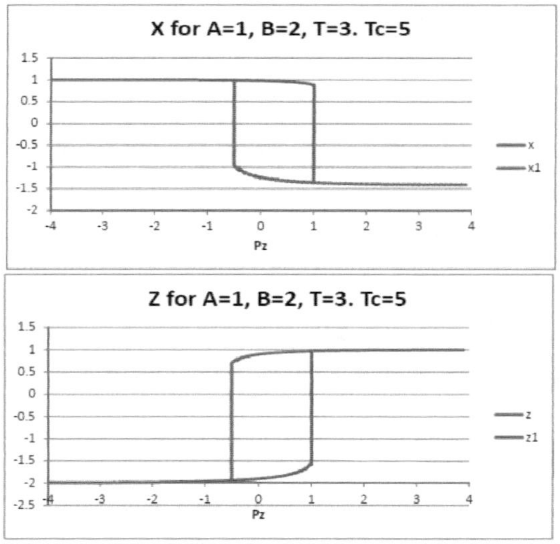

Figure 6. Hysteresis curve calculated from the 3-d free energy surfaces similar to shown in figure 5.

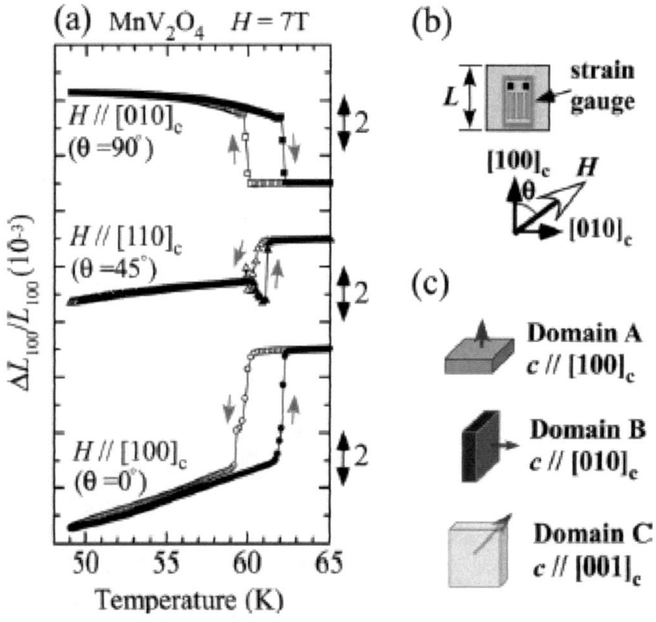

Figure 7. Temperature dependence of the magnetostriction at different orientation of magnetic field [13].

3. Jahn-Teller multiferroics as smart materials

A group of smart properties is found in Jahn-Teller multiferroics. These are crystals with simultaneous structural, magnetic, and/or electric (ferro-, antiferrotype) ordering. They all contain at least one sublattice of ions (structural units) with orbital electronic degeneracy or pseudo degeneracy. Some examples of ordered structures in Jahn-Teller multiferroics were discussed earlier [6, 16]. A smart property – magnetoelectricity – was also widely discussed [16, 17] for a Jahn-Teller ferroelastic, antiferromagnetic, antiferroelectric $TbPO_4$ crystal.

Among the multiferroics, the vibronic magnets [18] are of a special interest. The Jahn-Teller vibronic magnets are characterized by magnetic moments whose magnitude is linearly proportional to the vibronic (electron-phonon) coupling

$$M_z = g_z \beta \bar{S}_z$$

$$M_z = \frac{g_z\beta}{Z_0}\frac{2A\bar{P}H_x}{E_0^2}\left(e^{\Delta/kT} + \frac{1}{3}e^{\Delta/kT\left(1+\frac{2E_0^2}{\Delta^2}\right)} + \frac{E_0^2}{3E_0^2+\Delta^2}e^{-\frac{E_0^2}{\Delta kT}}\right)$$

$$Z_0 = e^{\frac{\Delta}{kT}}\left(1+e^{\frac{4E_0^2}{\Delta kT}}\right)+e^{-\frac{E_0^2}{\Delta kT}} \qquad (1)$$

$$M_z \equiv M_{zvibr} \sim A \sim \frac{V^2}{\hbar\omega}$$

Here the increase of symmetry strain $\bar{P}$ induces an increase of the magnetic moment. The constant *A* in (1) is the Jahn-Teller molecular field constant that is proportional to the square of the electron-phonon coupling. This effect is similar to the macroscopic effect of piezomagnetism which is one of the smart material properties. Some vibronic magnets could additionally have a ferro- or antiferroelectric ordering. In this case the corresponding smart properties are electrostriction and the dielectric response (susceptibility).

4. Applications of Jahn-Teller smart materials

One of the "traditional" smart properties of the Jahn-Teller crystals is the dependence of the ultrasound velocity upon temperature and external magnetic field. Examples of the measurements and calculations of these properties in rare earth and transition metal compounds at the low and room temperatures were discussed by Kaplan and Vekhter in reference [6].

Lately a new smart material property attracted much attention in connection with its numerous applications of self-assembly. This property could be considered as a special type of structural phase transition. It is not surprising that it was observed in the class of the Jahn-Teller crystals with spinel structure. An example of experimentally found self-assembly among the Jahn-Teller compounds is shown on the figure 8.

Overall, it should be mentioned that the idea of the Jahn-Teller smart materials had already been recognized in industry for a long time. That is why some applications of the Jahn-Teller compounds have been known for many years and are protected by various patents.

One of the first patents related to the Jahn-Teller effect was received by Bell Telephone Laboratories Inc. about 50 years ago – "Temperature-stable Sonic Transmission Elements Comprising Crystalline Materials Containing Jahn-Teller Ions" (inventors are D. Fraser, E. Gyorgy, R. Le Craw, F. Schnettler). The application of the materials is based on the properties discussed before in this presentation, the temperature dependence of the velocity of the acoustic waves in Jahn-Teller crystals that is proportional to the square-root of the modulus of elasticity ($V=(C/\varrho)^{1/2}$ and the possibility of changing this velocity by means of the external magnetic and fields.

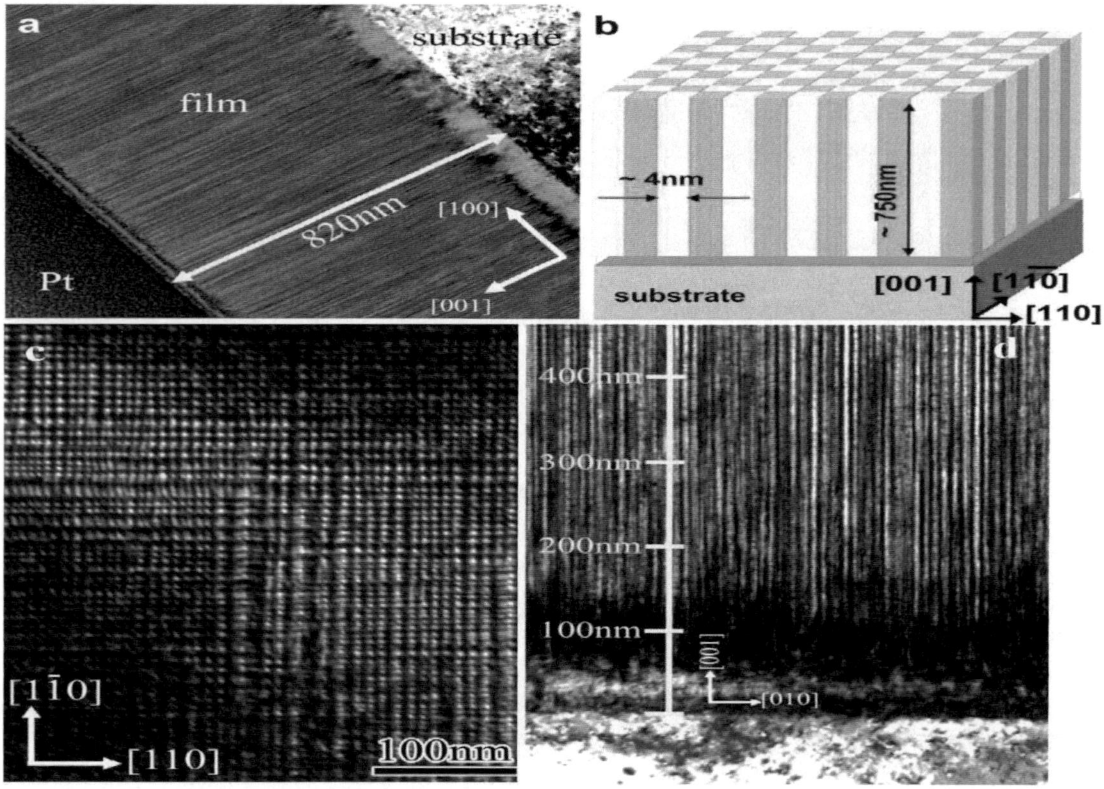

Figure 8. Self-assembly in Jahn-Teller crystals [19].

In the seventies a Patent US 3961289 was awarded to Matsushita Electric Industrial Co., Ltd for "Ultrasonic Delay Material" (inventor Y. Kino). The material under discussion is "a spinel-type oxide exhibiting a cooperative Jahn-Teller effect". This material shows an extraordinary reduction of sound velocity "there through at temperatures near the Jahn-Teller phase transition temperature."

M. Kaplan, B. Vekhter, I. Bersuker, M. Maizenberg, V. Moshkovich, V. Bobrov (Patent SU 1013837 A) in the eighties suggested a new method for measuring the external magnetic fields based on the dynamic magnetostriction of the Jahn-Teller crystals. This method has many different applications. One of them is related to the non-destructive testing of materials. The corresponding device is shown in figure 9.

In the nineteen nineties it was established that the Jahn-Teller effect is fundamentally important for the understanding of the mechanism of the working of the lithium batteries. For these batteries the best positive active electrode is a crystal with the cooperative Jahn-Teller effect. The first positive electrode material for the lithium batteries was patented by Japan Storage Battery Co, Ltd. (this material is LiMn2O4, inventors - K. Amine, H. Yasuda, Y. Fujita, 1997). The first Jahn-Teller materials for the lithium batteries were developed by J. Goodenough and his co-workers [21]. Today there are many different patented Jahn-Teller materials applicable as the electrodes for lithium batteries.

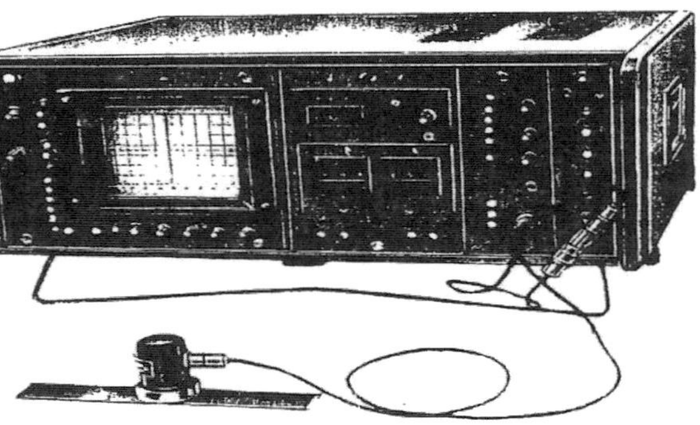

Figure 9. Acousto-electronic device for magnetic field measurements. The attached converter contains the Jahn-Teller crystal with big dynamic magnetostriction.

Lately a Patent WO 2006133129 A2 was issued to Rutgers University for "Nano-Scale Self Assembly in Spinels Induced by Jahn-Teller Distortion" (inventors – S.-W. Cheong, S.Yeo). It is about a method for making a self-assembled spinel having an ordered nanocrystal superlattice. Without using any organic liquids it is possible to fabricate a nanoscale checkerboard pattern of square rods. The ordered array is the result of the spinodal decomposition between Jahn-Teller-active and – inactive ions in $ZnMnGaO_4$ compounds. The picture of the ordered ion structure is shown on the figure 8.

5. Conclusion

In conclusion we want to emphasize and call attention to the fact that it should be accepted that a) the Jahn-Teller materials are mostly smart materials; b) in the search of a material with a required smart property the Jahn-Teller class of materials is very attractive; c) the Jahn-Teller multiferroics are especially promising type of smart materials [20]; d) new magnetostriction experiments with single domain crystals are necessary for elucidating the mechanism of the magnetic shape memory.

References

[1] Buehler W J, Gilfrich J W and Willey RC 1963 *J. Appl. Phys.* **34** 145
[2] Gehring G A and Gehring K A 1975 *Rep. Prog. Phys.* **38** 1
[3] Otsuka K and Wayman C M (eds.) 1998 *Shape Memory Materials* (Cambridge University Press)
[4] Ullakko K, Huang J K, Cantner C, Kokorin V V and O'Handley R C 1996 *Appl. Phys. Lett.* **69** 1966
[5] O'Handley R C., Murray S J, Marioni M, Hembach H and Allen S M. 2000 *J. Appl. Phys.* **87** 4712
[6] Kaplan M D and Vekhter B G 1995 *Cooperative Phenomena in Jahn-Teller Crystals* (New York, London: Plenum Press)
[7] Sokolov V I and Kazey Z A 1988 in *Physics of Magnetic Materials*, Eds. Gorzkowski W, Lahovich H K, Szymczak H (World Sci. Publ.) p 444
[8] Vekhter B G, Kaplan M D, Sokolov V I and Kazey Z A 1986 *JETP Lett.* **43** 369
[9] Kaplan M D 1982 *JETP Lett.* **35** 105

[10] Leask M J M, Troper A C and Wells M R 1981 *J. Phys.* C **14** 3481
[11] Kaplan M D 1982 *Phys. Stat. Solidi* B **112** 351
[12] Suzuki T, Katsumura M, Taniguchi K, Arima T and Katsufugi T 2007 *Phys. Rev. Lett.* **98** 127203
[13] Nii Y, Abe N, Taniguchi K and Arima T 2012 *Appl. Phys. Lett.* **100** 051905
[14] Nojiri H, Kaneko K, Motokawa M, Hirota K, Endoh Y and Takahashi K 1999 *Phys. Rev.* B **60** 4142
[15] Kaplan M D and Zimmerman G O 2008 *Phys. Rev.* B **77** 104426
[16] Kaplan M 2006 *Proc. of Int. Sugano Symp. on Left-Right Symmetry in Physics, Chemistry and Biology* (Tsukuba) p 10
[17] Bluck S and Kahle H G 1988 *J. Phys.* C **21** 5193
[18] Kaplan M D 2013 *J. Phys. Conf. Ser*. 428, 012034
[19] Park S, Horibe Y, Asada T, Wielinsky L S, Lee N, Bonano P L, O'Malley S M, Sirenko A A, Kazimirov A, Tanimura M, Gustafsson T and Cheong S W 2008 *Nano Letters* **8** 720
[20] Kaplan M D 2015 *BIT's 4th World Congress of Advanced Materials, Proc*. 081
[21] Mizushima K, Jones P C, Wiseman P J and Goodenough J B 1980 Mater. Res. Bull. **15** 781

IOP Conf. Series: Journal of Physics: Conf. Series **833** (2017) 012008 doi:10.1088/1742-6596/833/1/012008

Andrew Liehr and the structure of Jahn-Teller surfaces

Liviu F. Chibotaru and Naoya Iwahara

Theory of Nanomaterials Group, Department of Chemistry, Katholieke Universiteit Leuven, Celestijnenlaan 200F, 3001 Leuven, Belgium.

Abstract. The present article is an attempt to draw attention to a seminal work by Andrew Liehr "Topological aspects of conformational stability problem" [1, 2] issued more than half century ago. The importance of this work stems from two aspects of static Jahn-Teller and pseudo-Jahn-Teller problems fully developed by the author. First, the work of Liehr offers an almost complete overview of adiabatic potential energy surfaces for most known Jahn-Teller problems including linear, quadratic and higher-order vibronic couplings. Second, and most importantly, it identifies the factors defining the structure of Jahn-Teller surfaces. Among them, one should specially mention the minimax principle stating that the distorted Jahn-Teller systems tend to preserve the highest symmetry consistent with the loss of their orbital degeneracy. We believe that the present short reminiscence not only will introduce a key Jahn-Teller scientist to the young members of the community but also will serve as a vivid example of how a complete understanding of a complex problem, which the Jahn-Teller effect certainly was in the beginning of 1960s, can be achieved.

1. Introduction

After the publication of the Jahn-Teller theorem in 1937 [3, 4], there was a period of almost two decades during which the Jahn-Teller effect practically did not attract attention except via a few albeit very important publications [5, 6, 7, 8]. The burst of interest for Jahn-Teller effect begins with the works of Moffitt and Liehr [9, 10], Moffit and Thorson [11], Öpik and Pryce [12], Liehr and Ballhausen [13, 14] and Longuet-Higgins et al. [15] in the second half of 1950s. In this "golden" period, Andrew Liehr made seminal contributions which initiated several directions in Jahn-Teller research. He was one of the authors who first described theoretically the nuclear dynamics (dynamical Jahn-Teller effect) and the effect of quadratic vibronic coupling in Jahn-Teller systems [10]. He also was one of the first who identified the Jahn-Teller splitting of the spectroscopic bands [9]. However his most important contribution is related to the investigation of static Jahn-Teller problems and the identification of the main factors defining the structure of Jahn-Teller surfaces. His research in this domain, as well as of other authors at that time, was reviewed in a more than a hundred pages paper comprising the treatment of static Jahn-Teller (Part I) [1] and pseudo-Jahn-Teller (Part II) [2] problems. This work represents the first comprehensive review on Jahn-Teller effect and still remains the most complete overview of static Jahn-Teller effect. In this work, Andrew Liehr gives adiabatic potential energy surfaces (APES) for almost all known Jahn-Teller problems including linear, quadratic and higher-order vibronic couplings. He also identifies the factors influencing the structure of Jahn-Teller surfaces, among which the minimax principle is of paramount importance. In the present article we overview

IOP Conf. Series: Journal of Physics: Conf. Series **833** (2017) 012008 doi:10.1088/1742-6596/833/1/012008

and discuss these two basic aspects of the Liehr's work.

2. Adiabatic potential energy surfaces of Jahn-Teller problems

In the Part I [1], Liehr derived the analytical expressions for APES of almost all Jahn-Teller systems. His approach was to show specific systems and to visualize the APES and the pseudorotation of the Jahn-Teller distortions corresponding to nuclear motion at the bottom of the APES.

Liehr starts by considering the Renner-Teller effect in linear systems (C_2H_2 and triatomic molecule), and then investigates the vibronic problems of two-dimensional hydrocarbon molecules with axial symmetry (C_nH_n, n = 3, 4, ..., 8). Since the π electronic orbitals of these systems can be twofold degenerate, the $E \otimes e$ Jahn-Teller problems for $n = 3, 5, 6, 7$ and the $E \otimes (b_1 + b_2)$ Jahn-Teller problems for $n = 4, 8$ are considered. For these systems, not only linear vibronic coupling but also higher order vibronic coupling with respect to the Jahn-Teller and non-Jahn-Teller active modes was taken into account. As an example, well-known results for the triangular molecule are reproduced in figure 1 and figure 2. Figure 1 shows the APES of the $E \otimes e$ Jahn-Teller system with trigonal symmetry including the higher order vibronic coupling and Figure 2 shows the pseudorotations of the Jahn-Teller deformations.

After investigating the Jahn-Teller effect in the planar molecules, he moves to his study in three-dimensional regular polyhedra such as octahedron, cube and tetrahedron. He first discusses the $E \otimes e$ problem in a similar fashion as for the planar systems. Figure 3 reproduces the well-known picture of pseudorotation of Jahn-Teller deformations in a cube with doubly degenerate electronic state. Further he discusses the systems with triply degenerate T_1 and T_2 electronic states in detail without spin-orbit coupling. He concentrates, in particular, on the $T \otimes (e + t_2)$ Jahn-Teller problems including all quadratic terms. The latter involves the cross terms between e and t_2 vibrational modes as well as the coupling to the bilinear coordinates of e and t_2 vibrational modes. This Jahn-Teller system has two types of minima: the first one after e deformation only and the second, non-trivial one, after both e and t_2 deformations. The pseudorotation of the latter is reproduced below figure 4). Turning to systems with the spin-orbit coupling, Liehr considers the ^2T electronic term which splits into Kramers doublet Γ_7 and four-fold degenerate Γ_8 spin-orbit multiplet. He treats the vibronic problem including the couplings between Γ_7 and Γ_8 multiplets, and then focuses on the Γ_8 quadruplet. The latter displays conical intersection at octahedral or tetrahedral symmetry point.

At the end of his analysis of the regular polyhedra, the Jahn-Teller effect in the icosahedral systems (icosahedron and dodecahedron) is also briefly mentioned. At variances to other problems, the form of APES is not derived, only the symmetry of expected extrema after Jahn-Teller active modes is discussed using the "adiabatic correlation" [16]. He predicted that the symmetry of the system becomes as high as D_{5d} or D_{3d} for all degenerate electronic states. Thus he writes:

> "The extrema of icosahedral (I_h) potential energy surfaces lie at the six equivalent D_{5d} geometries, the ten equivalent D_{3d} geometries, and the fifteen equivalent C_{2h} geometries..."

Furthermore, he mentioned the possibility that the system can further deform due to the double degenerate electronic states at D_{5d} and D_{3d} points, an assumption which doesn't prove to be true because the ground adiabatic state is already non-degenerate at these symmetry points [17, 18].

Finally, he investigated the irregular polyhedra (mono- and bipyramidal structures). The most important result here is the influence of the spin-orbit coupling on the Jahn-Teller effect. Contrary to the case of regular polyhedra, the spin-orbit coupling lifts the degeneracy of the electronic states, which modifies the radial positions of the minima and make the minima of

IOP Conf. Series: Journal of Physics: Conf. Series **833** (2017) 012008 doi:10.1088/1742-6596/833/1/012008

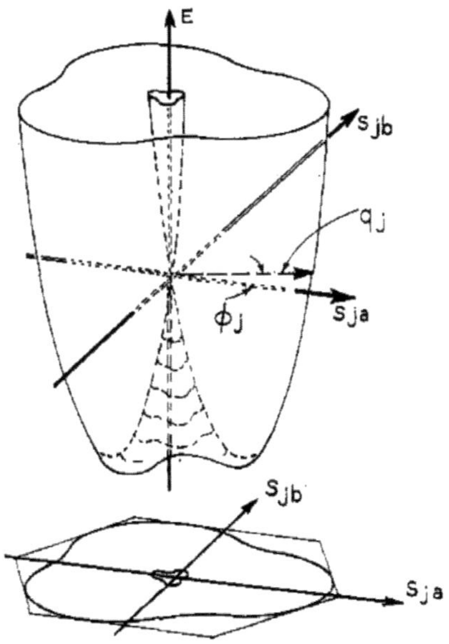

Figure 1. A part of figure 2d from Ref. [1]. The APES of the $E \otimes e$ Jahn-Teller system. S_{ja} and S_{jb} are the $e\theta$ and $e\epsilon$ mass-weighted vibrational coordinates, angle ϕ is the polar coordinates in the space of the e vibrational mode [19].

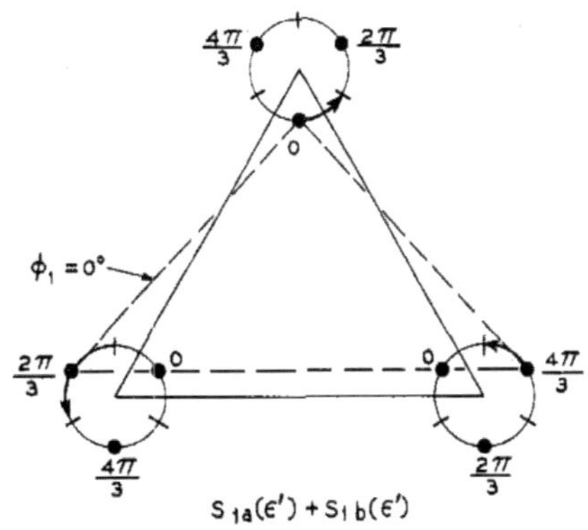

Figure 2. A part of figure 2b from Ref. [1]. Pseudorotation of the Jahn-Teller distortions of $E \otimes e$ Jahn-Teller system with triangular structure. The angles $\phi = 0$, $2\pi/3$, $4\pi/3$ stand for the position of the minima in the APES.

APES inequivalent (figure 5). The lift of the degeneracy can be predicted based on the group theoretical analysis, the analytical expression giving us better insight into the structure of the APES.

3. Factors defining the structure of Jahn-Teller surfaces

As a summary of his analysis of APES in various Jahn-Teller problems, Andrew Liehr formulates fourteen insights (rules, laws) governing their structure. Many of them remain to be of fundamental importance for the understanding of the Jahn-Teller effect (the numbers below correspond to those in the abstract of Ref. [1]). Below we quote some of them.

The first conclusion concerns the basic group-theoretical technique (descent of symmetry) describing the Jahn-Teller effect:

1) *Principle of mathematical inheritance*: the Jahn-Teller-Renner behavior of a polyatomic group theoretic system is completely determined by that of its elemental subgroups.

Further he gives conclusions directly emerging from the group-theoretical analysis:

2) *The formation of Jahn-Teller family*: isomorphous point groups exhibit isomorphous Jahn-Teller deportments.

As long as the irreducible representations of electronic wave functions and vibrational modes are the same, we obtain the same Jahn-Teller Hamiltonian and the same type of Jahn-Teller

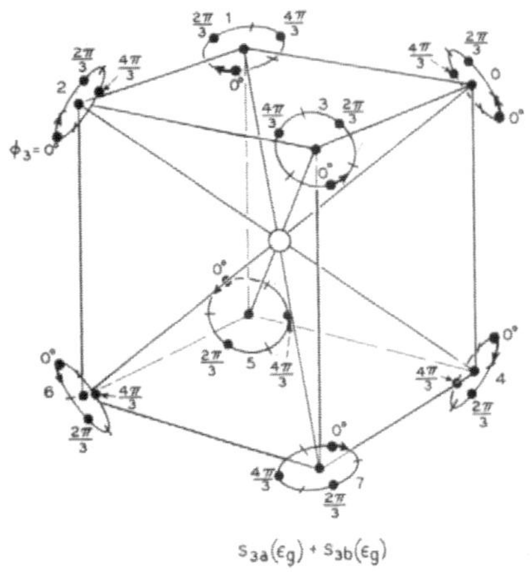

Figure 3. Figure 15c from Ref. [1]. Pseudorotation of the Jahn-Teller distortions of the $E \otimes e$ Jahn-Teller system with cubic structure.

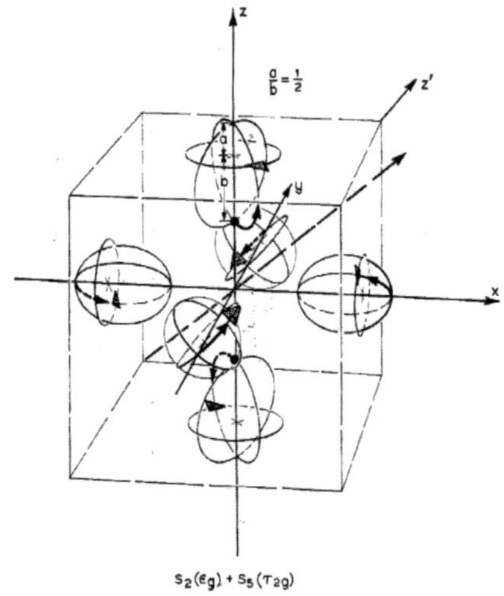

Figure 4. Figure 30a from Ref.[1]. Pseudorotation of the Jahn-Teller distortions of $T \otimes (e \oplus t_2)$ Jahn-Teller system with octahedral structure.

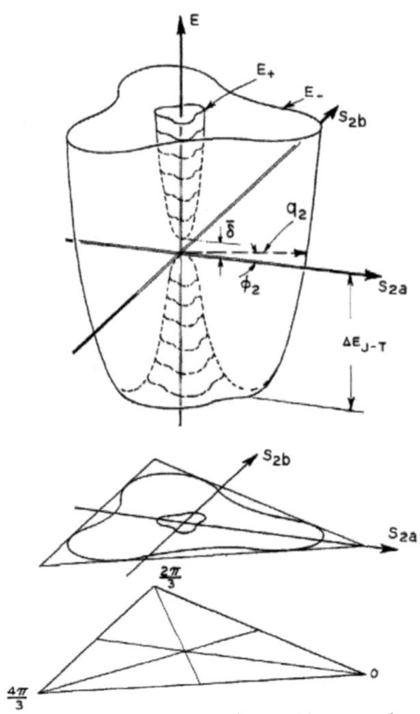

Figure 5. A part of figure 46 from Ref. [1]. The effect of the spin-orbit coupling on the adiabatic potential energy surface of bipyramidal system.

IOP Conf. Series: Journal of Physics: Conf. Series **833** (2017) 012008 doi:10.1088/1742-6596/833/1/012008

surfaces. Further, as the result of the thorough study of the planar systems, he reaches the next conclusion:

> 3) *Law of prime numbers*: Whereas the prime number groups produce a single unique Jahn-Teller-Renner topography, the non-prime number groups produce all these topographies which are required by the principle of mathematical inheritance (1).

Besides static Jahn-Teller problem, he also addresses general properties of vibronic wave functions:

> 4) *Symmetry transcendence*: The dynamical quantization and the topography of the Jahn-Teller problems are completely specified by group theoretic and permutational symmetry precepts.

Liehr observes that simple analytical solutions for APES are obtained at the symmetry points:

> 6) *Factorization theorem*: the Jahn-Teller-Renner energy resolvants factor only at locations of high nuclear symmetry.

Using these solutions, he draws the information about the shape of the entire APES in several specific cases:

> 12) *Canon of dimensional variability*: Doubly degenerate electronic states possess continuous Jahn-Teller-Renner electronic energy surfaces[1], but triply degenerate electronic states possess disjoint electronic energy surfaces in certain directions.

The complete development of the rule 12) was given later by Pooler [20, 21].

Liehr also discusses the effect of spin-orbit coupling on the APES:

> 9) *Spin-orbit law*: spin-orbit forces remove cuspidal Jahn-Teller radial electronic energy singularities for geometries less regular than the cube.

As mentioned in Sec. 2, he compares the spin-orbit Γ_8 states in cubic system with ^2E state in bipyramidal system, and finds that the spin-orbit coupling removes the degeneracy at high-symmetry point in the latter. Such splitting can be understood from the group theoretical analysis, whereas the importance of this statement is due to the analytical expressions which give additional insight.

4. The minimax principle

A special attention deserves the principle put forward by Andrew Liehr as the minimax rule (or minimax law) [1]:

> 5) *Minimax rule*: Although not mathematically required, the symmetry of the stable Jahn-Teller conformation is always the highest symmetry which is yet compatible with the loss of the initial inherent symmetry.

In the summary of the review he formulates his finding in the form of a theorem [2]:

> *Theorem 2 (Conjectured)*: The group theoretical symmetry of the stable Jahn-Teller conformation is the highest which is yet compatible with the loss of initial electronic degeneracy.

He emphasizes that this theorem is actually a conjecture made on the basis of available solutions for Jahn-Teller problems and not a rigorous demonstration, at variance, e.g., with the Jahn-Teller theorem (Ref. [1], p. 457):

[1] Liehr uses here the term "continuous" in the sense of "equipotential", i.e., speaks about the trough at the bottom of the APES.

> "...there exist no symmetry arguments which will prevent a molecular system from assuming a structure whose eurythmic elements are those of a subgroup of the total (original) molecular group, which subgroup is of a *lower* order than the minimal order necessary to lift the initial electronic degeneracy. However, from the analytical considerations outlined in this article, it is seen *that a free molecule in a degenerate electronic state will descend in symmetry only as far as the nearest point group which will remove the degeneracy.* This statement can be recast into the following form: *a Jahn-Teller molecule tends to preserve as much symmetry as possible while acting to remove its inherent electronic degeneracy.*"

One should outline clearly the statement contained in the Liehr's minimax principle. It emerges clearly from the text (e.g., his analysis of the $T \otimes (e \oplus t_2)$ problem as mentioned in Sec. 2) that his minimax principle refers to *the lowest order* of the vibronic coupling sufficient to create minima on the lowest APES of *degenerate* electronic state. Then one should consider two types of Jahn-Teller problems. The first type are the Pooler's Jahn-Teller problems (or problems with a trough) which arise when the symmetrized square (antisymmetrized square for double valued representation) of electronic irreducible representation (irrep) Γ_0 contains only one single nuclear irrep Γ besides the total symmetric one, A ($[\Gamma_0^2] = A \oplus \Gamma$) [21]. These systems have a trough in the ground APES and the minima can be obtained only by considering quadratic vibronic coupling (eaxmples: $E \otimes e$, $T \otimes h$ and several other Jahn-Teller problems cited in the Pooler's paper. The second type of Jahn-Teller problems are those in which the symmetrized square of Γ_0 contain more than one low-symmetry nuclear irreps (active nuclear modes): $[\Gamma_0^2] = A \oplus \Gamma_1 \oplus \Gamma_2 \oplus \cdots$. In such systems the minima appear in the linear order of vibronic coupling (examples: $E \otimes (b_1 \oplus b_2)$, $T \otimes (e \oplus t_2)$, etc).

In the second type of Jahn-Teller problems, e.g., $T \otimes (e \oplus t_2)$, the minima are realized (depending on the strength of Jahn-Teller coupling to each set of active modes) either in the e-set or t_2-set of active nuclear vibrations. We emphasize that there are no minima mixing nuclear coordinates of different subsets (e.g., e and t_2) when the linear vibronic coupling only is considered. This is actually a general feature of linear Jahn-Teller problems with several different sets of active nuclear coordinates, which always allows to consider the Jahn-Teller problem with individual subsets when the minima of APES are investigated (e.g., $T \otimes e$ and $T \otimes t_2$ instead of $T \otimes (e \oplus t_2)$). This means that we always deal with Jahn-Teller problems with one single set of nuclear vibrations when investigate the minima of APES, to which the Liehr's minimax principle perfectly applies.

As mentioned above, the Liehr's principle is only concerned with the loss of initial electronic degeneracy by the minimal order of vibronic coupling which already provide minima on the APES. Further symmetry lowering due to higher-order vibronic coupling ($T \times (e \oplus t_2)$ problem mentioned in Section 2) including pseudo-Jahn-Teller effect (**figure** 7.31 in Ref. [22]) will affect only slightly the already existing equilibrium distortion because the lowest-order vibronic couplings (when non-zero) usually give a dominant contribution. The Liehr's principle concerns precisely these dominant contributions of equilibrium Jahn-Teller distortions.

Despite being conjectured more than half a century ago, this fundamental principle remains valid till our days and became reinforced by new examples of Jahn-Teller problems. One exception from this general rule is the icosahedral $T \otimes h$, where the minima on the lowest APES can correspond to either the highest subgroup D_{5d} (in full accord with the minimax principle) or to the next one, D_{3d}, depending on the sign of the parameter defining the warping of the trough due to quadratic vibronic coupling [23, 24]. The Liehr's minimax rule has broad connection with the Curie's principle [25, 26, 27], where the latter states that the greatest common subgroup of the symmetries of the system and perturbations characterizes the resulting effect of the symmetry. Liehr's minimax rule provides further information on the symmetry of the structure at the minimum of the APES. Thus, the importance of the Liehr's minimax principle can hardly be overstated. The distortion of a high-symmetry nuclear configuration can proceed, according to the Jahn-Teller theorem, after any of the non total symmetric nuclear coordinates $Q_{\Gamma\gamma}$ (or their linear combinations) which are contained in $[\Gamma_0^2]$: $\Gamma \subset [\Gamma_0^2]$. As a rule, Γ corresponds to a degenerate irreducible representation and even there can be several different Γ's entering the $[\Gamma_0^2]$ [19]. This is the reason why the space of Jahn-Teller active nuclear distortions is multidimensional in most Jahn-Teller problems. As a consequence, the possible Jahn-Teller distortions towards a stable electronic configuration represents a multidimensional continuum. The power of the Liehr's principle is that it identifies in this multidimensional space a few directions (corresponding to the highest possible symmetry of the distorted molecule), which *a priori* correspond to stable energy points. The overwhelming fulfilling of the minimax principle in Jahn-Teller problems is of course not accidental and should have a deep physical reason. Its unraveling represents a challenging task for the Jahn-Teller community.

5. Discussion and conclusions

Andrew Liehr was an outstanding scientist pioneering many fundamental aspects of Jahn-Teller effect and its manifestation in real molecules. His major contribution was the investigation of a large variety of static Jahn-Teller and pseudo-Jahn-Teller problems and the identification of the factors defining the basic properties of Jahn-Teller surfaces. Among them, the minimax principle stands out since it represents a statement the importance of which is only overweighed by the Jahn-Teller theorem. Indeed, if the latter proves the structural instability of high-symmetry molecular configurations in orbitally degenerate electronic states, the Liehr's minimax principle tells us how the high-symmetry molecular configurations become distorted. According to Liehr's finding they always tend to preserve the highest symmetry consistent with the loss of orbital degeneracy.

Unfortunately, the research carrier of Andrew Liehr lasted for only one decade. Despite such a short period of time, he was able to make contributions of fundamental importance which deserve to be remembered, especially what concerns the investigation of Jahn-Teller surfaces, a domain to which Andrew Liehr contributed more than any other.

Acknowledgments
We would like to thank Isaac Bersuker, Victor Polinger and Janette Dunn for useful discussions. N. I. is a postdoctoral researcher under Japan Society for the Promotion of Science (JSPS) Postdoctoral Research Fellowships for Research Abroad and also gratefully acknowledges financial support from GOA grant from the KU Leuven.

References

[1] Liehr A D 1963 *J. Phys. Chem.* **67** 389–471 (Part 1)
[2] Liehr A D 1963 *J. Phys. Chem.* **67** 471–494 (Part 2)
[3] Jahn H A and Teller E 1937 *Proc. R. Soc. Lond. A* **161** 220–235
[4] Jahn H A 1938 *Proc. R. Soc. Lond. A* **164** 117–131
[5] Van Vleck J H 1939 *J. Chem. Phys.* **7** 61–71
[6] Van Vleck J H 1939 *J. Chem. Phys.* **7** 72–84
[7] Van Vleck J H 1940 *Phys. Rev.* **57** 426–447
[8] Abragam A and Pryce M H L 1950 *Proc. Phys. Soc. A* **63** 409
[9] Liehr A D and Moffitt W 1956 *J. Chem. Phys.* **25** 1074–1074
[10] Moffitt W and Liehr A D 1957 *Phys. Rev.* **106**(6) 1195–1200
[11] Moffitt W and Thorson W 1957 *Phys. Rev.* **108**(5) 1251–1255
[12] Öpik U and Pryce M H L 1957 *Proc. R. Soc. Lond. A* **238** 425–447
[13] Liehr A D and Ballhausen C J 1957 *Phys. Rev.* **106**(6) 1161–1163
[14] Liehr A D and Ballhausen C J 1958 *Ann. Phys.* **3** 304 – 319
[15] Longuet-Higgins H C, Öpik U, Pryce M H L and Sack R A 1958 *Proc. R. Soc. Lond. A* **244** 1–16
[16] Liehr A D 1957 *J. Chem. Phys.* **27** 476–477

[17] O'Brien M C M 1969 *Phys. Rev.* **187** 407–418
[18] Khlopin V P, Polinger V Z and Bersuker I B 1978 *Theor. Chim. Acta* **48** 87–101
[19] Bersuker I B and Polinger V Z 1989 *Vibronic Interactions in Molecules and Crystals* (Berlin: Springer-Verlag)
[20] Pooler D R 1978 *J. Phys. A: Math. Gen.* **11** 1045
[21] Pooler D R 1980 *J. Phys. C: Solid St. Phys.* **13** 1029
[22] Bersuker I B 2006 *The Jahn-Teller Effect* (Cambridge: Cambridge University Press)
[23] Dunn J L and Bates C A 1995 *Phys. Rev. B* **52** 5996–6005
[24] Chancey C C and O'Brien M C M 1997 *The Jahn-Teller Effect in C_{60} and Other Icosahedral Complexes* (Princeton: Princeton University Press)
[25] Curie P 1894 *J. Phys. Theor. Appl.* **3** 393–415
[26] Shubnikov A 1956 *Usp. Fiz. Nauk* **59** 591–602 (English translation: 1988 *Comput. Math. Applic.* **16** 357 - 364)
[27] Bersuker I B 2016 *Spontaneous Symmetry Breaking in Matter Induced by Degeneracies and Pseudodegeneracies* (John Wiley & Sons, Inc) pp 159–208

IOP Conf. Series: Journal of Physics: Conf. Series **833** (2017) 012009 doi:10.1088/1742-6596/833/1/012009

Distorted allotropes of bi-benzene: vibronic interactions and electronic excitations

V Krasnenko, V Boltrushko and V Hizhnyakov

Institute of Physics, University of Tartu, W. Ostwaldi Str 1, 50411, Tartu, Estonia

E-mail: hizh@ut.ee

Abstract. Bi-benzene - chemically bound two benzene molecules in stuck position is studied both analytically and numerically. There are several allotropes of bi-benzene having different geometry. The reason of the existence of sundry distorted structures is the pseudo-Jahn-Teller effect. The parameters of vibronic couplings causing distortions are found. For the calculation of these parameters both, the vibronic coupling of carbon atoms in different C_6 rings and the vibronic coupling in the rings are considered. The contribution of the distortion of C_6-planes to the latter coupling is also found. The energies of all the electronic states of π-electrons in all bi-benzene allotropes are determined by using the calculated vibronic interaction parameters.

1. Introduction

Benzene is one of the basic petrochemicals with a simple molecular formula C_6H_6. The carbon ring C_6 in this molecule is almost identical to those in graphene, graphite and carbon nanotubes. The properties of benzene, unlike saturated hydrocarbons, are essentially determined by the presence of the conjugated 6 π-electron cloud of D_{6h} symmetry [1], [2], [3], [4]. For example, this cloud makes the benzene molecule planar [5], although the planar structure is not very stiff; see e.g. numerical studies [6], [7], [8] of the stability of this structure with respect to several distortions. In usual conditions benzene is a liquid with predominantly weak van-der Waals forces acting between the molecules [9], [10], [11], [12], [13], [14]. However, recently in our theoretical study [7] it was found that at a short distance these molecules may interact chemically forming metastable compounds of benzene molecules stacked on each other at the distance ~ 0.16 nm in a hollow position, called as bi-benzene. Analogous bound states have been found for benzene and graphene [8], [15] and for two graphene sheets [16]. Moreover graphene nanoribbons (GNRs) and carbon nanotubes (CNTs) may also form chemically bound states [17]. In all these cases the bondings take place between adjoined C_6 rings. An examination shows that the reason of this chemical bonding is in 6 π-electrons: the inter-bonds of these electrons, connecting the adjoined C_6 rings, are of σ-type and they are stronger than the intra π-type bonds in C_6 rings. Therefore for a short distance it is energetically favorable to switch-off the intra-π bonds and to form stronger inter-σ bonds.

An important property of the adjoined C_6 rings discovered in Refs. [7], [15], [17] is that not all intra-π bonds can be switched off to inter-σ bonds. As a consequence the C_6 rings may be strongly distorted. This rearrangement of the bonds followed by distortions of adjoined C_6 rings opens new possibilities for various applications of these structures including the opportunity to fabricate All-Carbon-Devices.

IOP Conf. Series: Journal of Physics: Conf. Series **833** (2017) 012009 doi:10.1088/1742-6596/833/1/012009

In Ref. [7] we have shown that distortions of the adjoined C_6 rings in bi-benzene may be considered as the result of the pseudo-Jahn-Teller effect (PJTE) caused by the length-dependence of the inter-bonds of carbon atoms in adjacent C_6 rings. This effect works in bi-benzene because the HOMO-LUMO energy difference in this system is rather small ($\sim 2eV$ Ref. [7]) - much smaller than in benzene ($\sim 5eV$). We have also found that PJTE is responsible for distortions of adjoined C_6 rings in conjoined structures of benzene, GNRs and CNTs [17], [7], [15]. PJTE in adjoined C_6 rings has some analogy with considered in Refs. [18], [19], [20], [21], [22] PJTE in benzene. It may also be compared with the Berry softening of the in-plane b_{2u} vibration (1306 cm^{-1} mode) which connects the symmetric D_{6h} structure of the benzene molecule with a Kekulé-like structure of D_{3h} symmetry: the observed reduction of the mode frequency by $\sim 300\text{cm}^{-1}$ is a result of the vibronic interaction stemming from the dependence of the π-type bond of $2p_z$-electrons on its length [20], [23]. This interaction is insufficient for making D_{3h} symmetry of a single benzene molecule to be energetically favorable. However it still shows-up as a remarkable decrease of the frequency of the b_{2u} mode. Note also that the account of PJTE appeared to be essential in a number of cation and anion radicals of benzene [20].

In the present communication we investigate the vibronic interactions causing distortions of bi-benzene bound states. We consider that only π-electrons of conjoined C_6 rings remarkably contribute to the vibronic interactions responsible for the formation of different conjoined structures of these molecules. In our work the combined analytical and numerical (density functional theory, DFT) description of the problem is given. To this end we use simple tight binding model allowing one to treat the vibronic interactions analytically. To take these interactions into account we consider that the parameters of the model depend on the lengths of the bonds and on the distortions of C_6 planes. A comparison of the theoretical results with the DFT calculations of the ground state for properly chosen configurations allows us to determine the numerical values of vibronic interaction parameters. The knowledge of these parameters makes possible to analytically consider all states of the π-electrons for any configuration of the system. Here we have found these states for the allotropes of bi-benzene.

2. Allotropes of bi-benzene

The different rearrangements of the bonds of π-electrons, which are followed by subsequent different distortions of benzene rings, lead to different allotropes of bi-benzene. Four valence electrons of carbon atoms occupying almost degenerate $2S$ and $2P$ states, form in a benzene molecule three sp^2 hybrid orbitals. One of these orbitals makes a bond with hydrogen and two residuary orbitals establish the σ-bonds with the nearest carbon atoms. The residual $2S^1$ electron is promoted to the empty $2p_z$ state, making possible the formation of π-type bonds with the nearest carbon atoms. As a result, twelve $2p_z$ states of both spins in benzene remain half-occupied. The σ-bonds are stronger than π-bonds. Therefore, if to place two benzene molecules on a short distance in hollow position, then, as it was shown in Ref. [7] it will be energetically favorable to switch intra-π bonds to inter-σ bonds forming bi-benzene. At that different intra-π bonds may switch to the inter-σ bonds differently. As a result bi-benzene may form different allotropes. In Ref. [7] four possible allotropes were found. Schematically the shape of these allotropes is given in Fig. 1: the symmetric allotrope and three distorted allotropes. The differences in the geometry of all these allotropes are remarkable, they are caused by differences in the chemical bonds, which are reflected in the changes of electronic densities.

The calculations in Ref. [7] were made using CASTEP of *Material Studio* program module [24]. In these calculations the generalized gradient approximation (GGA) with the Perdew-Burke-Ernzerhof functional [25] was used. The cutoff energy of the plane waves, which determines the size of the basis set, was taken as 400 eV. The Monkhorst-Pack k-points grids were set as $2 \times 2 \times 4$ for the Brillouin zone integration.

In the symmetric structure (see Fig. 2A) six $2p_z$-electrons of every C_6 ring are removed from

IOP Conf. Series: Journal of Physics: Conf. Series **833** (2017) 012009 doi:10.1088/1742-6596/833/1/012009

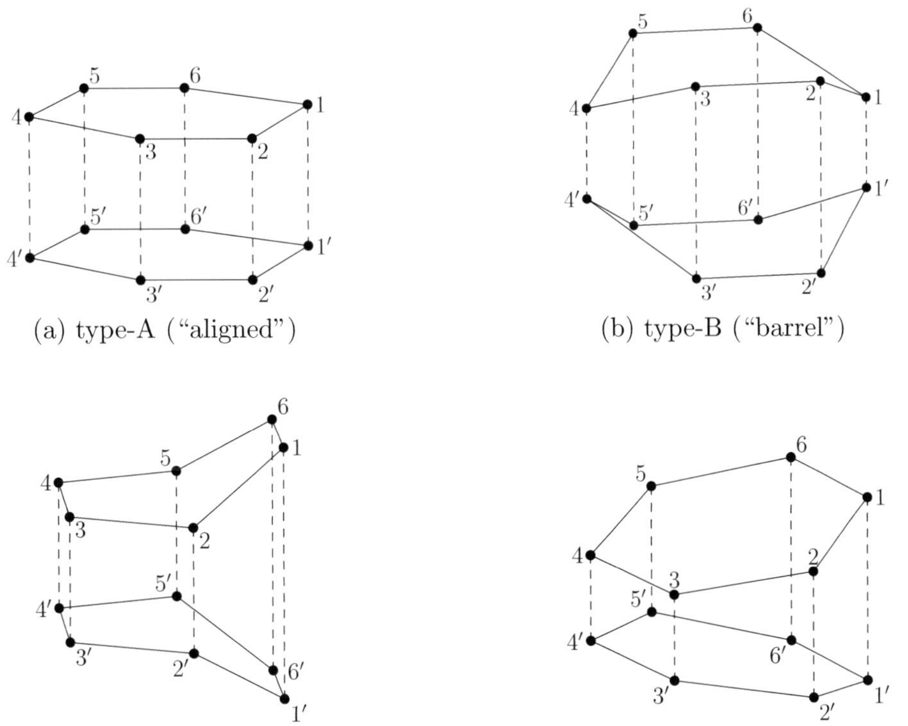

(a) type-A ("aligned") (b) type-B ("barrel")

(c) type-C ("croc") (d) type-D ("duck")

Figure 1. Schematic structures of bi-benzene

the $\sigma - \pi$ bonds in C_6 rings and these are used to form six σ-bonds between the C_6 rings. As a result, all nearest carbon atoms in the symmetric bi-benzene become chemically bound by the σ-bonds of practically the same length 1.55 Å (see Table 1). In the distorted structures (Fig. 2 B, C and D) a redistribution of the inter- and intra-bonds of $2p_z$ electrons takes place, resulting in the weakening of some inter-σ bonds and the formation of double $\pi - \sigma$ intra-bonds in C_6 rings. For example, in the structure B four inter-bonds are elongated from 1.55 Å to 2.69 Å. Simultaneously, two opposing each other intra-bonds in every C_6 ring are shortened due to the forming double bonds of the length 1.33 Å. In the C structure two inter-σ bonds are weakened and elongated from 1.55 pm to 3.50 Å. This is followed by the formation of one $\sigma - \pi$ bond of the length 1.33 Å in each C_6 ring. All other bonds change only slightly. In the third, most stable D structure only two adjoining inter-bonds remain practically unchanged. Two nearby inter-bonds are weakening moderately, while the remaining two inter-bonds are weakening strongly. Correspondingly, two opposing each other intra-σ bonds in every C_6 ring remain practically unchanged, two intra-σ bonds are increased due to formation of $\sigma - \pi$ intra-bonds with the length 1.34 Å and two remaining intra-bonds are gaining strength only slightly by forming σ-bond with the partial π-bond of the length 1.48 Å.

3. Vibronic Hamiltonian of bi-benzene

Here we consider the vibronic interactions of $2p_z$ electrons of conjoined C_6 rings contributing to different allotropes of bi-benzene. We take into account that electron-electron correlation and spin-orbital interaction of these electrons in benzene are relatively weak and may be excluded from consideration of vibronic interaction. The Hamiltonian of $2p_z$ electrons in bi-benzene can be taken as the sum of the Hamiltonians of two benzene molecules and to include there

Table 1. Characteristic lengths of *in-plane* bonds ($C_n - C_{n+1}$), z-shifts (transverse components of $C_n - C_{n+1}$), *out-of-plane* bonds ($C_n - C'_n$) and intra/inter-plane coupling parameters (β, β') for different structures of bi-benzene. Lengths are given in Å, coupling parameters β, β' in eV.

	type-B						type-C						type-D					
atom, n	1	2	3	4	5	6	1	2	3	4	5	6	1	2	3	4	5	6
in-plane	1.52	1.33	1.52	1.52	1.33	1.52	1.49	1.56	1.55	1.56	1.49	1.34	1.34	1.49	1.57	1.49	1.34	1.46
z-shift	-0.55	0	0.55	-0.55	0	-0.55	0.69	0.03	0	-0.03	-0.69	0	0.59	0.70	0	-0.70	-0.59	0
out-of-plane	1.64	2.73	2.73	1.64	2.73	2.73	2.98	1.61	1.56	1.56	1.61	2.98	4.07	2.96	1.58	1.58	2.95	4.06
β	1.5	2.5	1.5	1.5	2.5	1.5	1.7	1.35	1.4	1.35	1.7	2.5	2.5	1.7	1.3	1.7	2.5	1.8
β'	3.5	2.5	2.5	3.5	2.5	2.5	0.88	2.83	2.95	2.95	2.83	0.88	0.30	0.9	3.63	3.63	0.9	0.3

additionally the interaction between the adjoined carbon atoms of C_6 rings. In the tight-binding approximation this Hamiltonian gets the form:

$$H = \sum_{n=1}^{6} \left[\alpha a_n^\dagger a_n + \alpha a_n'^\dagger a'_n + \left(\beta_{n,n+1} a_{n+1}^\dagger a_n + \beta_{n,n+1} a_{n+1}'^\dagger a'_n + \beta'_n a_n^\dagger a'_n + Hc \right) \right] \quad (1)$$

Here $a_n^\dagger$ and a_n are the creation and destruction operators of electrons on atom n, a'_n and $a'^\dagger_n$ are the destruction and creation operators of the valence electrons in the second benzene molecule, α is the one-site energy, $\beta_{n,n+1}$ are the transfer (overlapping) integrals of the π-bond of nearest atoms in the C_6 rings which depend on the lengths and on the orientation of the corresponding bonds with respect to the symmetry plane (their signs are not important here, we chose them to be positive), Hc is the Hermitian conjugated term. For benzene the $\pi - \pi$ exchange interaction equals $\beta = 2.5\,\text{eV}$. In Ref. [7] it was found that the length of the σ-bonds in symmetric bi-benzene is practically the same as the length of the intra-σ bonds in benzene (0.16 nm). Therefore, we can take for exchange interaction of the adjacent carbon atoms in different C_6 rings (the $\sigma - \sigma$ interaction) at this distance $\beta' \approx 3.7\,\text{eV}$ which is the strength of the σ-bonds in a benzene molecule. Consequently, in symmetric state of bi-benzene $\beta' \approx 1.5\beta$. Note that β' is remarkably larger than β. Due to this ratio it is energetically more favorable to switch at short distance the intra-π bonds into the inter-σ bonds and to form bi-benzene. This fact is of primary importance for the formation of bi-benzene [7].

The Hamiltonian (1) includes into consideration the dependence of the exchange interactions on the distance between the carbon atoms in the C_6 rings and on the relative shift of them in normal direction. This allows us to take into account the difference in the lengths of intra-bonds in the C_6 rings of the bi-benzene allotropes and the distortion of the plane symmetry of these rings, not considered in previous analytical treatment of bi-benzene allotropes. These factors are essential to be taken into account for the description of the energies of the excited states of the distorted allotropes of bi-benzene.

In symmetric bi-benzene the stationary states of π-electrons are very simple: they have the form $|\Psi_k\rangle = 6^{-1/2} \sum_{n=1}^{6} e^{i\pi kn/3} |\Phi_n\rangle$, where $|\Phi_n\rangle$ are two-component states of two adjoining carbon atoms. In the k-basis the Hamiltonian H_2 of bi-benzene reads $H_{2,kk'} = (E_{1k} \cdot I + \beta\sigma_x)\,\delta_{kk'}$, where σ_x is the Pauli matrix. The eigenvalues of this matrix equal ± 1.

IOP Conf. Series: Journal of Physics: Conf. Series **833** (2017) 012009 doi:10.1088/1742-6596/833/1/012009

Therefore the energies of the levels of the undisturbed bi-benzene in the used approximation are $E_k = 2\alpha + 2\beta_0 \cos(\pi k/3) \pm \beta'$.

The energy difference of LUMO and HOMO in symmetric bi-benzene is $\Delta E_{LH}^{bibz} = 2(\beta' - \beta) \approx$ 2 eV, i.e. essentially smaller than the energy difference of LUMO and HOMO in benzene $\Delta E_{LH}^{bz} \approx 5.2$ eV. This causes the change of the order of the HOMO and LUMO when decreasing the distance between the adjoined C_6 rings, which is precondition of the existence of the chemically bound state of two benzene molecules [7].

There are 18 bonds between the nearest carbon atoms in bi-benzene: 12 intra-bonds and 6 inter-bonds, which all may have different lengths and adjoined angles. However we are interested here only in the structures with mirror symmetry, i.e. with identical C_6 rings. These structures have twelve different lengths of bonds and six different angles of intra-bonds. In this case the vibronic Hamiltonian of bi-benzene can be taken in the form:

$$H = \left(2E_0 + 2\sum_{n=1}^{6} u_0(Q_n) + \sum_{n=1}^{6} u'_0(q_n)\right) \cdot I + 2V, \tag{2}$$

where $u_0(q)$ and $u_0'(q)$ are the potential energy of the single intra- and inter-σ bonds, respectively,

$$V = \begin{pmatrix} m_1 & m_2 \\ m_2 & m_1 \end{pmatrix} \tag{3}$$

is the vibronic energy, m_1 and m_2 are following matrices:

$$m_1 = \begin{pmatrix} 0 & \beta_1 & 0 & 0 & 0 & \beta_6 \\ \beta_1 & 0 & \beta_2 & 0 & 0 & 0 \\ 0 & \beta_2 & 0 & \beta_3 & 0 & 0 \\ 0 & 0 & \beta_3 & 0 & \beta_4 & 0 \\ 0 & 0 & 0 & \beta_4 & 0 & \beta_5 \\ \beta_6 & 0 & 0 & 0 & \beta_5 & 0 \end{pmatrix}, \qquad m_2 = \begin{pmatrix} \beta_1' & 0 & 0 & 0 & 0 & 0 \\ 0 & \beta_2' & 0 & 0 & 0 & 0 \\ 0 & 0 & \beta_3' & 0 & 0 & 0 \\ 0 & 0 & 0 & \beta_4' & 0 & 0 \\ 0 & 0 & 0 & 0 & \beta_5' & 0 \\ 0 & 0 & 0 & 0 & 0 & \beta_6' \end{pmatrix}, \tag{4}$$

the factors two in Eq. 2 stands for two spin values. For a weakly distorted structure

$$\beta_n = \beta - kQ_n - \kappa z_n^2, \qquad \beta'_n = \beta' - k' q_n,$$

k, κ and k' are the positive constants of vibronic interactions inside the C_6 rings and between the C_6 rings, respectively, Q_n and the differences of the in-plane lengths of carbon bonds from their equilibrium value in the C_6 rings, z_n are the difference of the z-coordinates (perpendicular to symmetry plane) of atoms n and $n+1$; q_n are the differences of the lengths of the inter-bonds from their equilibrium value in the C_6 rings of symmetric bi-benzene. The minus signs in the equations for β_n and β'_n account for the fact that the strength of bonds decreases with the increasing of the distance between the carbon atoms. The same holds for the distortion terms of the plane.

Here we take into account that the small distortions of C_6 planes can be considered in terms of the appearance of small additional non-orthogonality of the basic states of the planar structure. This non-orthogonality results in the appearance of the additional matrix of overlapping integrals (see, e.g. [26]). In our concrete case these terms take into account the reduction of the overlapping integrals of the nearest π-functions when they move apart in z-direction. We include these terms into consideration by adding the $\propto \kappa$ terms to the exchange interaction parameters β_n. In the case of large Q_n, or z_n or q_n we apply the exponential reduction of the inter- and intra-bonds taking

$$\beta_n = \beta e^{-kq_n - \kappa z_n^2}, \qquad \beta'_n = \beta' e^{-k' Q_n}.$$

Note that the ground state corresponds to the double occupancy of the six lowest levels. Therefore the vibronic energy of this state for any values of q_n, Q_n and z_n is determined by a double sum of six lowest eigenvalues of the Hamiltonian for the given Q_n, q_n and z_n. This energy considered as a function of the coordinates determines the shape of the multi-dimensional potential surface of the ground state.

The energies of the electronic interactions $\beta \approx 2.5$ eV and $\beta' \simeq 3.7$ eV are known [7]. The parameter of the vibronic interaction k' determining the dependence of the vibronic interaction of the $\sigma - \sigma$ inter-bonds on their lengths (this interaction is responsible for the dimers formation) was found in Ref. [7] by combining the analytical and numerical (DFT) calculations of the vibronic contributions to the energy of the ground state of the system for its different configurations. This allows one to exclude the contribution of other interactions in numerical considerations. The calculated value of this parameter is $k' = 3.75$ eV/Å.

The remaining vibronic interaction parameters k and κ can be found by applying the same method. For example, to find k we compare the result of the calculations of the

(1) totally symmetrically distorted C_6 rings with $Q_n = Q \neq 0$ (in this case $\beta_n = \beta - kQ$);

(2) the rings elongated along the C_2 axis connecting the opposite corners with $Q_2 = Q_5 = Q\sqrt{3}$ and $Q_1 = Q_3 = Q_4 = Q_6 = 0$ (here $\beta_2 = \beta_5 = \beta - \sqrt{3}kQ$ and $\beta_1 = \beta_3 = \beta_4 = \beta_6 = \beta$).

In this case the difference of the energies of the elastic distortions for small Q comes solely from the vibronic interaction under consideration. Performing the analytical calculation of the vibronic energies of the ground state, we have found

$$E_{symm} - E_{dist} \approx 7(kQ)^2 e^{0.21k/\beta} / 9\beta. \tag{5}$$

Here the exponential factor $\exp(-0.21k/\beta)$ accounts for the reduction of the exchange energy β in C_6 rings on the increasing of the $C-C$ distance in symmetric bi-benzene by 0.21 Å. According to our DFT calculation $E_{symm} - E_{dist} \approx 30\,Q^2$ eV (for Q in Å). This gives $k \approx 7.5\,\mathrm{eV/Å}^2$ which is almost 2 times larger than k'. This was to be expected: the longitudinal size of π-functions is smaller than their transversal size, which makes corresponding bond more strongly dependent on the length of the $\pi - \pi$ bond. Besides, in case of $\pi - \pi$ vibronic interaction both "petals" of the π-functions give a contribution to the vibronic coupling, while only one of them contributes to the inter-σ coupling.

To find the parameter κ, describing the vibronic effect of the transversal distortions of the C_6 rings, we consider two following structures:

(1) the structure with two bent opposite angles (boatlike structure [6]); in this case $z_1 = z_4 \neq 0$ and $z_2 = z_3 = z_5 = z_6 = 0$;

(2) the structure with all six alternatively bent angles with $z_1 = -z_2 = z_3 = -z_4 = z_5 = -z_6$.

If in both these cases the absolute value of the difference of z-coordinates of the nearest atoms is taken to be the same, then the non-vibronic elastic energy of the second distortion would be three times larger than that of the first distortion. The contribution of this energy to the energy of the distorted structures will be extractable, when the total elastic energy of the second structure from the triple distortion energy of the first structure. Our analytical calculations of the vibronic energies of the ground state of the C_6 ring with these distortions give $32\kappa z^2/3$ (first structure) and $8\kappa z^2$ (second structure). Therefore,

$$\kappa = (3E_{2,z} - E_{3,z})/24z^2. \tag{6}$$

According to our calculations $3E_{2,z} - E_{3,z} \approx 14.0 \cdot z^2$ eV (for z in Å) which gives $\kappa \approx 0.60$ eV/Å^2. Consequently, the vibronic interaction stemming from the distortion of the C_6 planes is rather weak. This agrees well with the known fact that the planar structure of benzene is not very stiff [6].

IOP Conf. Series: Journal of Physics: Conf. Series **833** (2017) 012009 doi:10.1088/1742-6596/833/1/012009

4. Energies of electronic states

The established values of the vibronic parameters k, k' and κ allows us to calculate the exchange integrals β_n and β'_n $(n = 1, 2, ..., 6)$ for any configuration of carbon atoms of bi-benzene. The knowledge of these parameters gives a possibility to find analytically all the 12 states of all valence (π) electrons for different configurations of bi-benzene without complete calculations of the excited states. To this end one should find the eigenstates and eigenvalues of the vibronic matrix V for the chosen coordinates of carbon atoms, which can be done analytically. Note that in our calculations of bi-benzene vibronic parameters we apply the DFT method only for ground state calculations. Therefore, our theory is free from the usual difficulties of the DFT with the calculations of excited states, e.g. when ensuring orthogonality in the ground-state wave function for the optimization of excited states [27].

The possibility to find analytically all states of valence electrons for any configuration of bi-benzene is an advantage of our method. For example, it gives one a possibility in the frame of unified approach to explore in detail the multi-configurational potential surfaces of all electronic states of bi-benzene and to find relaxation passes and conical intersections important for different processes (see in this connection Ref. [15] where we have found several conical intersections in the excited states of benzene bonded to graphene).

Table 2. Vibronic energies of bi-benzene allotropes (in eV).

	⟵ Occupied energy levels						Unoccupied energy levels ⟶					
Level nr.	1	2	3	4	5	6 (HOMO)	7 (LUMO)	8	9	10	11	12
Type-A	-8.70	-6.20	-6.20	-1.30	-1.20	-1.20	1.20	1.20	1.30	6.20	6.20	8.70
Type-B	-6.36	-5.09	-4.23	-1.89	-0.62	-0.45	0.45	0.62	1.89	4.23	5.09	6.36
Type-C	-6.94	-5.04	-3.48	-3.22	-1.21	0.08	0.43	1.72	3.73	3.99	5.55	7.45
Type-D	-5.46	-3.40	-3.40	-2.86	-1.18	-0.93	0.93	1.18	2.86	3.40	3.40	5.46

In the Table 1 the values of exchange integrals β and β'_n $(n = 1, 2, ..., 6)$ for configurations corresponding to the distorted allotropes B,C and D are presented (the values of these parameters for undistorted allotrope A are known [7]: $\beta = 2.6$ eV, $\beta' = 3.7$ eV). Using these values we calculated energies of electronic states for all four allotropes of bi-benzene. These energies are given in Table 2. From this table one can easily find the energies of all 36 electronic transitions of π-electrons in every structure. Note that according to this table the A structure should absorb light from 2.4 eV (green) to far ultraviolet (UV, 17 eV) and should not absorb in red and infrared, while the structures B, C and D should absorb light starting from near infrared till far ultraviolet (UV, 12-14 eV).

5. Conclusion

In the present study we investigated the vibronic interactions of valence (π) electrons in bi-benzene. These interactions plays crucial role in this system causing the appearance of four different allotropes: the symmetric allotrope A and three distorted allotropes B, C and D (see Fig. 1). The micro-mechanism of distortions in this system is the pseudo-Jahn-Teller effect – the lowering of symmetry due to the vibronic mixing of ground and excited states. In bi-benzene this mixing has a strong effect due to rather small value of the LUMO-HOMO energy difference. Here we have found the energies of the occupied and unoccupied state of π-electrons of all four allotropes.

In the appearing structures both, the different inter-bonds and the different intra-bonds of π-electrons have different lengths. Besides, the C_6 planes of B and C allotropes have out-of-plane distortions. To take these effects into account we considered the vibronic interactions stemming from the length dependence of the exchange interactions of both inter- and intra-bonds. We also take into account the dependence of these interactions on the C_6 plane distortions. To find the parameters of these interactions, we applied our method of combined analytical and numerical (DFT) considerations. In the analytical study we used for π-electrons the tight binding model. We compared the theoretical results with the DFT calculations of the ground state (to avoid the DFT difficulties with the calculations of excited states) for the properly chosen configurations that enabled us to determine the numerical values of the vibronic interaction parameters. The knowledge of these parameters gives a possibility to find analytically all 12 states of π-electrons for any configuration of the system.

Finally we note that the possibility to have analytical description of all electronic states for any configuration of the system is an advantage of the used method. Among other tasks, which may be solved by using this method, one can mention a possibility to explore within the framework of unified approach the multi-configurational potential surfaces of all electronic states of bi-benzene and to find relaxation passes and conical intersections important for different processes.

Acknowledgements

The research was supported by the Estonian institutional research funding project IUT2-27, by the Ministry of Education and Research of Estonia, Project PUT430 and by the European Union through the European Regional Development Fund (projects 3.2.0101.11-0029 and EU49695).

References

[1] Cooper D L, Gerratt J and Raimondi M 1986 *Nature* **323** 699-701
[2] Pauling L 1987 *Nature* **325** 396
[3] Messmer R P and Schultz P A 1987 *Nature* **329** 492
[4] Harcourt R D 1987 *Nature* **329** 491-2
[5] Moran D *et al* 2006 *J. Am. Chem. Soc.* **128** 9342-3
[6] Feixas F, Matito E, Poater J and Solà M 2007 *J. Phys. Chem.* **111** 4513-21
[7] Boltrushko V, Krasnenko V and Hizhnyakov V 2015 *Chem. Phys.* **460** 90-6
[8] Krasnenko V, Kikas J and Brik M 2012 *Physica B* **407** 4557-61
[9] Arunan E and Gutowsky H S 1993 *J. Chem. Phys.* **98** 4294-6
[10] Lee E C *et al* 2007 *J. Phys. Chem.* **111** 3446-57
[11] Tachikawa H 2012 *RSC Adv.* **2** 6897-904
[12] Schnell M *et al* 2013 *Angew. Chem. Int. Ed.* **52** 5180-3
[13] Meijer E J and Sprik M 1996 *J. Chem. Phys.* **105** 8684-9
[14] Strutyński K, Melle-Franco M and Gomes J A N F 2013 *J. Phys. Chem. A* **117** 2844-53
[15] Krasnenko V, Boltrushko V and Hizhnyakov V 2016 *J. Chem Phys.* **144** 134708
[16] de Andres P, Ramrez R and Vergés J 2008 *Phys. Rev. B* **77** 045403-1–045403-5
[17] Krasnenko V, Boltrushko V, Klopov M and Hizhnyakov V 2014 *Phys. Scr.* **89** 044008
[18] Goode J G, Hofstein J D and Johnson P M 1997 *J. Chem. Phys.* **107** 1703-16
[19] Barlukova M M *et al* 2005 *J. Phys. Chem. A* **109** 4404-09
[20] Blancafort L and Solà M 2006 *J. Phys. Chem. A* **110** 11219-22
[21] Baldea I, Franz J and Köppel H 2007 *J. Mol. Struct.* **838** 94-9
[22] Mondal T and Mahapatra S 2009 *Phys. Chem. Chem. Phys.* **11,** 10867-80
[23] Berry R 1961 *J. Chem. Phys.* **35** 2253
[24] Accelrys Software Inc 2012 *Material Studio Release Notes, Release 6.1* (San Diego: Accelrys Software Inc.)
[25] Perdew J, Burke K and Ernzerhof M 1996 *Phys. Rev. Lett.* **77** 3865-8
[26] Newman D J 1970 *J. Phys. Chem. Solids* **31** 1143-9
[27] Young D 2001 *Computational Chemistry: A Practical Guide for Applying Techniques to Real World Problems* (New York: John Wiley & Sons)

Pseudo Jahn-Teller effect in control and rationalization of chemical transformations in two-dimensional compounds

N N Gorinchoy[1], **I B Bersuker**[1,2]

[1]Institute of Chemistry, Academy of Sciences of Moldova, Kishinev, Republic of Moldova
[2]Institute for Theoretical Chemistry, The University of Texas at Austin, Austin TX 78712, USA

ngorinchoy@yahoo.com

Abstract. We show that the pseudo Jahn-Teller effect (PJTE) is instrumental in predicting and rationalizing structural changes in chemical transformations of two-dimensional (2D) molecular systems by means of analyzing the symmetries and electron occupation of the ground and lowest excited electronic states and the energy gap between them, subject to their PJT coupling along the main distortion coordinates. Special attention is paid to rationalizing the PJTE origin of non-planarity of 2D compounds and to the restoration of their planar configurations. Examples of two series of 1,2- and 1,4-dithiin containing tricyclic compounds (carbon sulfide, thianthrene, and antracene and their derivatives) are used to demonstrate in detail the mechanism of (1) enhancement and suppression of the PJTE distortions (puckering) in redox processes, and (2) PJTE induced symmetry breaking and restoration of the planar configuration by chemical substitutions.

1. Introduction

The pseudo Jahn-Teller effect (PJTE) was shown to be instrumental in revealing the spontaneous symmetry breakings (SSB) in molecular systems and solids in nondegenerate states, and related to these SSB properties (see, e.g., [1] and references therein). In application to two-dimensional (2D) or quasi 2D systems the PJTE explains the origin of their main structural features, the out-of-plan distortions, puckering (puckering as a PJTE was first formulated in [2]). 2D systems may have unique applications due to a variety of their planar-surface-specific physical and chemical properties. The best known example of this kind is graphene with its important properties for electronics. Similar properties are sought for in silicene, boron nitride, zinc oxide, phosphorene, gallium nitride, as well as in other attractive subjects like 2D transition metal dichalcogenids and oxides, and there is an unlimited number of other organic and inorganic 2D and quasi-2D systems with expected similar properties (see the review [3]). A series of such cyclic and heterocyclic compounds was studied recently with regard to PJTE in the origin of their puckering (see in [3].

The present paper is devoted to the next step in the development of this trend: if the cyclic system is out-of-plan distorted, how can we restore the planar configuration with minor other changes? If we know the detailed mechanism of distortion of the planar configuration via the PJTE and we revealed the active excited state in this mechanism, we can try to influence the system by means of external perturbations that violate the PJTE condition of instability of the ground state. In principle, there are four possibilities to do this: (1) increasing just the energy gap between the electronic states mixed by the PJT vibronic coupling, e.g. by weak coordination with outside systems, (2) changing the symmetry of the active excited states by removing or adding electrons (which may change all the parameters of the PJTE including the energy gaps), (3) making appropriate substitutions that preserve the 2D structure restoring its planarity, and (4) by spectroscopic excitations. The possibility (1) was demonstrated first in the mechanism of hemoglobin oxygenation [4] in which the (driven by the PJTE) out-of-plane displacement of the iron atom from the porphyrin ring returns into the plane by oxygenation due to the increase of the energy gap to the active excited state, thus suppressing the PJTE. Examples of realization of the mechanism (2) were demonstrated recently [5-7]. For instance, by coordination of two chloride anions to nonplanar (distorted by the PJTE) chair structures of Si_6X_{12} (X = Cl, Br) to form dianions of the type $[Si_6Cl_{14}]^{2-}$, the latter were proved to have a planar Si_6-ring structure, in which the PJTE is suppressed because the orbitals of the active excited states of the ring become occupied by electrons from the Cl^- ions [5]. The substitution possibility to restore the planarity of cyclic compounds is demonstrated here for the first time. Spectroscopic excitations in this respect were not studied as yet.

We explore the enhancement and suppression of the PJTE distortions in redox processes and the PJTE induced symmetry breaking by chemical substitutions using two sets of 1,2- and 1,4-dithiin containing tricyclic compounds as examples. In the first set two tricyclic species were studied, the neutral carbon sulfide C_6S_8 and its dianion $C_6S_8^{2-}$ containing two S-atoms in the 1,2-positions of the central six-membered ring and one thione (C=S) bond in the five-membered rings on its either side. Both compounds and their structure were studied experimentally and confirmed by single-crystal X-ray diffraction [8]. Crystallographic data show that the neutral compound is not planar, it has a puckered structure of C_2 symmetry with a twisted 1,2-dithiin core, and its reduction has resulted in cleavage of the S-S bond and rotation of the two peripheral rings about the central C-C bond [8]. We show that in both systems the out-of-plane distortions are due to the PJT coupling between the ground A_1 and two excited electronic states of A_2 symmetry, and that by oxidation the PJTE in $C_6S_8^+$ is weakened being completely suppressed in $(C_6S_8)^{2+}$.

The second set of studied tricyclic systems contains thianthrene, or dibenzo-1,4-dithiin molecule, and anthracene, and their derivatives. According to crystal structure determinations the thianthrene molecule is bent at the S-S axis by 127°- 128°, and it has a low energy flapping vibration about the S–S axis [9, 10], while the oxidized thianthrene has a planar nuclear configuration with D_{2h} symmetry [11, 12]. Besides, if the C atoms at 1, 4, 6, and 9 positions are substituted by nitrogen atoms, the dithiin core becomes planar [13]. On the other hand, the anthracene molecule (An) is planar with the nuclear configuration of D_{2h} symmetry, whereas the 9,10-dihydroanthracene (9,10-H_2An) has a folding geometry of C_{2v} symmetry with a dihedral angle of about 150-160° [14].

All these structural features are fully controlled by the PJTE, its mechanism being demonstrated here in detail. For all the considered molecules the geometry optimization and vibrational frequency analysis in the planar high-symmetry and equilibrium nuclear configurations were carried out at the B3LYP level of DFT method [15]. All the calculations were performed with the GAUSSIAN09 package [16]. The Pople's 6-31+G(d,p) split basis sets [17] were utilized in all steps of the calculations. The potential energy profiles along the normal distortion coordinates were calculated with the CISD method. The applied active space for the C_6S_8 and $C_6S_8^{2-}$ species was composed of eight electrons and nine active orbitals, while that for antracene, thiantrene and their derivatives included six electrons and ten active molecular orbitals. The numerical values of the vibronic coupling constants were estimated by means of fitting the solutions of the secular equations to the ab initio calculated energy profiles as shown below.

2. Enhancement and suppression of PJTE distortions in redox processes.

As mentioned in the introduction, we demonstrate the change of the PJTE in redox processes using as example two tricyclic species, the neutral carbon sulfide C_6S_8 and its dianion $C_6S_8^{2-}$.

IOP Conf. Series: Journal of Physics: Conf. Series **833** (2017) 012010 doi:10.1088/1742-6596/833/1/012010

Calculations of the neutral carbon sulfide and its dianionic form in the planar configuration of C_{2v} symmetry show that both systems are unstable with respect to the out-of-plane distortions of a_2 symmetry, and imaginary frequency corresponding to this vibrational mode for the dianion is more than two times greater than for the neutral molecule (figure 1). The optimized structures show C_2 symmetry for the nuclear configurations. Calculated geometrical parameters are presented in table 1 together with the available experimental data.

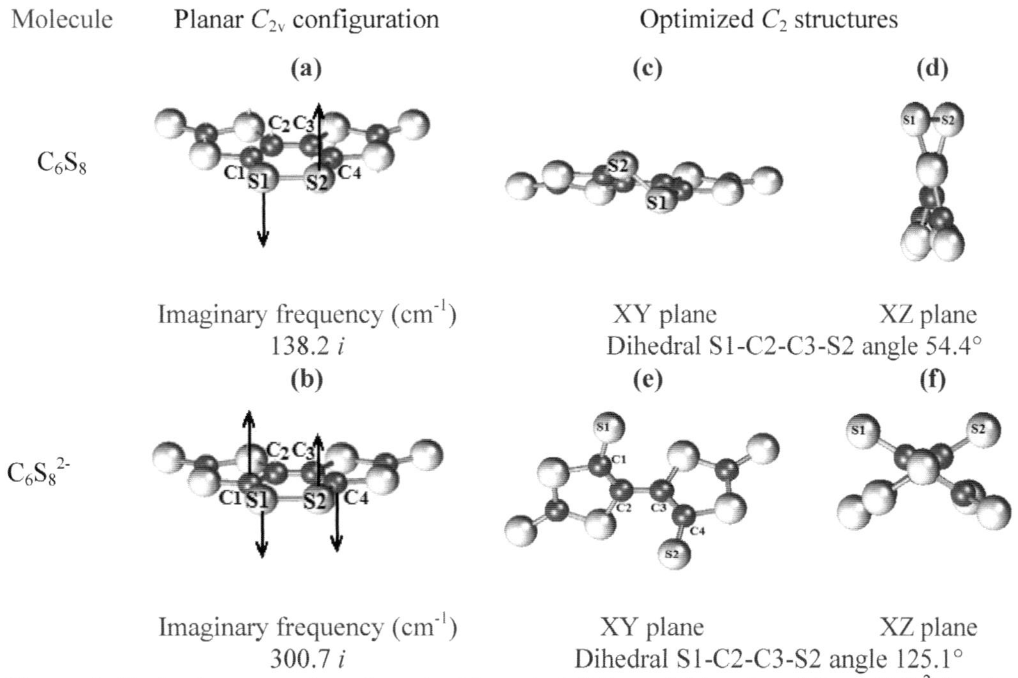

Figure 1. Puckering normal mode a_2 in (a) neutral C_6S_8 and (b) reduced form $C_6S_8^{2-}$ in planar C_{2v} configurations; optimized structures (C_2 configurations) are shown in two projections for C_6S_8 (c-d) and $C_6S_8^{2-}$ (e-f).

Table 1. Selected distances (Å) and angles (deg) for C_6S_8 and $C_6S_8^{2-}$ in the planar and equilibrium nuclear configurations.

Geometry parameters		**Molecules**					
		C_6S_8			**$C_6S_8^{2-}$**		
		Planar (C_{2v})	Equilibrium (C_2)	Exp. [8]	Planar (C_{2v})	Equilibrium (C_2)	Exp. [8]
Bond length (Å)	S1-S2	2.10	2.06	2.05	2.10	-	-
	S1-C1	1.77	1.77	1.76	1.78	1.73	1.72
	C1-C2	1.33	1.33	1.35	1.35	1.36	1.38
	C2-C3	1.47	1.47	1.46	1.44	1.47	1.45
Angle (Degree)	S2-S1-C1	104.8	97.2	97.9	106.2	-	-
	S1-C1-C2	129.8	123.1	122.5	126.6	131.7	130.3
	C1-C2-C3	125.3	123.2	124.5	127.2	127.3	126.2
Dihedral angle (Degree)	C1-S1-S2-C4	0.0	54.4	52.0±2	0.0	117.1	180
	C1-C2-C3-C4	0.0	27.8	-	0.0	125.6	180
Imaginary frequency a_2(cm^{-1})		138.2	-	-	300.7	-	-
Normal modes in Cartesian coordinates x		+0.11988 x_{S1} -0.01175 x_{C1} -0.01639 x_{C2} -0.02110 x_{S9} -0.02522x_{S11}	-	-	+0.09686 x_{S1} -0.12660 x_{C1} -0.01270 x_{C2} -0.01142 x_{S9} -0.00517x_{S11}	-	-

Consider first the MO energy level scheme and cross sections for the adiabatic potential energy surfaces (APES) of the ground state and the low-lying excited states of A_2 symmetry along the a_2 distortion mode for the neutral compound (figure 2). Of the seven low-lying excited electronic states, two of them possess A_2 symmetry. The first one is mainly formed by one-electron excitation from the occupied b_2 MO to the unoccupied b_1 MO, while the second excited state corresponds to the excitation from the HOMO of a_2 symmetry to the unoccupied a_1 MO. Each of them can undergo vibronic mixing with the ground A_1 state along the a_2 distortion mode.

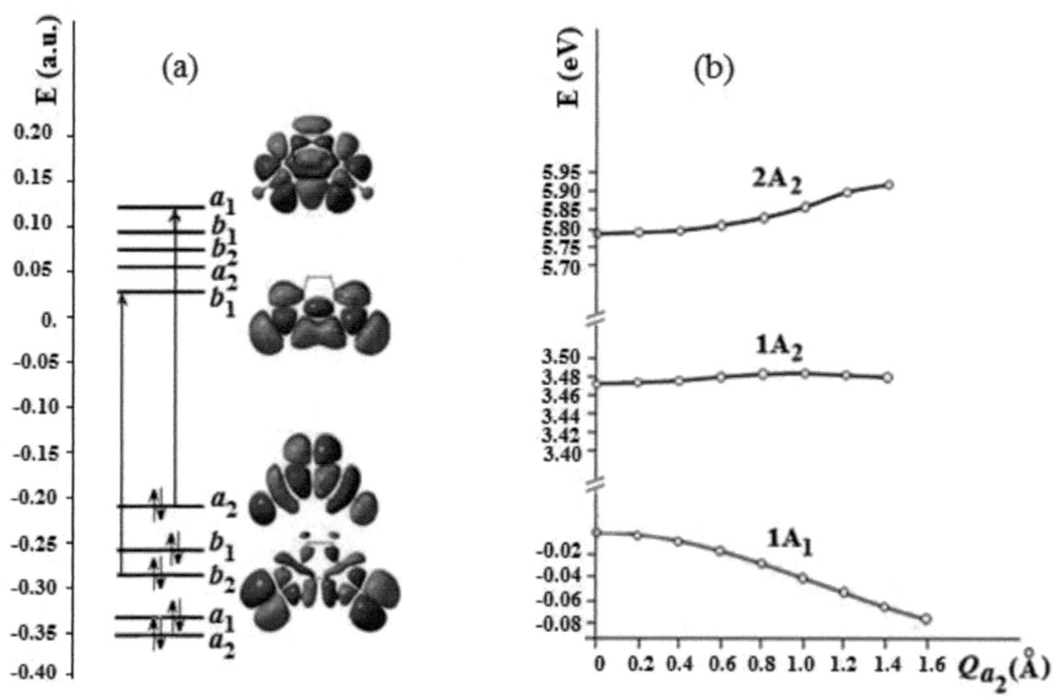

Figure 2. (a) MO energy levels scheme for the neutral C_6S_8 in the ground A_1 state with indication of the one-electron excitations to A_2 terms; (b) *ab initio* calculated energy profiles along the a_2 puckering coordinate for the ground A_1 and two excited A_2 electronic terms.

Therefore, in this case we have the three-level (A_1+ $1A_2$+ $2A_2$)$\otimes a_2$ PJTE problem [1] with the matrix equation (for simplicity we ignore here the second order perturbation corrections induced by the higher excited states)

$$\begin{vmatrix} \frac{1}{2}K_0Q^2 - \varepsilon & F_{01}Q & F_{02}Q \\ F_{01}Q & \frac{1}{2}K_1Q^2 + \Delta_{01} - \varepsilon & 0 \\ F_{02}Q & 0 & \frac{1}{2}K_2Q^2 + \Delta_{02} - \varepsilon \end{vmatrix} = \begin{vmatrix} a-\varepsilon & f & g \\ f & b-\varepsilon & 0 \\ g & 0 & c-\varepsilon \end{vmatrix} = 0 \ , \tag{1}$$

where K_0, K_1, and K_2 are the primary force constants for the ground A_1 and two low-lying excited A_2 states, respectively,

$$K_0 = \langle A_1|(\partial^2 H/\partial Q_{a2}^2)_0|A_1\rangle, \tag{2}$$

(and similar for K_1 and K_2 of the two excited states), Δ_{01}, and Δ_{02} are the energy gaps between the ground and the two excited states, and F_{01} and F_{02} are the corresponding vibronic coupling constants,

$$F_{01} = \langle A_1|(\partial H/\partial Q_{a2})_0|1A_2\rangle \quad \text{and} \quad F_{02} = \langle A_1|(\partial H/\partial Q_{a2})_0|2A_2\rangle. \tag{3}$$

The constants $a, b, c, f,$ and g are simplifying denotations.

Solving for the 3×3 secular determinant, one can get the following equation for the energies of the three states (1 ground state and 2 excited states) along the a_2 distortion mode:

$$\varepsilon^3 - (a+b+c)\varepsilon^2 + (ab+ac+bc-f^2-g^2)\varepsilon - abc + cf^2 + bg^2 = 0 \tag{4}$$

Using the well-known algebra relations between the products and sums of the three roots of this secular equation

$$\varepsilon_1 + \varepsilon_2 + \varepsilon_3 = a + b + c \tag{5}$$

$$\varepsilon_1\varepsilon_2 + \varepsilon_2\varepsilon_3 + \varepsilon_1\varepsilon_3 = ab + ac + bc - f^2 - g^2 \tag{6}$$

$$\varepsilon_1\varepsilon_2\varepsilon_3 = abc - cf^2 - bg^2 \tag{7}$$

we can estimate the values of the vibronic coupling constants by fitting these equations to the corresponding calculated ab initio energy profiles. The values of thus estimated PJT parameters are presented in table 2.

Table 2. Energy gaps Δ_{0i} (in eV), PJTE coupling constants F_{0i} (in eV/Å), and primary force constants K_i (in eV/Å^2) for the neutral carbon sulfide molecule C_6S_8 and its fully reduced form $C_6S_8^{2-}$ obtained by numerical fitting Eqs. (5)-(7) to the calculated energy profiles in Fig. 2.

Molecule	K_0	Δ_{01}	F_{01}	K_1	Δ_{02}	F_{02}	K_2
C_6S_8	0.42	3.47	0.57	0.11	5.79	0.98	0.15
$C_6S_8^{2-}$	0.49	1.86	0.61	0.09	4.65	0.98	0.15

At small values of Q the curvature of the APES for the ground state can be approximately estimated as $K \approx K_0 + K_{v1} + K_{v2}$, where $K_{v1} \approx -2F_{01}^2/\Delta_{01}$ = -0.187 eV/Å^2, and $K_{v2} \approx -2F_{02}^2/\Delta_{02}$ = -0.332 eV/Å^2. Substituting parameters from table 2 we obtain K = -0.098 eV/Å^2.

One can see that the vibronic contribution of the first excited A_2 state to the curvature of the AP of the ground state is much smaller than of the second excited state. Qualitatively, this can be explained by the fact that both mixing occupied b_2 and vacant b_1 MOs are localized on the peripheral cycles, whereas the strongest distortion occurs in the central ring (figure 1). The main negative contribution to the curvature, and therefore to the instability of planar nuclear configuration comes from the second excited state of A_2 symmetry which is formed by the one-electron excitation from the HOMO of a_2 symmetry to the vacant a_1 MO (figure 1).

When the carbon sulfide molecule is oxidized to its radical cation form, this a_2 MO becomes single-occupied. The removal of an electron from the HOMO of a_2 symmetry reduces the negative PJTE contribution to the curvature of the APES produced by the second A_2 excited state. Indeed, because the excited states are formed by one-electron excitations, in the MO approximation $F_{01}=\sqrt{2}f_1$ and $F_{02}=\sqrt{2}f_2$, where f_1 and f_2 are the orbital vibronic coupling constants [1],

$$f_1 = \langle b_2|(\partial H/\partial Q_{a2})_0|b_1\rangle, \qquad f_2 = \langle a_2|(\partial H/\partial Q_{a2})_0|a_1\rangle. \tag{8}$$

For the neutral molecule we have $K_{v1} \approx -4f_1{}^2/\Delta$ and $K_{v2} \approx -4f_2{}^2/\Delta$. For the cation $(C_6S_8)^{2+}$ the value of K_{v2} becomes equal to $K_{v2}(cat) \approx -2f_2{}^2/\Delta$. With the PJT parameters from Table 2 the resulting value of the curvature of the AP of $(C_6S_8)^{2+}$ is equal to $K_{cat} \approx -0.043$ eV/Å^2. Hence the oxidation of the system leads to the weakening of the PJTE, although the molecule is still distorted (imaginary frequency is equal to 80.5 *i*).

Further oxidation to the dication $(C_6S_8)^{2+}$ leads to the complete suppression of the PJTE and restoration of planar nuclear configuration. This confirms once again that the source of instability in this system is the PJT interaction with mainly one most active excited state formed by one-electron excitation from the HOMO.

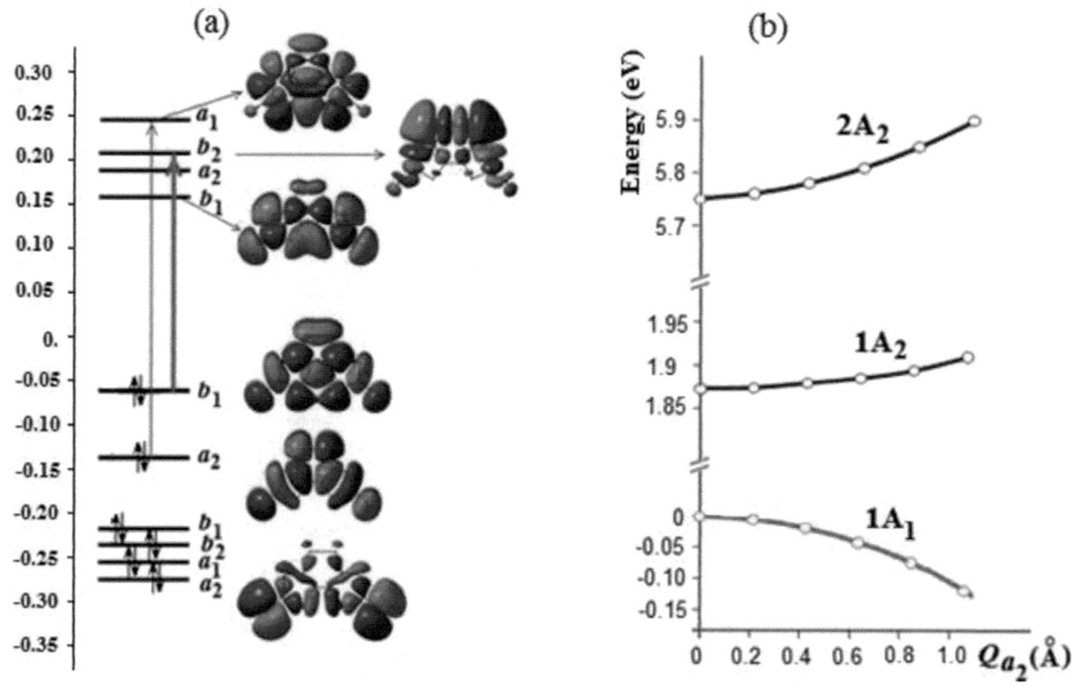

Figure 3. (a) MO energy levels scheme for the neutral $C_6S_8{}^{2-}$ in the ground A_1 state with indication of the one-electron excitations to A_2 terms; (b) *ab initio* calculated energy profiles along the a_2 puckering coordinate for the ground A_1 and two excited A_2 electronic terms.

As mentioned above, the totally reduced compound $(C_6S_8)^{2-}$ is much more distorted from the initial planar structure of C_{2v} symmetry than the neutral molecule. Analysis of the MO energy levels schemes of the two systems shows that in the dianion form the unoccupied in the neutral molecule MO b_1 becomes occupied by two electrons (figure 3). Its structure is changed, so that this MO covers not only the peripheral rings but also the central cycle. This leads to the occurrence of one more low-lying active A_2 state (formed by one-electron excitation from this MO to the vacant MO of b_2 symmetry). So in this case we have the full three-level PJTE problem. Following the same procedure as in the previous case, and using calculated APES cross-sections for the dianion, we can estimate the values of the PJT parameters for $(C_6S_8)^{2-}$ (table 2). At Q=0 the resulting value of the curvature of the APES is equal to $K \approx K_0 - 2F_{01}{}^2/\Delta_{01} - 2F_{02}{}^2/\Delta_{02} = -0.24$ eV/Å^2. We see that the absolute value of the curvature for the dianion estimated via the PJTE is almost three times larger than that for the neutral molecule. Thus the reduction of neutral carbon sulfide enhances significantly the PJTE induced distortion.

3. PJTE induced symmetry breaking and its suppression in substitution reactions

Using two other tricyclic compounds (thiantrene and antracene) and their derivatives as examples we show how the PJTE induces the symmetry breaking and how the latter can be suppressed by substitutions or oxidation. As mentioned in the Introduction, the thianthrene molecule, as the other 1,4-dithiin containing compounds [18, 19] is bent at the S-S axis [9, 10]. On the other hand a crystal structure determination shows that the oxidized thianthrene molecule has a planar nuclear configuration with D_{2h} symmetry [11, 12]. Besides, if the C atoms at 1, 4, 6, and 9 positions are substituted by nitrogen atoms, the dithiin core becomes planar [13].

The second example, the anthracene molecule (An) is planar with the nuclear configuration of D_{2h} symmetry, whereas the substituted 9,10-dihydroanthracene (9,10-H_2An) has a folding geometry of C_{2v} symmetry [14]. Optimized structures for thianthrene, anthracene and 9,10-dihydroanthracene molecules are shown in figure 4.

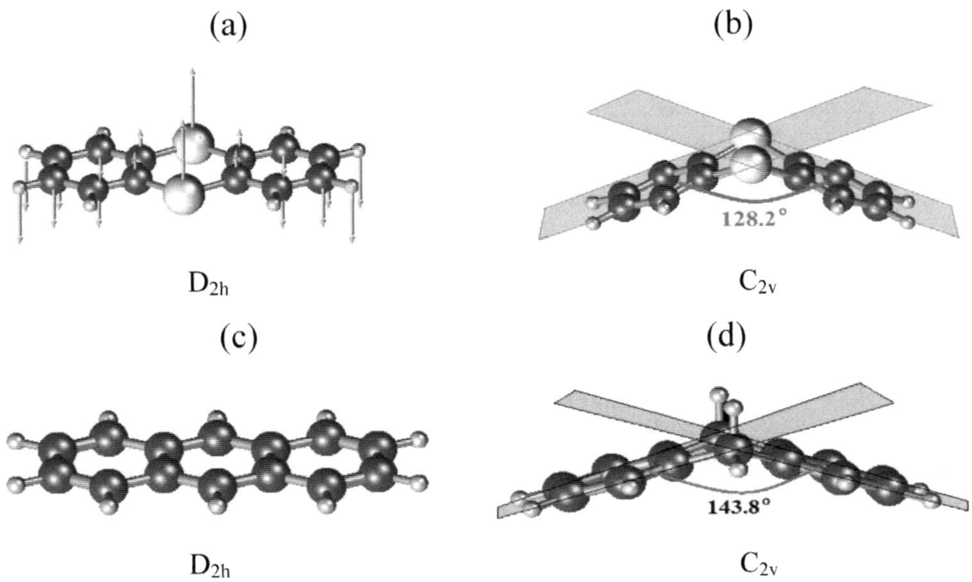

Figure 4. (a) Puckering b_{1u} normal mode of the thianthrene molecule in the planar D_{2h} configuration; (b) calculated optimized structure, C_{2v} configuration; (c) and (d) show the optimized structures of anthracene and 9,10-dihydroanthracene.

Involving the PJTE we can rationalize such different behavior of the molecules in question with respect to folding analyzing their electronic structure, namely, the presence or absence of excited states that control their PJT coupling to the ground state along the distortion coordinates. The ground electronic state of all these molecules in the highest possible D_{2h} symmetry configuration is the ^1A_g one. Since a *butterfly-like* distortion transforms according to B_{1u} irreducible representation of the D_{2h} symmetry group, only the excited states of B_{1u} symmetry are involved in the vibronic mixing with the ground state.

For thiantrene, electronic structures calculations in this configuration indicate the presence of one imaginary frequency (70.11 *i*) corresponding to the out-of-plane distortion of b_{1u}-symmetry. Figure 5 shows the APES cross sections for the ground state and the low-lying excited states of various symmetries along the b_{1u} distortion mode. It is seen that of the nine low-lying excited electronic states in the considered energy domain there is only one fairly high-lying excited state of B_{1u} symmetry which is formed by one-electron excitation from the HOMO of b_{1u}-symmetry to the unoccupied a_g MO (figure 6).

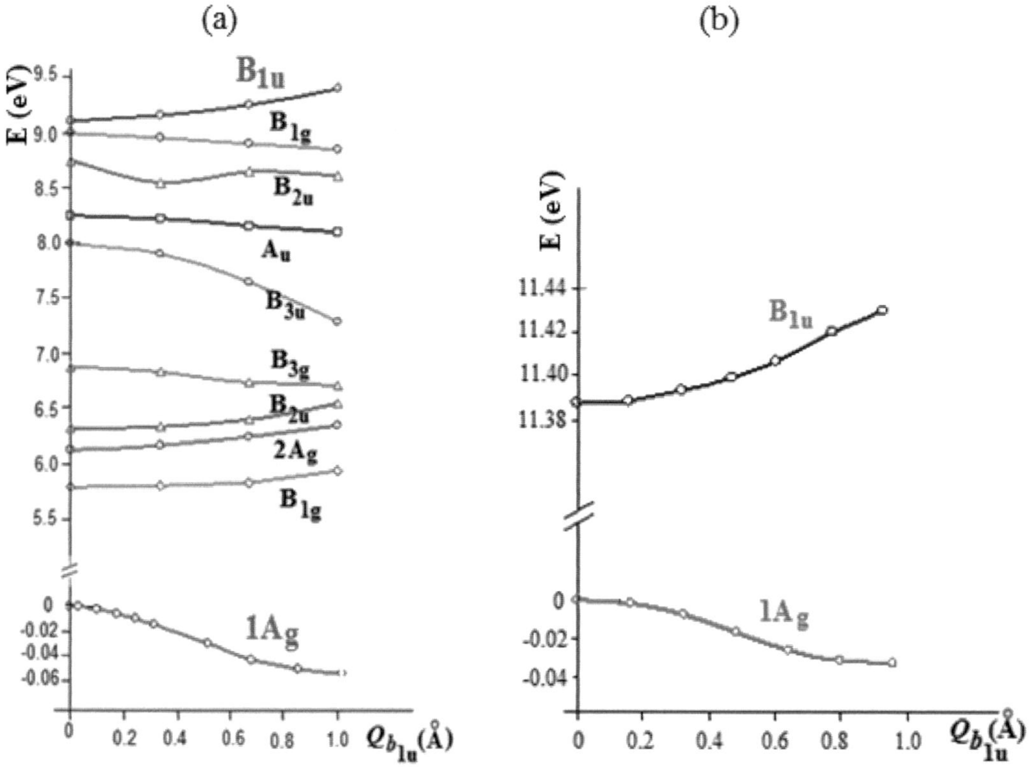

Figure 5. APES cross sections for the ground and excited states along the b_{1u} distortion mode for (a) thianthrene $C_{12}H_8S_2$ and (b) 9,10-H_2An $C_{12}H_8$. As the Q_{b1u} coordinate we used the z-coordinate difference between the atom of sulfur (or carbon in 9,10-H_2An) and the most distant carbon atom in the peripheral ring.

Thus, in this case we have a two-level $(A_{1g}+B_{1u})\otimes b_{1u}$ PJT problem with the following secular equation:

$$\begin{vmatrix} \frac{1}{2}K_0Q^2 - \varepsilon & F_{01}Q \\ F_{01}Q & \frac{1}{2}K_1Q^2 + \Delta_{01} - \varepsilon \end{vmatrix} = 0 \ . \quad (9)$$

The constants K_0, K_1, F_{01} and Δ_{01} in the Eq. (9) have the same meaning as in equation (1).

The solution of this 2×2 equation is straightforward:

$$\varepsilon_{1,2} = \frac{1}{4}(K_0 + K_1)Q^2 + \frac{\Delta}{2} \pm \frac{1}{2}\sqrt{\left[\frac{1}{2}(K_0 - K_1)Q^2 - \Delta\right]^2 + 4F^2Q^2}. \quad (10)$$

Fitting the *ab initio* calculated ground state APES cross-section along the b_{1u} coordinate of instability to the obtained analytical expressions provides for the estimation of the vibronic coupling parameters: K_0 = 1.12 eV/Å^2, K_1 = 0.96 eV/Å^2, F_{01} = 2.47 eV/Å, Δ_{01} = 9.17 eV. At small values of Q the curvature K of the APES for the ground state can be approximately estimated as $K \approx K_0 - 2F_{01}^2/\Delta_{01}$ = - 0.21 eV/Å^2.

As mentioned above, the oxidized thianthrene molecule has a planar nuclear configuration with D_{2h} symmetry. This fact can also be rationalized in the framework of the PJTE theory. Since the excited B_{1u} state is formed by the one-electron excitation from the HOMO b_{1u} to the unoccupied a_g MO (figure 6), the removal of an electron from this orbital reduces by half the PJTE negative contribution K_v to the curvature of the APES compared with a neutral molecule: $K_v \approx -4f^2/\Delta$ for the neutral molecule *versus* $K_v \approx -2f^2/\Delta$ for the radical cation, where $f = \langle b_{1u}|(\partial H/\partial Q_{b1u})_0|a_g\rangle$. Therefore, with a much smaller PJTE distortion, the thiantrene cation has a positive value of the curvature, $K = +\ 0.58$ eV/Å^2, and hence the radical cation has a planar nuclear configuration of D_{2h} symmetry. Thus, we can conclude that 1) the folding of the neutral thianthrene molecule is due to the PJT coupling of the ground A_g and the excited B_{1u} states, and 2) restoration of planar nuclear configuration of this system upon oxidation is directly related to the decrease in the orbital vibronic coupling between these states, and hence, suppression of the PJTE.

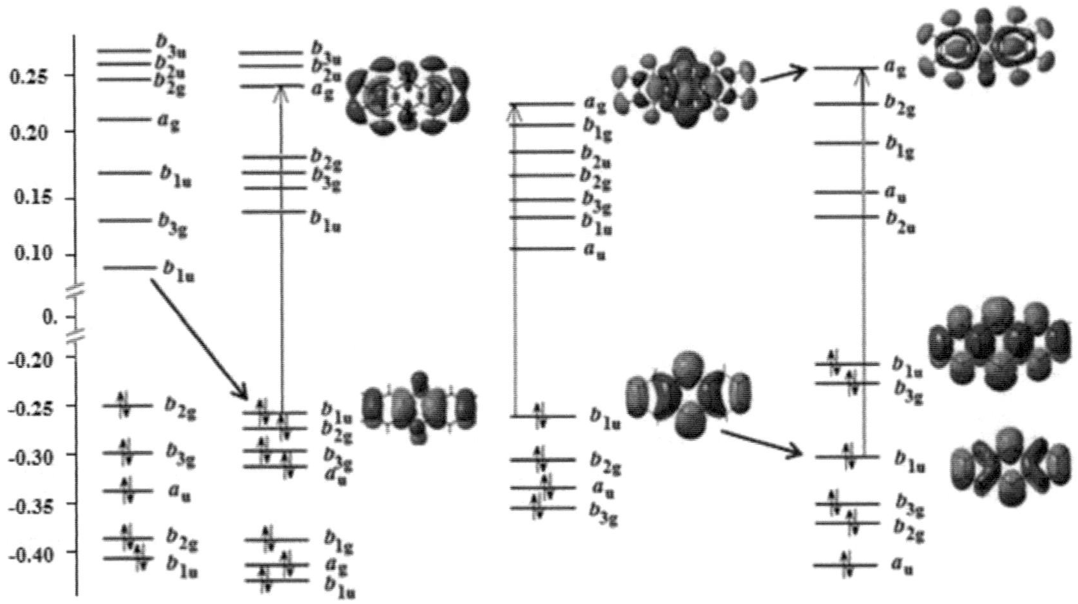

Figure 6. MO energy levels scheme for the An, 9,10-H_2An, $C_{12}S_2H_8$, and $C_{12}S_2N_4H_8$ molecules in the ground A_g state with indication of the one-electron excitations to B_{1u} terms.

The suppression of the PJTE and restoration of the planar configuration takes place also by substitution of the carbon atoms at the 1, 4, 6, and 9 positions by the more electonegative nitrogen atoms. From comparison of MO energy level schemes of these two systems, $C_{12}S_2H_8$ and $C_{12}S_2N_4H_8$ (figure 6), it is seen that the energy gap between the mixing b_{1u} and a_g MOs, which produce the PJTE instability, increases significantly when moving from the former to the latter, reducing the PJTE. In addition, the emergence of two new upper occupied MOs (b_{3g} and b_{1u} in figure 6) increases the value of K_0. Calculations of this system in the D_{2h} nuclear configuration do not show any imaginary frequency, this molecule is planar in the equilibrium configuration.

As for anthracene, there are no low-flying excited states of B_{1u} symmetry that are active in the PJTE. In comparison, in 9,10-dihydroanthracene the b_{1u} MO becomes double-occupied (figure 6). This leads to the occurence of an excited B_{1u} state formed by the one-electron excitation $b_{1u} \rightarrow a_g$, just as in thianthrene. Hence in this case we have the two-level PJT problem, too. As in the case of thianthrene, calculating the APES cross sections (figure 5b) and fitting them to analytical formula for the two-level PJT problem, we have estimated the parameters of the PJTE. Calculated PJT parameters are: $K_0 = 0.68$ eV/Å^2, K_1=0.85 eV/Å^2, F_{01} =2.17 eV/Å, Δ_{01} =11.38 eV. At small values of Q, $K \approx K_0 - 2\ F_{01}^2/\Delta_{01} = -0.14$ eV/Å^2. The results allow us to conclude that in this case the folding of the planar configuration of 9,10-dihydroanthracene is also induced by the PJTE coupling between the ground and one excited state.

4. Conclusions

Using a series of tricyclic molecular compounds and their derivatives as examples, it is shown in detail how the PJTE produces instability and distortions of their planar 2D configuration, and how oxidation, reduction, and substitutions influence the PJTE resulting in either additional buckling of the system or in the restoration of the planar configuration. Summing up the results we can draw some more specific conclusions relevant to the MO presentation:

(1) The out-of-plane distortions of all the considered tricyclic systems are due to the PJTE; the main negative contribution to the curvature of the APES of the ground states, and therefore the instability of their planar nuclear configurations, comes from the excited states formed by one-electron excitation from the HOMO to the unoccupied MO with appropriate symmetry.

(2) If the HOMO state is the mainly one responsible for the PJT interaction with the active excited states leading to distortion of the planar configuration, oxidation of the systems by removing electrons from this MO leads to the suppression of the PJTE and may result in restoration of planar configuration; similar suppression of the PJTE and restoration of the planar configuration may take place by adding electrons to the active excited state.

(3) In the systems under consideration, substitution of any atoms by more electronegative atoms suppresses the PJTE; substitution by more electropositive atoms enhances the PJTE due to donation of additional electrons to the system, and can result in the PJTE induced symmetry breaking.

References

[1] Bersuker I B 2013 *Chem. Rev.* **113** 1351; Bersuker I B 2006 *The Jahn-Teller Effect* (Cambridge UK: Cambridge University Press)
[2] Bersuker I B 2001 *Chem. Rev.* **101** 967
[3] Bersuker I B, this issue, doi:10.1088/1742-6596/833/1/012001
[4] Bersuker I B and Stavrov S S 1988 *Coord. Chem. Rev.* **8** 1
[5] Sergeeva A P and Boldyrev A I 2010 *Organometallics* **29** 3951; Pokhodnya K, Olson C, Dai X, Schulz D L, Boudjouk P, Sergeeva A P and Boldyrev A I 2011 *J. Chem. Phys.* **134** 1; Ivanov A S, Bozhenko K V and Boldyrev A I 2012 *Inorg. Chem.* **51** 8868
[6] Jose D and Datta A 2012 *J. Phys. Chem.* C **116** 24639
[7] Ilkhani A R, Gorinchoy N N and Bersuker I B 2015 *Chem. Phys.* **460** 106
[8] Chou J H, Rauchfuss T B and Szczepura L F 1998 *J. Am. Chem. Soc.* **120** 1805
[9] Schaefer T, Sebastian R, Beaulieu C 1991 *Canadian Journal of Chemistry* **69** 927
[10] Amato M E, Grassi A, Irgolic K J, Pappalardo G C, Radics L 1993 *Organometallics* **12** 775
[11] de Vondele J V, Lynden-Bell R, Meijer E J, Sprik M 2006 *Journal of Physical Chemistry B* **110** 3614
[12] Ito A, Ino H, Ichiki H, Tanaka K 2002 *Journal of Physical Chemistry A* **106** 8716
[13] Joule J A 1990 *Advances in Heterocyclic Chemistry* **48** 301
[14] Moffat John B 1990 *Theoretical Aspects of Heterogeneous Catalysis* Van Nostrand Reinhold New York p. 270

[15] Polly R, Werner H J, Manby F R, Knowles P J 2004 *Mol. Phys.* **102** 2311
[16] Frisch M J, Trucks G W, Schlege H B, et al. 2009 *Gaussian 09, Revision B.01* Gaussian, Inc.: Wallingford, CT
[17] Hehre W J, Ditchfield R, Pople J A 1972 *J. Chem. Phys.* **56** 2257
[18] Ilkhani A R, Hermoso W, Bersuker I B 2015 *Chem. Phys.* **460** 75
[19] Pratik S Md, Chowdhury Ch, Bhattacharjee R, Jahiruddin Sk, Datta A 2015 *Chem. Phys.* **460** 101

Pseudo Jahn-Teller effects and optical activities of negatively curved hydrocarbons

M Hatanaka

Graduate School of Engineering, Tokyo Denki University, 5 Senju-Asahi-cho, Adachi-ku, Tokyo 120-8551, Japan

E-mail: mhatanaka@xug.biglobe.ne.jp

Abstract. As a fundamental study on energetics of negatively curved structures, puckering energetics of [*n*]circulene is studied in view of vibronic interactions. In [7]circulene, the D_{7h} geometry is subject to pseudo Jahn-Teller effect (PJT) and unstable with respect to the E_1'', E_2'', and E_3'' vibrational modes. It is distorted into C_s or C_2 geometries to form valleys of the potential surface ca. 9 kcal/mol below the D_{7h} geometry. The C_2 structure is slightly stable than C_s one, which is consistent with experiments. The resultant C_2 geometry is optically active, and the CD (circular dichroism) and ORD (optical rotatory dispersion) spectra are well simulated. This type of optical activities is interesting in that they should be proportional to the PJT-distortion. PJT-induced optical activity in [2*m*+1]circulenes should be mainly originated from vibronic coupling constants in the initial and final states, which leads to helical rule characteristic of C_2 molecules.

1. Introduction

Curved structures are often found in large-size hydrocarbons, fullerenes, nanotubes, and modified graphenes [1, 2]. The Gauss curvatures K of these materials are classified into three groups: 1) positive curvature ($K > 0$) in bowed structures, 2) zero curvature ($K = 0$) in 'flat' structures, and 3) negative curvature ($K < 0$) in saddled structures. Small circulenes such as [5]circulene (corannulene) and many fullerenes belong to the first group. Normal nanotubes, graphene sheets belong to the second group, in which the local structures are regarded as 'flat'. Materials classified into the third group are often called negatively curved structures, in which the Gauss curvatures are negative due to the saddle structures. Intuitively, the saddled structures with negative curvatures come from different signs of the curvatures along the principal axes. Large circulenes, some jointed or branched nanotubes, and warped graphenes [2] are classified into the third group. The chemical and physical properties of such negatively curved structures have not been scrutinized yet, because of difficulty in the synthesis and characterization [1].

Of all the negatively curved structures, saddled geometries induced by seven-membered rings are essentially important in that the strain energy is not so large and the conformations are not so complex to grasp the fundamental features of the saddled carbons. In this sense, [7]circulene (figure 1) is the most important molecule to grasp the theoretical aspects of the puckering energetics and to predict chemical properties characteristic of the distorted geometries. [7] circulene was first synthesized by Yamamoto et al. in 1983 [3], and the structure at room temperature was characterized by X-ray diffraction. The stable structure was found to have a C_2 geometry, and the geometry was theoretically reproduced by Shen et al., of which stability relative to the D_{7h} structure was estimated to be ca. 8 kcal/mol [4]. Though their vibration analysis suggested that there were very low-frequency vibration

IOP Conf. Series: Journal of Physics: Conf. Series **833** (2017) 012011 doi:10.1088/1742-6596/833/1/012011

modes, the origin of the puckering energetics was not pursued in detail. Intuitively, degenerate low-frequency vibrations seem to induce shallow valleys in the potential surface through the pseudo rotations around the D_{7h} geometry. Then, the degenerate puckering vibrational modes cause two principal stable geometries with C_2 and C_s symmetry. Until now, however, the C_s geometry of [7]circulene has not been found. Thus, it seems meaningful to scrutinize the puckering energetics of [7]circulene in view of the vibronic interactions, that is, pseudo Jahn-Teller effect (PJT). In particular, specification of the active electronic states and the active vibrational modes are of interest in relation to the resultant geometries. Moreover, chemical properties induced by the PJT is worth analysing for characterization of the molecular structures and for applications. Above all, the optical activities are important, because the C_2 and C_s geometries are thereby distinguished experimentally. Indeed, di-bromo and di-methyl derivatives of [7]circulene are optically active, and the one of the enantiomers afforded CD (circular dichroism) spectra [3]. The ORD (optical rotatory dispersion) spectra should also be obtained by direct observation or Kramers-Kronig transformation of the CD spectra.

The theoretical calculations of the CD/ORD spectra will serve as a foundation for developing optical active carbons. The simulated spectra are applicable to grasp the optical activities of more extended systems with a few saddle points such as defective graphenes, warped graphenes, and higher circulenes. In this study, geometries and optical activities of higher circulenes are also analysed in view of the amplitude pattern of the orbitals and puckering vibrational modes. General aspects of PJT-induced optical activities and some applications are also presented.

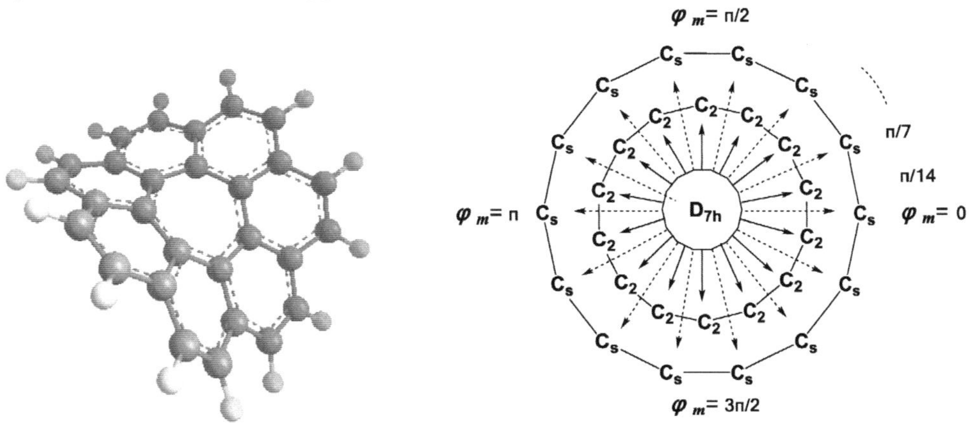

Figure 1. [7]circulene (C_2 geometry) and the pseudo rotation in D_{7h} point group.

2. Pseudo Jahn-Teller effect of [7]circulene

The origin of the instability in non-degenerate polyatomic systems is vibronic contribution only [5]. Puckering of planar molecules has also been rationalized to PJT, and the advances in general theory have been reviewed in detail [6]. Recent studies cover not only small molecules but also two-dimensional thin films such as inorganic graphene analogues [7]. Therefore, structural analysis along the PJT scheme is quite reasonable in [7]circulene. The subjects of interest are active vibrational modes, active excitation states, active orbitals (orbital pattern), the stable structure (point group), and energy gaps between the high and low symmetries. In the present case, the resultant chemical properties, e. g., optical activities are also of interest.

The PJT-active vibrational modes are determined by analysing the Hertzberg-Teller expansion of the Hamiltonian H:

$$H = H_0 + \left(\frac{\partial H}{\partial Q_i}\right)_0 Q_i + \frac{1}{2}\left(\frac{\partial^2 H}{\partial Q_i^2}\right)_0 Q_i^2, \qquad (1)$$

IOP Conf. Series: Journal of Physics: Conf. Series **833** (2017) 012011 doi:10.1088/1742-6596/833/1/012011

$$E(Q_i) \cong E_0 + \langle \psi_0 | \left(\frac{\partial H}{\partial Q_i} \right)_0 | \psi_0 \rangle Q_i + \frac{1}{2} \left\{ \langle \psi_0 | \left(\frac{\partial^2 H}{\partial Q_i^2} \right)_0 | \psi_0 \rangle + 2 \sum_{n \neq 0} \frac{\left| \langle \psi_n | (\partial H / \partial Q_i)_0 | \psi_0 \rangle \right|^2}{E_0 - E_n} \right\} Q_i^2, \tag{2}$$

where H_0 and E_0 correspond to the non-distorted structures, and ψ_0 is the original wavefunction of the high-symmetry geometry. When the state ψ_0 is non-degenerate, the second term in Eq. (2) is zero except for the total symmetric modes. Two terms in the curly brackets determine the curvature of the potential surface. They consist of the classical force constant and relaxability. The former is always positive, and the latter is always negative. When the last term is dominant, PJT is induced and the curvature of the potential surface becomes negative. From the non-vanishing condition of the last term, the active vibrational modes and the excited states should belong to the same irreducible representation.

Strictly speaking, the potential curve of the ground-state energy is obtained from a secular equation. In the two-level framework, it becomes [5]:

$$E(Q_i) \cong \frac{1}{2} K_0 Q_i^2 - \sqrt{\Delta^2 + F^2 Q_i^2} + const., \tag{3}$$

where 2Δ is energy interval between the ground state Γ and the most active excited state Γ', $K_0 = \langle \Gamma | (\partial^2 H / \partial Q_i^2)_0 | \Gamma \rangle$, $F = \langle \Gamma | (\partial H / \partial Q_i)_0 | \Gamma' \rangle$. Of course K_0 is the classical force constant, and F is the vibronic coupling constant. This expression can be expanded with respect to Q_i, and the approximate form within the second-order terms is essentially identical to Eq. (2).

Let us consider a regular j-sided polygon and the puckering displacement of the j-th moiety of the system, z_j. The displacement z_j is described by Cremer-Pople coordinate:

$$z_j = \sqrt{\frac{2}{N}} \cdot \sum_m^{\frac{1}{2}(N-1)} q_m \cos\{\varphi_m + 2\pi m(j-1)/N\} \qquad (N = \text{odd number}), \tag{4}$$

where q_m is amplitude of the m-th puckering normal modes, φ_m is the phase angle which determines the symmetry of the vibrational modes. In the present case, $N = 7$. The larger m becomes, the more the nodes appears. Thus, E_1'', E_2'', and E_3'' puckering modes in the D_{7h} point group correspond to $m = 1$, 2, and 3, respectively. The pseudo rotation of the present D_{7h} system is schematically shown in figure 1. A D_{7h} geometry is distorted into C_2 and C_s point groups at $n\pi/14$ (n; integer, not the ring size). C_2 and C_s geometries are alternatively appeared according to the increase of φ_m. Between the C_2 and C_s points, the geometry belongs to C_1 representation.

To grasp the active orbitals for the PJT, orbital diagram of the [7]circulene is shown in figure 2 [8]. For this purpose, very accurate calculations are unnecessary. Instead, qualitative description by Kohn-Sham orbitals at B3LYP/3-21G level of theory is given. There are active e'' (e_1'', e_2'', and e_3'') orbitals in the occupied space, and e' (e_1', e_2', and e_3') orbitals in the virtual space. The one-electron excitation (typical examples are shown by arrows) affords the PJT-active electronic states E'' (E_1'', E_2'', and E_3''), according to the non-trivial direct products, e. g.:

$$e_2'' \otimes a_1' = e_2'' \qquad e_3'' \otimes a_1' = e_3'' \qquad e_1'' \otimes a_1' = e_1''$$

$$e_2'' \otimes e_1' = e_1'' + e_3'' \qquad e_3'' \otimes e_1' = e_2'' + e_3'' \qquad e_1'' \otimes e_1' = a_1'' + a_2'' + e_1''$$

$$e_2'' \otimes e_2' = a_1'' + a_2'' + e_3'' \qquad e_3'' \otimes e_2' = e_1'' + e_2'' \qquad e_1'' \otimes e_2' = e_1'' + e_3''.$$

We see that one-electron excitations from e'' to e' (or a_1') orbitals afford the PJT-active E'' electronic states. Figure 2 shows the amplitude patterns of the active e'' (e_1'', e_2'', and e_3'') orbitals. These orbitals are delocalized. In the edge region, the amplitudes are not trivial, and should be coupled with the active puckering modes. HOMO (e_2'') is very important for the present PJT distortion. Figure 2 also shows e' (e_1', e_2', and e_3') orbitals. These virtual orbitals are also localized at the edge region. These are coupled with puckering modes including the peripheral C-H bending displacement.

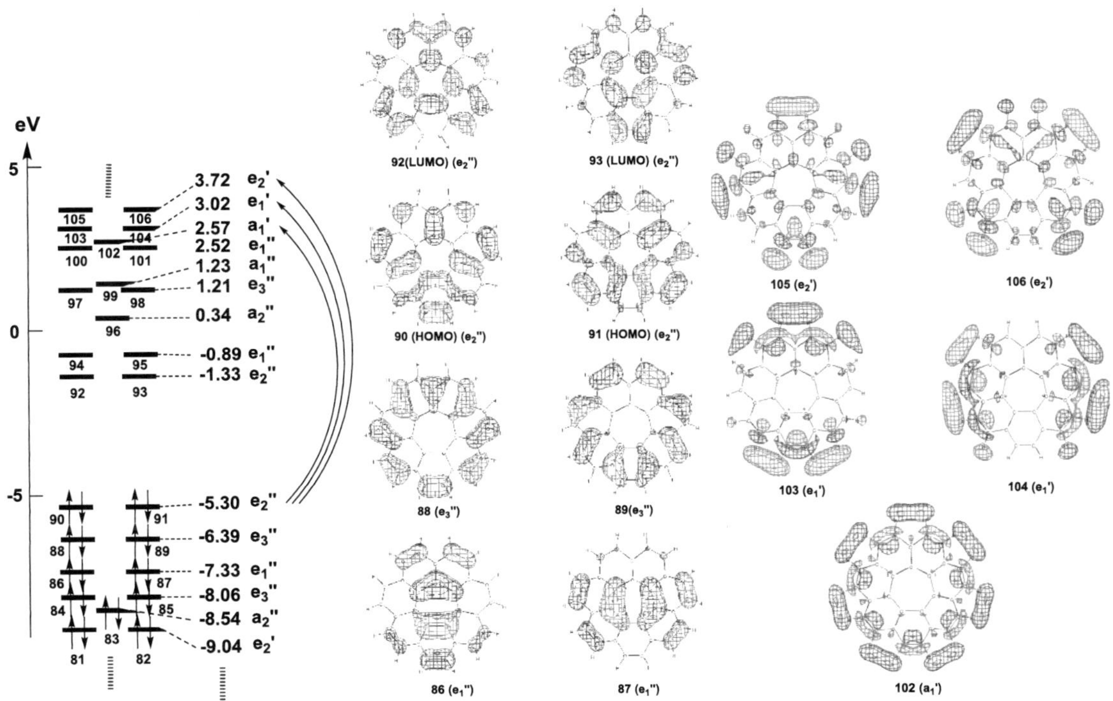

Figure 2. Molecular orbitals of D_{7h}-[7]circulene (B3LYP/3-21G//B3LYP/3-21G). Reprinted with permission from M. Hatanaka, *J. Phys. Chem. A*, **120**, 1074 (2016). Copyright (2016) American Chemical Society.

Figure 3 shows the electronic state diagram for D_{7h}-[7]circulene. The state diagrams including the excited states are obtained by configuration interactions (CI), TDDFT, MCSCF, and other post-Hartree-Fock calculations. Here, the semi-empirical single CI results at CIS(23,7)/PM3//PM3 level is shown, because only qualitative description is needed. The PJT-active states are E' (E_1', E_2', and E_3') states, which lie ca. 6 eV above the ground state.

From the vibration analysis of the D_{7h}-[7]circulene, it was found that the lowest E_2'' and E_3'' vibrational modes are PJT active, though whether each frequency is real or imaginary depends on the basis sets. The frequencies became imagine under some basis sets, in which the absolute values are below 100 cm^{-1}, but real values below 100 cm^{-1} were also obtained at the other basis sets [8]. The basis-set impact shows that the magnitude of the classical force constant and the relaxability is nearly the same order, and thus, the resultant vibrational frequencies become very small. However, this does not means that the vibronic interaction itself is trivial. The displacements around the peripheral carbon and hydrogen atoms were significant, which suggests the vibronic interactions occur through interactions between the peripheral amplitudes of the active orbitals and the peripheral displacements of the active vibrational modes.

Energetically, the C_2 geometry was most stable, consistent with the X-ray results [3,8]. The energy difference between C_2 and C_s geometries was small (ca. 0.05 kcal/mol). The gap between D_{7h} and C_2 was ca. 9 kcal/mol. We note that C_2 structures are optical active if one of the enantiomers is optically resolved, because there are two asymmetric conformers with *M* (minus) and *P* (plus) helicities. This formal classification is akin to those of helicenes, which are famous optically active molecules. In particular, hexahelicene and heptahelicene are good comparative examples of the C_2-[7]circulene. Figure 4 shows the optimized structures of the [7]circulenes, hexahelicene, and heptahelicene. In view of the structural resemblance, optical activities of [7]circulene with C_2 symmetry are worth analysing. Given that the optical activities of *M*- and *P*-[7]circulene are considered to be induced by vibronic interactions, role of the PJT in the resultant optical rotatory strength is theoretically described later.

IOP Conf. Series: Journal of Physics: Conf. Series **833** (2017) 012011 doi:10.1088/1742-6596/833/1/012011

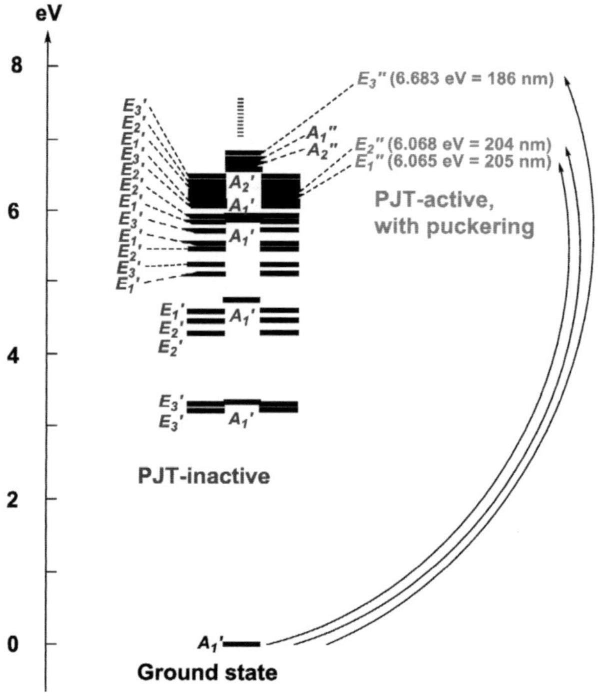

Figure 3. State diagram (CIS(23,7)/PM3//PM3) of D_{7h}-[7]circulene.

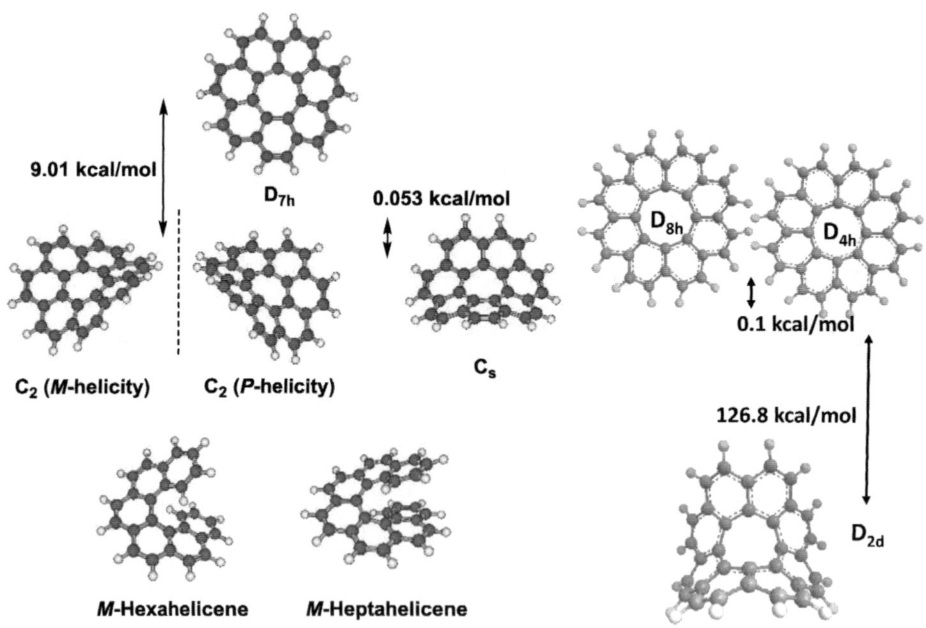

Figure 4. Optimized conformers of [7]circulene and [8]circulene (B3LYP/3-21G//B3LYP/3-21G). The structural resemblance between [7]circulene and helicenes is also shown.

3. [8], [9], and higher circulenes

[8]circulene derivatives have D_{2d} skeletons. Possible PJT scheme is considered to be $D_{8h} \rightarrow D_{4h} \rightarrow D_{2d}$, similar to cyclooctatetraene (COT). DFT calculations by B3LYP/3-21G//B3LYP/3-21G revealed that the most stable structure has a D_{2d} geometry, which was 127 kcal/mol below the D_{4h} structure. The D_{8h} structure is unstable, of which energy is 0.1 kcal/mol above the D_{4h} structure. The energetics is schematically shown in figure 4. These are optically inactive. In general, [4*n*]circulenes are optically inactive even if they are subject to any PJT. This is easily shown by amplitude pattern of the puckering vibrational modes or orbital pattern in D_{8h} (or D_{4h}) point group, as depicted in figure 5. The ground state of a [4*n*]circulene is non-degenerate, and PJT active states are given by one-electron excitation from b_{1u} (or b_{2u}) to a_{1g} (or a_{2g}), in which one of them is vacant. Thus, the PJT active states and puckering vibrational modes belong to B_{1u} or B_{2u}, which lead to D_{2d} geometries.

[9]circulene has not been synthesized yet. However, by the semi-empirical PM3 method, I found C_2 and C_s geometries below a D_{3h} geometry. The C_2 and C_s geometries have almost same energy (the difference is below 0.1 kcal/mol). Though the detail of energetics and vibrational modes are not clear at present, the C_2 geometry is optically active in principle. In general, [2*m*+1]circulenes (odd-circulenes) should be optical active due to the C_2-distorted structures. This is realized through anti-symmetric amplitude pattern of the vibrational modes.

[4*n*+2]circulenes such as [10]circulene may also be subject to PJT. Though optically active C_2-distorted structures are possible in principle, they are probably unstable, because the adjacent atomic sites should be twisted. Twisted-saddled structures predicted by Christoph et al. [1] are probably not originated from PJT, but from re-relaxation of saddled structures.

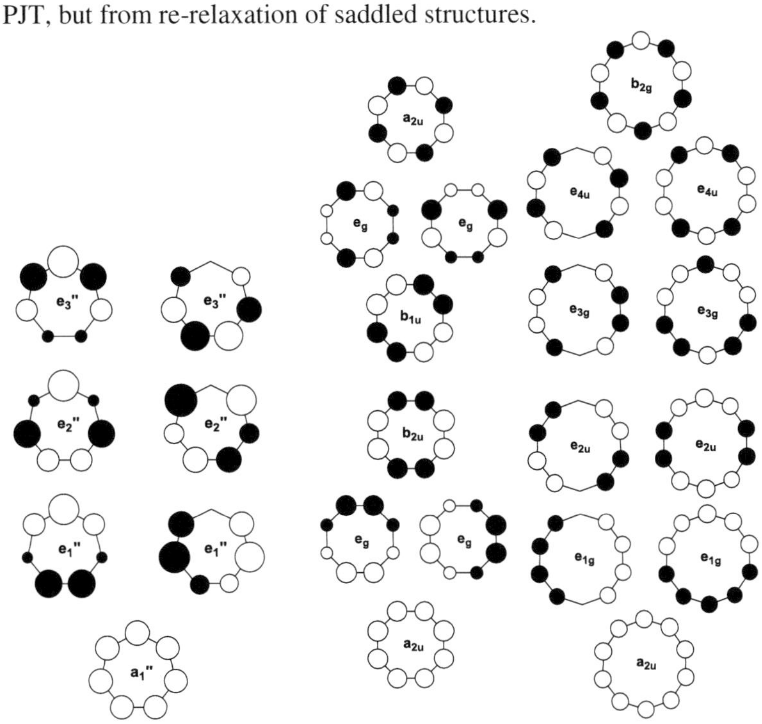

Figure 5. Schematic vibrational modes in 7-, 8-, and 10-membered ring systems.

3. PJT-induced optical activities

While the C_s conformers are optically inactive, C_2 conformers are optically active. How strong the intensity is, compared with helicenes? Does the so-called helical rule of Cotton effect hold? Optical rotatory strength $R_{I\text{-}F}$ is calculated based on Rosenfeld formula;

$$R_{I-F} = \mathrm{Im}\left\{\langle I|\sum_n \mu_e(n)|F\rangle\cdot\langle F|\sum_{n'}\mu_m(n')|I\rangle\right\}, \tag{5}$$

$$[R_{I-F}] = \frac{100R_{I-F}}{\mu_D\mu_B} = 1.08\times10^{40}R_{I-F}(\mathrm{cgs}), \tag{6}$$

$$[R_{\Psi_G-\Psi_E}] \cong -\frac{7313}{E_E-E_G}\sum_{a,r}\sum_{b,s}c_a^r c_b^s\{\langle a|\nabla|r\rangle\cdot\langle s|\boldsymbol{r}\times\nabla|b\rangle\}, \tag{7}$$

where I and F mean the initial and final states, $\boldsymbol{\mu}_e$ and $\boldsymbol{\mu}_m$ are electronic and magnetic transition moment operators, and the sum in Eq. (5) is taken for all the electrons. R_{I-F} is often reduced to the second formula, in which μ_D and μ_B are Debye unit and Bohr magneton, respectively. The third formula is the extended formula for the configuration interactions (CI), in which the energy and coordinates are in units of eV and Å. We calculated $[R_{I-F}]$ by this expression by using PM3-CASCI wavefunctions [8]. The simulated CD/ORD spectra are shown in figure 6. In the long-wavelength region, minus-sign Cotton effects are seen in all cases due to anti-symmetric B excited state, which indicates the helical rule. The intensity of spectra in [7]circulene is much smaller than that in helicenes, because of the nearly degenerate electronic states.

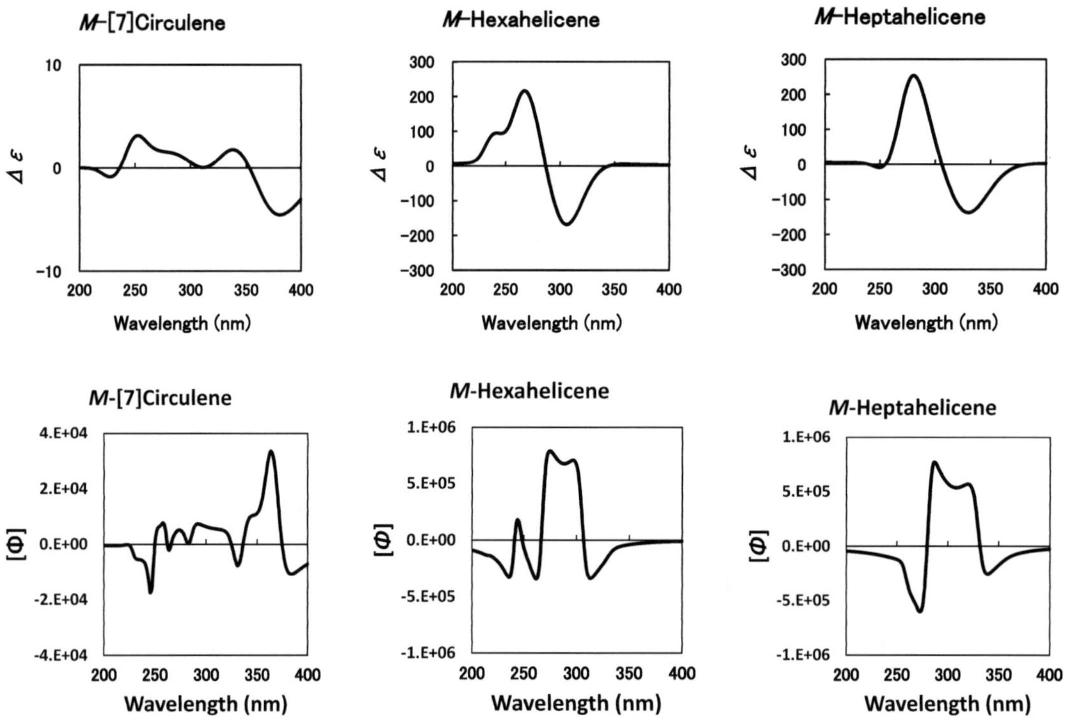

Figure 6. CD (circular dichroism) and ORD (optical rotatory dispersion) spectra of [7]circulene, hexahelicene, and heptahelicene simulated by PM3-CASCI level. Reprinted with permission from M. Hatanaka, *J. Phys. Chem. A*, **120**, 1074 (2016). Copyright (2016) American Chemical Society.

The optical rotatory strength induced by the PJT is estimated by perturbation theory. When change in the magnetic transition moment is neglected, the first-order approximation affords;

$$R_{I-F} \cong \mathrm{Im}\left\{ \begin{array}{l} \displaystyle\sum_{M_0 \neq I_0} \frac{\left\langle I_0 \left| \left(\frac{\partial H}{\partial Q_i} \right)_0 \right| M_0 \right\rangle}{E_{I_0} - E_{M_0}} Q_i \left\langle M_0 \left| \sum_n \mu_e(n) \right| F_0 \right\rangle \cdot \left\langle F_0 \left| \sum_{n'} \mu_m(n') \right| I_0 \right\rangle \\ \displaystyle + \sum_{N_0 \neq F_0} \frac{\left\langle N_0 \left| \left(\frac{\partial H}{\partial Q_i} \right)_0 \right| F_0 \right\rangle}{E_{F_0} - E_{N_0}} Q_i \left\langle I_0 \left| \sum_n \mu_e(n) \right| N_0 \right\rangle \cdot \left\langle F_0 \left| \sum_{n'} \mu_m(n') \right| I_0 \right\rangle \end{array} \right\}, \quad (8)$$

where the subscript 0 means the D_{7h} geometry, index M_0 and N_0 run ground and excited states except for I_0 and F_0. The integrals $<I_0|(\partial H/\partial Q_i)_0|M_0>$, etc., are vibronic coupling constants between I_0 and M_0, etc. The direct product between I_0 and M_0 should identical to the representation of $(\partial H/\partial Q_i)_0$. In the present case, for the excitation from the ground state (I_0 is A_1' state), M_0 should be E'' (E_1'' or E_2'' or E_3'') states in the first term. Similaly, N_0 in the second term should also be limited so that the direct product between F_0 and N_0 contains one of the E'' states. Anyway, only low-lying excited states contribute to the optical rotatory strength. Then, R_{I-F} is approximately proportional to the displacement Q_i, and is an odd function of Q_i. This situation corresponds the minus and plus sign of R_{I-F}. It goes without saying that the higher-order terms should be the third-order, fifth-order, and so on. Though the precise estimation of R_{I-F} is complex, it is notable that this formalism is akin to charge transfer model of the so-called octant rule, and leads to the conventional helical rule for C_2-symmetry molecules.

Roughly estimating, magnitude of R_{I-F} is approximately proportional to (relaxability/ΔE)$^{0.5}$ of the initial and final states, where ΔE (= 2Δ) is the energy gap to the lowest PJT active state. Given that relaxability is related to strain energy and the strain is determined nearby the irregular polygons only, optical activities of higher circulenes probably increase with the ring size. As applications to extended systems, optical activities of the heptagon-embedded graphenes are probably no so different from that of [7]circulene. Furthermore, in warped graphenes, contribution of the each heptagon to R_{I-F} is probably additive.

4. Concluding remarks

Puckering energetics of [*n*]circulene was analysed in view of vibronic interactions. It was shown that the origin of the C_2 and C_s geometries of [7]circulene is attributed to pseudo Jahn-Teller effect (PJT), and modern DFT calculations supported the experimental results. The optical activities in such C_2 geometries were deduced by perturbation theory. While in [2*m*+1]circulenes PJT-induced optical activities are possible, in [4*n*] and [4*n*+2]circulenes, any PJT does not induce optical activities ([4*n*]) or results in unstable twisted geometries ([4*n*+2]). PJT-induced optical activities will be useful for fundamental analyses of CD/ORD spectra in negatively curved structures.

Acknowledgments
Thanks are due to ACS publications for permission to reprint the part of figures.

References

[1] Christoph H, Grunenberg J, Hopf H, Dix I, Jones P G, Scholtissek M, Maier G 2008 *Chem. Eur. J.* **14** 5604
[2] Kawasumi K, Zhang Q, Segawa Y, Scott L T, Itami K 2013 Nature Chemistry, 5 739
[3] Yamamoto K, Harada T, Nakazaki M, Naka T, Kai Y, Harada S, Kasai N 1983 *J. Am. Chem. Soc.* **105** 7171; Yamamoto K, Harada T, Okamoto Y, Chikamatsu H, Nakazaki M, Kai Y, Nakao T, Tanaka M, Harada S, Kasai N 1988 *J. Am. Chem. Soc.* **110** 3578
[4] Shen M, Ignatyev I S, Xie Y, Schaefer III H F 1993 *J. Phys. Chem.* **97** 3212
[5] Bersuker I B 2013 *Chem. Rev.* **113** 1351
[6] Bersuker I B 2001 *Chem. Rev.* **101** 1067
[7] Jose D, Datta, A 2014 *Acc. Chem. Res.* **47** 593
[8] Hatanaka M 2016 *J. Phys. Chem.* A **120** 1074

IOP Conf. Series: Journal of Physics: Conf. Series **833** (2017) 012012 doi:10.1088/1742-6596/833/1/012012

Pseudo Jahn-Teller effect in permittivity of ferroelectric perovskites

V Polinger[1] and I B Bersuker[2]

[1]Department of Chemistry, University of Washington, 109 Bagley Hall, Seattle, WA 98195, USA
[2]Institute for Theoretical Chemistry, University of Texas at Austin, 105 E. 24th St. Austin, TX 78712, USA

E-mail: vpolinger@msn.com

Abstract. Microscopic origin of dielectric susceptibility and permittivity of cubic ferroelectric perovskites ABO_3 ($BaTiO_3$, $KNbO_3$, etc.) in their high-symmetry paraelectric phase is explored. Ferroelectricity in these crystals is due to the strong pseudo Jahn-Teller effect (PJTE) resulting in off-centre instability of the transition metal ion B. In addition to the noticed earlier big enhancement of the permittivity by the local PJTE, it is found that the latter induces also anisotropy, contrary to widespread expectation for isotropic polarization in cubic crystals. The temperature and field dependence of polarization, dielectric susceptibility, and permittivity, as well as their angular dependence, are revealed as resulting from the local PJTE instability by numerical diagonalization of the Hamiltonian including the vibronic coupling and temperature-averaged interaction with the external electric field.

1. Introduction

We consider the dielectric susceptibility and permittivity of ferroelectric perovskites ABO_3, like $BaTiO_3$, $KNbO_3$, etc., in which the polar instability of the elementary-cell octahedron $[BO_6]$ is due to the pseudo Jah-Teller effect (PJTE) [1-3]. The adiabatic potential energy surface (APES) of this octahedron was shown to have eight equivalent wells situated along the four trigonal axes C_3 (figure 1), in which the atom B is off-center shifted making a local dipole moment, with potential barriers between the wells. At higher temperatures, in the paraelectric phase the local polar displacements in different centers have arbitrary (incoherent and dynamic) orientations that make the macroscopically averaged crystal seen as cubic. At a lower temperature $T = T_C$, the ferroelectric phase transition results in orientation ordering of the local dipoles (partial or full) resulting in a polarized crystal, and the multi-valley APES explains the temperature-dependent formation of all the four ferroelectric phases [1].

A detailed presentation of this vibronic coupling theory is given in a series of papers [1, 2] and books [3] including a recent more elaborate theoretical description based on the Green's functions approach [1]. Its main conclusions are confirmed in dozens of experimental observations forming a pattern (see one of them in [4] and many others cited in [1-3]). Recently the theory was expanded to also include the possibility of coexisting PJTE-induced ferroelectric and open-shell magnetic properties in perovskites ABO_3, and a table was given listing the B atoms with electron configurations

$d^1 - d^{10}$ that can produce such a multiferroic [2]. A possibility of magnetic- ferroelectric crossover is predicted there as well.

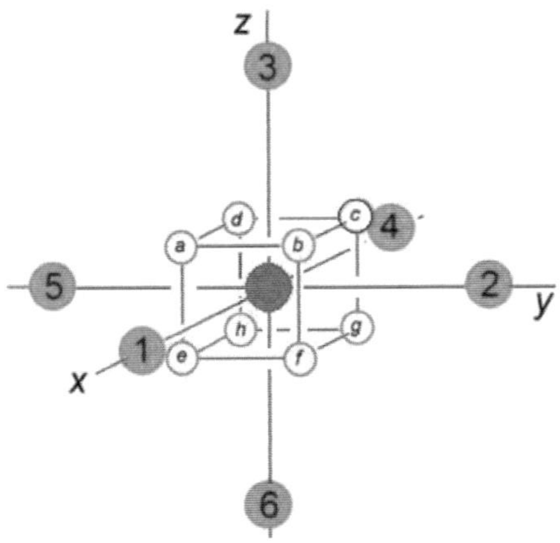

Figure 1. In the paraelectric phase of a cubic perovskite ABO_3 the transition metal atom B (in red) is deemed to be in the center of the octahedron of oxygen atoms (in blue). Due to the PJTE there are eight off-center shifted (shown not to scale) symmetry-equivalent positions, indicated *a, b, c, ...*, of the atom B along trigonal axes (after [1]).

For $BaTiO_3$ in the paraelectric phase numeric estimates [5] show that the energy barrier between the minima of the APES is rather low, of the order of 200 cm^{-1}, but the vibrational frequency in the direction to the barrier is also low, so the zero-vibrations level is at about half the barrier height, at ~100 cm^{-1}. Hence we may expect the tunneling transitions between the minima at low temperatures being overwhelming with more intensive over-the-barrier transitions at higher temperatures. Accordingly is moving the vector of the dipole moment formed by the Ti^{4+} ion positioned in the minima. For an ensemble of dipolar molecules oriented by an external electric field and disoriented by thermal motions the dielectric constant ε was estimated by Debye [6]. It includes two contributions. One is the *displacive* term. It takes into account the polar distortion (polarization) of the molecule by the external electric field. The other one, the *orientation* contribution, comes from the orientation of the dipole moment by the field. The classical Debye-Curie-Weise-Langeven theory takes into account both these contributions to permittivityε, as well as its dependence on the electric-field frequency ω:

$$\varepsilon(\omega) = \varepsilon_0 \left(1 + \frac{\chi_d}{1+\omega^2\tau_d^2} + \frac{\chi_o}{1+\omega^2\tau_o^2} \right) \tag{1}$$

Here ε_0 is the electric constant (also known as the permittivity of the vacuum), χ_d and χ_o are the electric susceptibilities of displacive and orientation contributions, respectively, and τ_d и τ_o are the corresponding relaxation times.

Usually the displacive contribution χ_d is relatively small; in molecular systems and ionic crystals in static electric fields ($\omega = 0$) it varies within the limits $\chi_d \sim 0.1 \div 10$. The orientation contribution χ_o for systems with a proper dipole moment is much larger (for water $\chi_d = 5$, $\chi_o = 75$), but there is no orientation effect in nonpolar molecules and in crystals without the JTE or PJTE. In $BaTiO_3$ the permittivity in the paraelectric phase reaches the value of $\varepsilon/\varepsilon_0 \approx 20{,}000$, and as was recently shown [7], in estimates by orders of magnitude that almost all of it comes from the orientation contribution of the PJTE-induced rotating dipole moments in the Ti centers. Without the PJTE, there is no other explanation of this giant permittivity and a similar strongly enhanced flexoelectricity [7].

Note that for these giant effects to take place in matter, the only requirement is the presence of local centers with polar instability in several (or infinite number of) equivalent directions with free or hindered rotations of the dipoles (the case of free rotation of the dipoles is similar to a polar liquid). In the perovskites ABO_3 under consideration such polar centers may occur at B sites (in the clusters $[BO_6]$) due to the PJTE. In the most studied case of $BaTiO_3$ the polar instability is strong with an eight-minimum APES, but relatively shallow barriers between the wells. Therefore, the rotation of the local dipoles (either by tunneling or by over-the barrier transitions) is hindered. In other perovskites such a possibility may occur subject to the local ground state being sufficiently close in energy to excited electronic states (bands) with the same spin multiplicity but opposite parity [1-3]. Similar conditions apply to any other atomic centers in matter. In all these cases free or hindered rotation of the dipoles may occur causing an enhanced response to external perturbations. The free rotation case comes about when, instead of the considered in this paper multi-minimum case, the JTE or PJTE results in a two-dimensional trough.

In anisotropic crystals the linear polarization **P** by the electric field **E** is a tensor, $P_j = \Sigma_k \chi_{jk} E_k$. Accordingly, the permittivity is a symmetric tensor of the second rank, $\varepsilon_{jk} = \varepsilon_0\ (1 + \chi_{jk})$. In cubic crystals the main three components of the tensor ε_{jk} coincide, and the permittivity is isotropic. On the other hand, in ferroelectric crystals with PJTE the orientation contribution to the susceptibility χ_o should depend on the direction of the electric field. Indeed, as mentioned above, each elementary cell of such a crystal is unstable with respect to the displacement of the central atom along trigonal axes C_3, hence the magnitude of the vector **P** is a maximum when the electric field vector **E** points along one of these axes. It follows that in these ferroelectric crystals the permittivity should have a *cubic anisotropy*, even in the cubic paraelectric phase where it is expected to be isotropic in the absence of the PJTE.

In the present paper we report temperature and field dependence of the polarization, dielectric susceptibility, and permittivity, as well as their anisotropy in cubic-perovskite ferroelectrics of the type ABO_3, subject to the PJTE in the B centers. In the alternating external electric field the frequency dependence $\varepsilon(\omega)$ includes the contribution of both the electrons and nuclei, but the contribution of the latter to $\varepsilon(\infty)$, due to their inertia, is negligible. Considering here the anisotropy of χ и ε due to nuclear displacements, we limit our analysis to the case of static electric field, meaning $\omega = 0$. The frequency dependence $\varepsilon(\omega)$ is shown in equation (1). In the paraelectric phase there is no spontaneous polarization, hence interaction between the active PJTE centers is not significantly strong. Therefore we ignore the cooperative effects and, *with respect to the local PJTE* properties we consider each elementary cell independently.

Specifics of the calculation presented in this paper relate to the most studied case of $BaTiO_3$. However, derived here qualitative conclusions apply to all ferroelectric perovskites ABO_3 as long as the PJTE is strong enough to provide instability of the on-center position of the atom B in the cubic phase. This is due to the overall cubic symmetry of the eight-minimum APES. Details, similarities, differences, and possible implications of the PJTE in other cubic perovskites in the variety of transition metal ions from d^0 to d^{10} are discussed in [2].

In the next section 2 we introduce the vibronic Hamiltonian of the octahedral cluster $[BO_6]$ of each elementary cell of the cubic-perovskite crystal, the eight-minimum APES induced by the PJTE, and the reduction of the Hamiltonian to the 8×8 matrix in the space of the 8 vibronic basis states localized in the eight trigonal minima. Section 3 shows the main results obtained by numerical calculations, while the conclusions in section 4 are devoted to some further discussion of the results.

2. PJT origin of the polar instability of the elementary octahedron $[BO_6]$. The eight-minimum model

2.1. Background

The origin of ferroelectric instability in cubic-perovskite $BaTiO_3$ as due to the local off-centre displacements of the Ti^{4+} ion in the octahedron $[TiO_6]$ induced by the PJTE was first introduced in 1966 [1]. In molecular orbital (MO) presentation the polar instability of this octahedron is due to the vibronic coupling between the highest occupied MO (HOMO), which are mostly oxygen $2p$-orbitals, with the lowest unoccupied MO (LUMO) which are mostly t_{2g}-type $3d$-orbitals of the Ti^{4+} (figure 2). In the absence of external perturbations and internal random fields the vibronic Hamiltonian of the octahedron $[TiO_6]$ can be presented in the matrix form as follows:

$$\mathbf{H}_0 = \Delta\mathbf{C}_0 + \frac{1}{2m^*}\left(p_x^2 + p_y^2 + p_z^2\right) + \frac{1}{2}K_0\left(q_x^2 + q_y^2 + q_z^2\right) + \mathbf{H}_{\mathrm{pJT}} \quad (2)$$

where p_α ($\alpha = x, y, z$) are the components of the linear momentum $\mathbf{p}$ conjugated with the coordinates q_α ($\alpha = x, y, z$), the components of the vector $\mathbf{q}$ transforming as T_{1u}, K_0 is the primary force constant, and m^* is the effective mass. The motion of the center of mass of the octahedron $[TiO_6]$ is separated out, so $\mathbf{q}$ includes off-center displacements of the ion Ti^{4+} and the opposite motion of the oxygen octahedron O_6. The 9 × 9-matrix $\mathbf{C_0}$ is the electronic part of the Hamiltonian presented in the basis of the nine one-electron states (figure 2) mixed by the three polar modes q_α ($\alpha = x, y, z$) active in the PJTE. The parameter 2Δ is the HOMO-LUMO energy gap and $\mathbf{H_{pJT}}$ stands for the Hamiltonian of the PJT vibronic coupling of the nine electronic states (HOMO and LUMO) to the three modes q_α. In the linear approximation we can set $\mathbf{H_{pJT}} = F(\mathbf{C}_x q_x + \mathbf{C}_y q_y + \mathbf{C}_z q_z)$ where F is the linear vibronic coupling constant and $\mathbf{C}_x$, $\mathbf{C}_y$ and $\mathbf{C}_z$ are matrices 9 × 9 acting in the basis of the nine HOMO-LUMO states. They are composed of the corresponding Clebsch-Gordan coefficients taken from [8]. In these matrices the coefficient 1/√2 is factored out and included in the vibronic constant F. The explicit form of the matrices $\mathbf{C}_0$, $\mathbf{C}_x$, $\mathbf{C}_y$ и $\mathbf{C}_z$, as well as the details of the solution of the corresponding PJTE problem are given in [1].

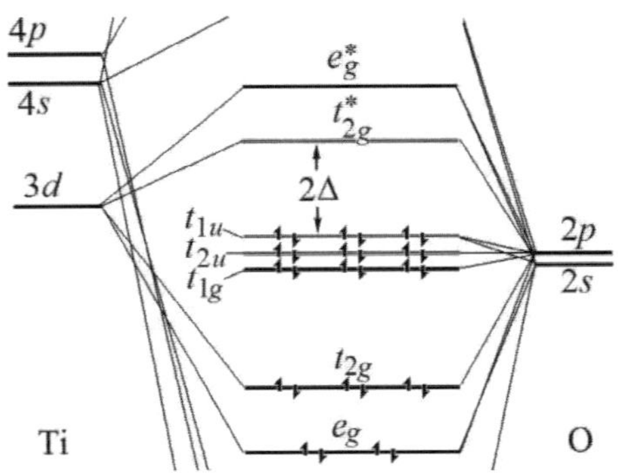

Figure 2. MO energy level diagram for the octahedral cluster $[TiO_6]^{8-}$ (after [1]). Shown in red are energy levels of the nine HOMO-LUMO states t_{1u}, t_{2u}, and t_{2g} coupled to the odd-parity modes q_x, q_y, and q_z active in the PJTE. 2Δ is the energy gap between LUMO and HOMO.

A more elaborate treatment of the problem going beyond the cluster model was given recently by means of the Green's function's approach [5]. It allowed taking into account the interaction of the B center with the whole crystal via the electronic and vibrational band structures. The Hamiltonian (2) was shown to be the effective Hamiltonian if the parameters K_0 and Δ are substituted by their weighted averages over the electronic and vibrational bands.

Note that the vibronic coupling makes the electronic states dependent on the nuclear coordinates $\mathbf{q} = \langle q_x, q_y, q_z \rangle$. It takes place in the first order of perturbation theory with respect to $\mathbf{H_{pJT}}$. For instance, one of the HOMO, the $t_{1u}(z)$, formed mostly from the oxygen orbitals $2p$, under the t_{1u} nuclear displacements acquires an admixture of the orbital $t_{2g}(yz)$, one of the three LUMO, the t_{2g} orbitals of the central atom B:

$$\psi(\mathbf{r},\mathbf{q}) = \psi_{T1u}^{(z)}(\mathbf{r},0) + F\sum_{\alpha=x,y,z}\frac{\left\langle \psi_{T1u}^{(z)}(\mathbf{r},0)\middle|\mathbf{C}_\alpha\middle|\psi_{T2g}^{(yz)}(\mathbf{r},0)\right\rangle}{E_{T1u} - E_{T2g}} q_\alpha \psi_{T2g}^{(yz)}(\mathbf{r},0) \quad (3)$$

were $\mathbf{r}$ is the electronic coordinate. Two important conclusions (often missed in publications) follow from this, at first sight simple expression:

(a) in the absence of the PJTE, when $F = 0$, the last term in equation (3) is zero and the wavefunction of the $t_{1u}(z)$ HOMO under consideration remains "frozen" atomic, $\psi_{T1u}(\mathbf{r}, 0)$, in spite of nuclear displacements. It follows that without the PJTE the HOMO orbitals are "bound" to the symmetrical positions of the nuclei and do not follow their displacements in the first order. In the last term in equation (3), the matrix elements in the numerator describe the emergent nonzero overlap of HOMO with LUMO which are orthogonal in the regular octahedron. Due to the PJTE symmetry breaking (figure 3) the symmetry restriction is lifted and the overlap becomes not equal to zero. Added overlap means added covalence bonding between the reference atoms. It becomes possible only due to the contribution of the $\mathbf{H}_{pJT}$, i.e. the PJTE. This conclusion is a part of the general statement that the only source of symmetry breaking in polyatomic systems is the PJTE in nondegenerate states and the JTE in degenerate states (see in [1-3]).

(b) The second term in equation (3) is proportional to the ratio $F/(2\Delta)$, where 2Δ is the noted above HOMO-LUMO energy gap (figure 2). The larger $F/(2\Delta)$, the stronger the added covalency and the PJTE, provided the stiffness of the system K_0 is the same. For a long time smallness of the energy gap 2Δ was believed to be the primary criterion of the efficiency of the PJTE. However, as was shown

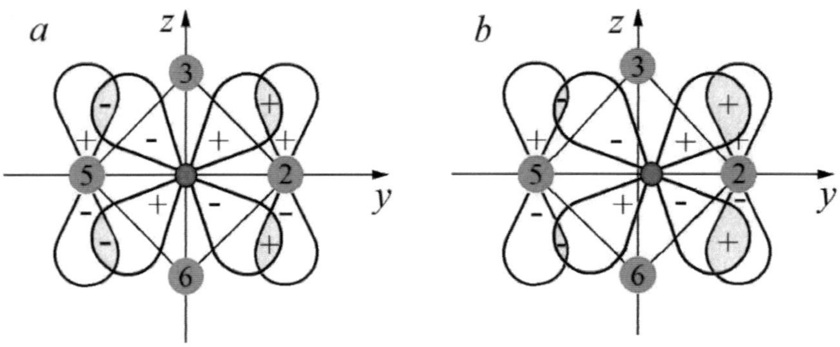

Figure 3. Illustration to the added covalency by the PJTE in the octahedron [BO_6] shown in the *yz* cross-section plane (after [5]). The numbering of the atoms follows figure 1 with the oxygen atoms in blue and the B atom red. ***(a)*** The B atom is in the center of the octahedron. The HOMO $t_{1u}(z)$ is composed of oxygen orbitals $2p_z$. The LUMO is the orbital $t_{2g}(yz)$ of the central ion B. As the positive contributions are cancelled by the negative ones, the total overlap (shown in yellow) is zero; the LUMOs are orthogonal to HOMOs. ***(b)*** The off-center displacement of the B ion lifts the orthogonality of the HOMO-LUMO orbitals. The nonzero overlap induces additional covalency that facilitates the polar distortion.

by numeric calculations [9], the PJTE instability may take place even if this gap is very large (~14 eV in the planar configuration of NH_3), provided the other two parameters in its criterion, $K_0 < F^2/(2\Delta)$ (introduced in [6], the first paper on this effect) are appropriately significant. In $BaTiO_3$ the energy gap 2Δ is of the order of the band gap, $2\Delta \approx 2.5$ eV. Actually, in the PJTE coupling to higher energy levels of the conduction band contribute as well, so the effective energy gap is larger. By numeric estimates of the relevant parameter values [5], it was shown that the condition $K_0 < F^2/(2\Delta)$ is true and, therefore, despite the relatively large energy gap the polar instability in this crystal is due to the PJTE.

2.2. *APES of the cluster* [BO_6] *in perovskite crystals* ABO_3

In search for the adiabatic potential energy we can ignore the kinetic energy term in equation (2). Diagonalization of the one-electron matrix Hamiltonian $\Delta\mathbf{C}_0 + \mathbf{H}_{pJT}$ gives nine adiabatic eigenvalues $\lambda_j(q_x, q_y, q_z)$ with $j = 1, 2, \ldots, 9$. In case when the central atom B is a transition metal ion with a d^0 shell, the lowest six ones correspond to HOMOs and the remaining three are LUMOs. Populating the HOMOs with 12 electrons (by 2 from each oxygen atom) we find the ground-state singlet adiabatic electron term [1], $\boldsymbol{U}(q_x, q_y, q_z) = ½K_0q^2 - 2\sqrt{[\Delta^2 + 2F^2(q^2 - q_x^2)]} - 2\sqrt{[\Delta^2 + 2F^2(q^2 - q_y^2)]} - 2\sqrt{[\Delta^2 + 2F^2(q^2 - q_z^2)]}$ with $q^2 = q_x^2 + q_y^2 + q_z^2$. Its extremes were analyzed in the first paper on this subject [1]. Under condition of a relatively strong PJTE, $8F^2/\Delta > K_0$, the ground-state APES $\boldsymbol{U}(q_x, q_y, q_z)$ was shown to have a maximum at the center, $\mathbf{q} = 0$, and, also, eight symmetry-equivalent minimum points where the atom B is off-center shifted along trigonal symmetry axes as shown in figure 1. The trigonal wells are separated by six tetragonal and 12 orthorhombic saddle points.

The overall landscape of the surface $\boldsymbol{U}(q_x, q_y, q_z)$ is shown in figure 4. The eight equivalent wells correspond to the central atom off-center shifted along one of the four trigonal axes (see also figure 1). The coordinates of the trigonal minimum points are $(\pm Q_0, \pm Q_0, \pm Q_0)$, where $Q_0 = ½(F/K_0)\sqrt{(64 - \delta^2)}$, and $\delta = \Delta K_0/F^2$ is the dimensionless energy gap Δ in units of F^2/K_0. The nuclear energy at these points is $\boldsymbol{U}(\pm Q_0, \pm Q_0, \pm Q_0) = -0.375(F^2/ K_0)\ (64 + 16\delta + \delta^2)$. Next, higher in energy are the mentioned above orthorhombic saddle points shown in figure 4 by open circles, and six tetragonal saddle-points along the three coordinate axes (not shown in the figure) that are higher than the orthorhombic points.

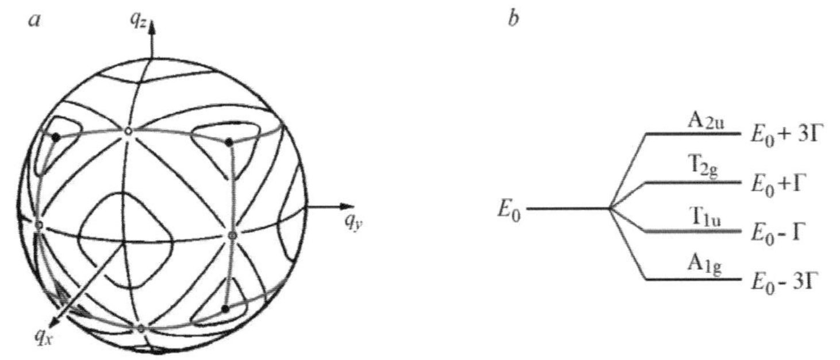

Figure 4. (a) APES of the cluster $[BO_6]$ in 3D-space of the coordinates q_x, q_y и q_z (after [1]). Bold black dots indicate trigonal minima, while open circles are the orthorhombic saddle points. Red curves are trajectories of either tunneling or over-the barrier transitions along the fastest steep-slope. **(b)** Tunneling splitting of the ground vibronic energy level

Their coordinates and energies are given in [1, 2, 5]. With the position, symmetry, and energy of the extremes of the APES we can estimate the sequence of order-disorder phase transitions by stepwise overcoming the corresponding barriers between the wells with temperature [1-3]. It explains the origin of all four phases in these crystals as following from the described above shape of the APES induced by the PJTE.

For the temperature of the paraelectric to ferroelectric phase transition in $BaTiO_3$ the estimate of the parameters of the APES yields [5]: $F \approx 4.7$ eV/Å, $K_0 \approx 97$ eV/Å^2, $\delta \approx 6.2$. For these parameter values trigonal wells are rather deep, ~0.2 eV, but the lowest potential barrier between them is much smaller, $\boldsymbol{U}[110] - \boldsymbol{U}[111] \approx 200$ см$^{-1}$. Similar values are obtained by DFT [5]. With the local vibrational frequency $\omega \sim 200$ cm^{-1} we get the zero vibrational level at about 100 cm^{-1}, meaning that the transitions between the wells can take place via tunnelling at low temperatures and by over-the-barrier activation at higher temperatures. The tetragonal barriers along [100], [010], *etc.*, are at about 560 cm^{-1}, meaning 2.75 higher than the orthorhombic barriers at 200 cm^{-1}. Therefore, practically the probability of direct tunnelling through them is zero. Over-the-barrier activation in this case becomes competitive at kT > 800K. At any reasonable temperature transitions via the regular cubic configuration at $\mathbf{q} = 0$ with the barrier height of about 0.2 eV is impossible. It follows that in these systems tunnelling and over-the-barrier transitions between trigonal wells take place only via the orthorhombic saddle-points between them (shown in figure 4*a* in red lines).

2.3. Tunnelling in the eight-minimum model

The model of an adiabatic potential energy surface with eight wells, irrelevant to their origin, was considered in [11], where it was assumed that, starting from a regular cube formed by the eight equivalent wells (see the cube *a, b, c, ...* in figure 1), there are three types of tunneling transitions between any two of them: along the edges, the large diagonal, and the small diagonal. Accordingly, in the basis of the eight vibrational states localized in the eight wells, the corresponding overlap integrals

$S_{kn} = \langle k \mid n\rangle$ and matrix elements of the Hamiltonian $H_{kn} = \langle k |\mathbf{H}_0| n\rangle$ can be reduced to three types, respectively: S, S', and S'', and H, H', and H''. The eight basis functions form a reducible representation of the cubic group that splits into four irreducible representations: $A_{1g} + T_{1u} + T_{2g} + A_{2u}$. The symmetry adapted wave functions, which are the linear combinations of the states localized in the wells, and the corresponding eigenvalues are given in [11]. For the transitions along the edge, long diagonal, and short diagonal of the cube, the parameters of tunneling splitting are Γ, Γ' and Γ'', respectively.

As mentioned above, the only tunnelling and over-the barrier transitions to be included are via the orthorhombic saddle points. This means $\Gamma' = \Gamma'' = 0$, $H' = H'' = 0$, and $S' = S'' = 0$. This assumption results in an equidistant tunneling energy level spectrum (figure 4*b*) with $E(A_{1g}) \approx E_0 - 3\Gamma$, $E(T_{1u}) \approx E_0 - \Gamma$, $E(T_{2g}) \approx E_0 + \Gamma$ and $E(A_{2u}) \approx E_0 + 3\Gamma$. For the sake of convenience, assume the ground state energy $E_0 = 0$. Then we get: $E(A_{1g}) \approx -3\Gamma$, $E(T_{1u}) \approx -\Gamma$, $E(T_{2g}) \approx \Gamma$ и $E(A_{2u}) \approx 3\Gamma$.

3. The effect of external electric field

3.1 The dipole moment of the cluster $[BO_6]$ *and polarization of the crystal in external electric fields.* In the trigonal wells of the APES the cluster $[BO_6]$ has a dipole moment $\mathbf{d}$. In the electric field $\mathbf{E}$ the energy of the cluster is $-\mathbf{E} \cdot \mathbf{d}$. In general, $\mathbf{d}$ includes nuclear and electronic contributions, $\mathbf{d} = \mathbf{d}_{nucl} + \mathbf{d}_{el}$. To estimate the nuclear contribution $\mathbf{d}_{nucl}$, we present the coordinate $\mathbf{q}$ as a linear combination of the nuclear displacements $\mathbf{R}_\kappa$ of each atom, $\mathbf{q} = \Sigma_\kappa c_k \mathbf{R}_\kappa$. We assume that the oxygen octahedron is of the mass $6m$, where $m = 16$ u is the mass of the oxygen atom, carries a charge of $6(-2e) = -12e$, and moves as a whole with respect to the atom B without distortion (in fact, there is a small distortion of the oxygen's octahedral configuration which we ignore). In this case the center of mass of the octahedron O_6 coincides with its geometric center. When this octahedron shifts to the point $\mathbf{R}$ and the ion B with the mass M moves to the point $\mathbf{R}_0$, the center of mass of the cluster $[BO_6]$ moves to the point $\mathbf{R}_{cm} = (6m\mathbf{R} + M\mathbf{R}_0)/(6m + M)$. Keeping the latter unchanged, $\mathbf{R}_{cm} = 0$, we get $6m\mathbf{R} + M\mathbf{R}_0 = 0$, or $\mathbf{R}_0 = -(6m/M)\mathbf{R}$. In particular, for $[TiO_6]$ we have: $M = 48$ u $= 3m$ and, therefore, $\mathbf{R}_0 = -2\mathbf{R}$. This means the displacement of Ti is twice those of the oxygen atoms, and in the opposite direction, its relative displacement being $\mathbf{q} = \mathbf{R}_0 - \mathbf{R}$. If we assume, as above, that the charge on B equals $+4e$, than $\mathbf{d}_{nucl} = (+4e)\mathbf{R}_0 + (-12e)\mathbf{R} = (4\mathbf{R}_0 - 12\mathbf{R})e$. With $\mathbf{R}_0 = = (2/3)\mathbf{q}$ and $\mathbf{R} = (-1/3)\mathbf{q}$ we have $\mathbf{d}_{nucl} = (20/3)e\mathbf{q}$.

The dipole moment of the electrons, $\mathbf{d}_{el} = -e\bar{\mathbf{r}}$, is proportional to the mean value $\bar{\mathbf{r}}$ of the electron coordinate $\mathbf{r}$. In the ground state of the undistorted cluster, when $\mathbf{q} = 0$, we have $\bar{\mathbf{r}} = 0$. Evidently, we can assume that $\bar{\mathbf{r}}$ grows with $\mathbf{q}$, and in the first approximation $\bar{\mathbf{r}} = \text{const} \cdot \mathbf{q}$. For the system in one of the minima, say, the one along [111], the main contribution to the dipole moment comes from the six HOMO. Ignoring the small contribution from the admixture of t_{2g} orbitals of the atom B, we find that $\bar{\mathbf{r}}$ is approximately equal to the displacement of the oxygen octahedron O_6, $\bar{\mathbf{r}} \approx \mathbf{R}$, and with $\mathbf{R} = (-1/3)\mathbf{q}$ we come to $\bar{\mathbf{r}} \approx -\mathbf{q}/3$. Accordingly, $\mathbf{d}_{el} = -e\bar{\mathbf{r}} \approx e\mathbf{q}/3$, and the total dipole moment of the cluster $\mathbf{d} = \mathbf{d}_{nucl} + \mathbf{d}_{el} \approx (20/3)e\mathbf{q} + (1/3)e\mathbf{q} = 7e\mathbf{q}$. Note that the nuclear contribution in the PJTE-induced dipole moment is 20 times larger than the electronic one. So, in a good approximation the latter can be neglected.

In an external electric field $\mathbf{E}$ the cubic-perovskite dielectric crystal ABO_3 becomes polarized. Induced by the field, the vector of the corresponding polarization $\mathbf{P}$ is linearly dependent of the electric field; its components are $P_j = \Sigma_k \chi_{jk} E_k$, where χ_{jk} are elements of the tensor of dielectric susceptibility of the system. Along symmetry axes vectors $\mathbf{E}$ and $\mathbf{P}$ are parallel and $\chi = P/E$. According to equation (1), in static electric field ($\omega = 0$) the permittivity ε is determined by both the displacive χ_d and the orientation χ_o contribution. However, as shown in [7], the latter is overwhelmingly greater, $\chi_o >> \chi_d$, and, therefore, $\chi = \chi_o + \chi_d \approx \chi_o$.

The polarization is the volume density of the average dipole moment induced by the electric field, $\mathbf{P} = \langle \mathbf{d}\rangle/a^3$, where a is the crystal lattice constant and $\langle\ldots\rangle$ means averaging over the quantum states of the system with their Boltzmann population factors at the given temperature T,

$$\langle \mathbf{d} \rangle = \frac{1}{Z}\sum_n \langle n | e^{-H/kT}\mathbf{d} | n \rangle = \frac{1}{Z}\sum_n e^{-E_n/kT}\langle n | \mathbf{d} | n \rangle, \quad Z = \sum_n \langle n | e^{-H/kT} | n \rangle = \sum_n e^{-E_n/kT} \tag{4}$$

Here $|n\rangle$ are the wavefunctions, and E_n are the eigenvalues of Hamiltonian $\mathbf{H} = \mathbf{H}_0 - \mathbf{E}\cdot\mathbf{d}$. In the basis of the eight tunneling states the matrix of this Hamiltonian is as follows [11, 12]:

$$\mathbf{H} = \Gamma \begin{array}{c} A_{1g} \\ T_{1u}x \\ T_{1u}y \\ T_{1u}z \\ T_{2g}\xi \\ T_{2g}\eta \\ T_{2g}\zeta \\ A_{2u} \end{array} \begin{pmatrix} -3 & \alpha A & \beta A & \gamma A & 0 & 0 & 0 & 0 \\ \alpha A & -1 & 0 & 0 & 0 & \gamma A & \beta A & 0 \\ \beta A & 0 & -1 & 0 & \gamma A & 0 & \alpha A & 0 \\ \gamma A & 0 & 0 & -1 & \beta A & \alpha A & 0 & 0 \\ 0 & 0 & \gamma A & \beta A & 1 & 0 & 0 & \alpha A \\ 0 & \gamma A & 0 & \alpha A & 0 & 1 & 0 & \beta A \\ 0 & \beta A & \alpha A & 0 & 0 & 0 & 1 & \gamma A \\ 0 & 0 & 0 & 0 & \alpha A & \beta A & \gamma A & 3 \end{pmatrix} \tag{5}$$

All the parameters in this matrix are given in units of Γ which is factored out of the matrix. On the diagonal are the obtained above energies of the eight tunnelling levels. The parameter of the electric field is denoted by $A = -Ed_0/(\Gamma\sqrt{3})$, where $d_0 = 7eQ_0\sqrt{3}$ is the dipole moment of the cluster $[BO_6]$ when nuclear motion is localized in one of the trigonal wells, $E = |\mathbf{E}|$ is the magnitude of the applied electric field, while α, β, and γ are the direction cosines of the electric field [13],

$$\alpha = \frac{E_x}{|\mathbf{E}|} = \cos\theta\cos\varphi, \quad \beta = \frac{E_y}{|\mathbf{E}|} = \cos\theta\sin\varphi, \quad \gamma = \frac{E_z}{|\mathbf{E}|} = \sin\theta \tag{6}$$

The results of numerical diagonalization of the matrix (5) and consequent calculations of the field-dependent tunneling energy levels, lattice polarizations **P,** dielectric susceptibility χ, and their angular and temperature dependence are shown in figures 5 – 7. The results of numerical calculations of the

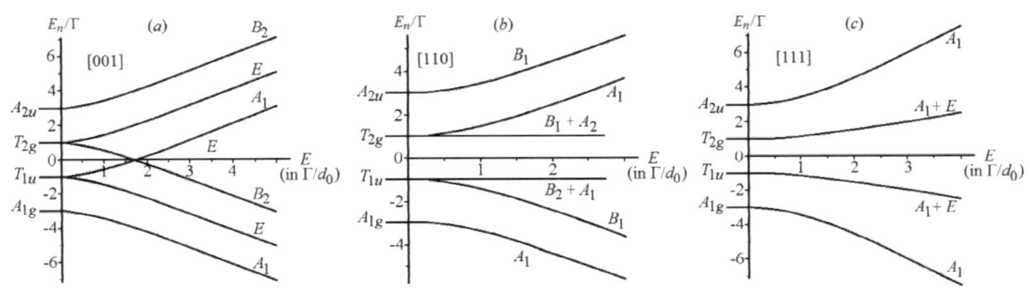

Figure 5. The variation of tunnelling energy levels E_n (in units of Γ) of the cluster $[BO_6]$ with the magnitude of the external electric field E in units of Γ/d_0. (*a*) The electric field is applied in the [001] direction. Energy levels are labelled by the irreducible representations of the reduced symmetry C_{4v}. (*b*) Electric field is applied in the [110] direction. The reduced symmetry is C_{2v}. (*c*) Electric field is applied in the [111] direction. The reduced symmetry is C_{3v}.

polarization of the crystal $P = \langle d \rangle / a^3$ with the vector $\langle \mathbf{d} \rangle$ after equation (4) are given in figure 6, where its dependence on both the magnitude and direction of the applied electric field **E**, as well as on temperature are shown. P is shown in units of $\boldsymbol{P}_m = d_0/a^3$ which is the maximum value of $\boldsymbol{P}$ reached at saturation along the direction [111] of the vector **E**, while the magnitude of the vector **E** is given in units of $E_0 = \Gamma/d_0$, where Γ, as above, is the tunneling parameter (figure 4b). From these data it follows, first, that there is a strong dependence of the orientation part of polarization on magnitude of the electric field. It should not be there without the PJTE. The displacive polarization of the ionic

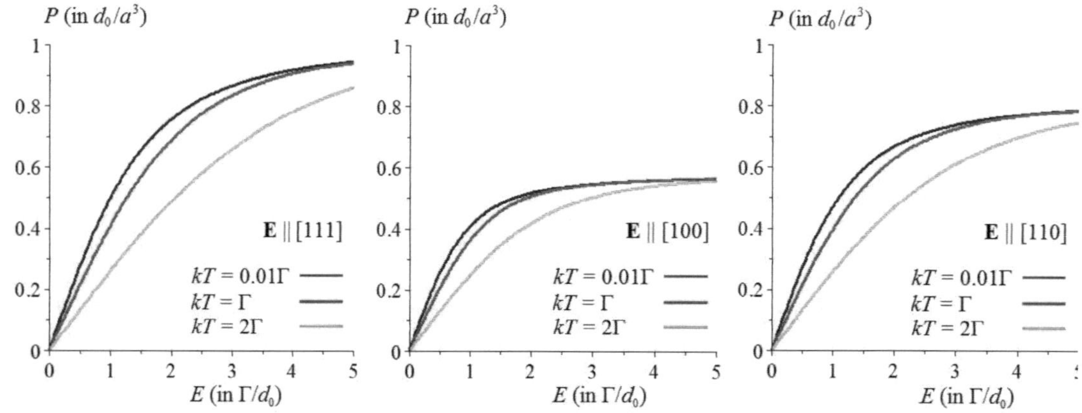

Figure 6. Variation of the polarization P (in units of d_0/a^3) of the cluster $[BO_6]$ with the external electric field E (in units of Γ/d_0) along the main symmetry axes [111], [100] and [110] at three temperatures.

lattice is by several orders of magnitude smaller. Also, all graphs in figure 6 show saturation of the polarization beginning from the value $E \approx 2\Gamma/d_0$. This means polarization of the crystal comes close to saturation when the energy of the dipole moments in the electric field reaches the order of magnitude of the tunneling splitting (or over-the-barrier transitions). At this point the dipoles become fully oriented along the applied electric field. This confirms the previous estimates of the orientation contribution to the magnitude of polarization given in [7].

Next, the differences in P values in the three directions, trigonal **E** ‖ [111], tetragonal **E** ‖ [100], and orthorhombic **E** ‖ [110] in figure 6 show that there is a significant angular dependence of the polarization (not considered in [11]) which, again, should not be there in a cubic crystal in the absence of the PJTE. Induced by the latter, the polarization is maximal along the trigonal axes, where the absolute minima are positioned, while the smallest P value emerges along the highest tetragonal saddle-points of the APES. The temperature and field dependent angular diagrams of the polarization **P** are shown in figure 7.

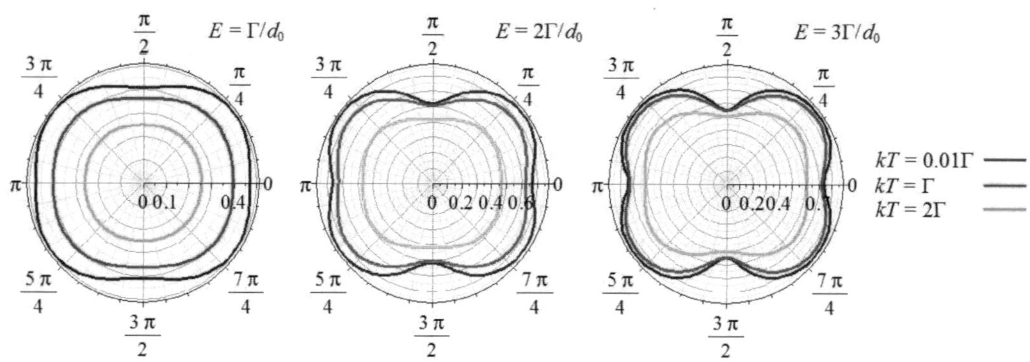

Figure 7. Angular dependence of the polarization **P** in the cross-section plane $q_x = q_y$, $\varphi = \pi/4$ (see figure 1) at different temperatures and electric field magnitude E. The magnitude of **P** is maximal along trigonal directions, $|q_x| = |q_y| = |q_z|$, lower in the orthorhombic, and minimal in the tetragonal directions.

A qualitative explanation of this orientation effect is illustrated in figure 8 in which the cross-section of the APES in the plane (110), $q_x = q_y$, is shown (cf. figure 4a). In sufficiently weak fields $E << \Gamma/d_0$, all the trigonal minima remain approximately equivalent (see the left-hand side of figure 8). In a stronger field, for instance, when $E >> \Gamma/d_0$ and E || [111], energy of the trigonal minimum with the coordinates (Q_o, Q_o, Q_o) is lowered, whereas the energy of the one at ($-Q_o$, $-Q_o$, $-Q_o$) is elevated, and there are similar differences between the other minima, which manifest pairwise correlation along and opposite to the direction of the applied electric field. As a result the nuclear motion is ultimately localized in the lowest well, at which point the orientation contribution to the dielectric susceptibility χ_o is frozen, and only a small contribution of the displacive one is left.

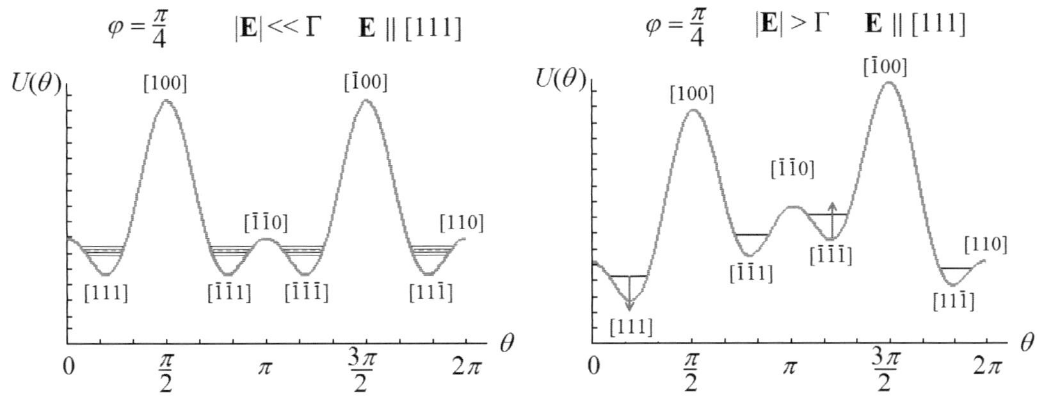

Figure 8. Qualitative cross-section of the APES of the cluster $[BO_6]$ in the plane (110) where $q_x = q_y$. In a weak electric field, $\boldsymbol{E}d_0 << \Gamma$, (left) all the trigonal wells are at the same energy, while in strong fields, $\boldsymbol{E}d_0 > \Gamma$ with the vector **E** || [111] (right) the well in this direction is significantly lower in energy, resulting in a significant localization of the nuclear motion in this well.

The cubic anisotropy of the polarization is due to the different magnitude of the average dipole moment at different direction of the vector **E.** For **E** || [111] in saturation all the dipoles are frozen in the lowest trigonal well along the field with the dipole moment d_0 of the unit cell. In the case when **E** || [110], two of the eight wells are lowest in energy. As they remain symmetry-equivalent in the field **E** || [110] the resonance conditions for tunneling are kept. The tunneling between them results in an average dipole moment of $(\sqrt{2/3})d_0$. If the applied electric field points along the tetragonal axes, **E** || [100], four wells remain equivalent in the field along the tetragonal axes, resulting in the average

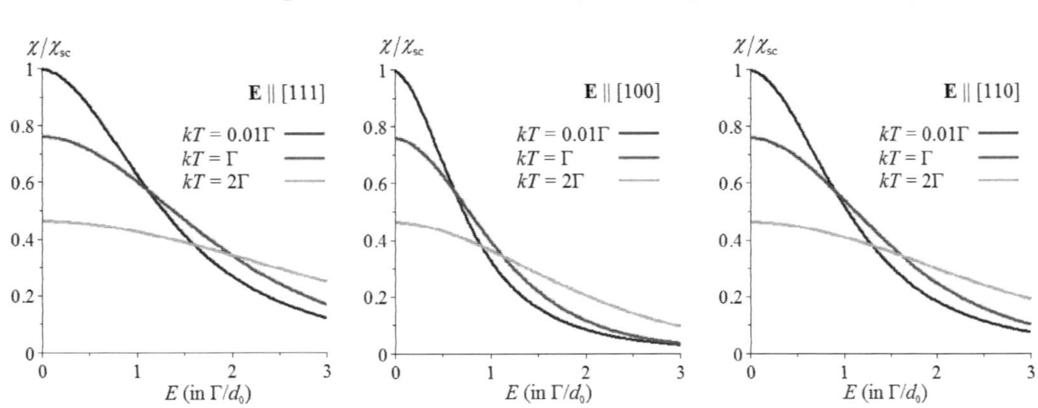

Figure 9. Dielectric susceptibility χ of the cluster $[BO_6]$ in electric fields **E** || [111], **E** || [100] and **E** || [110], at kT = 0.01Γ, Γ and 2Γ, shown in units $\chi_{sc} = d_0^2/(3\Gamma\varepsilon_0 a^3)$, where d_0 is the dipole moment of the unit cell with the B atom in the trigonal minimum and Γ is the tunneling parameter, while E is given in units Γ/d_0.

dipole moment equal to $(1/\sqrt{3})d_0$. The calculated values of P in figure 6 fully agree with these qualitative predictions.

3.2. Dielectric susceptibility and permittivity.

The variation of dielectric susceptibility χ with the magnitude of the applied electric field pointing in three directions, **E** || [100], **E** || [110]and **E** || [111], at three characteristic temperatures, is shown in figures 9 and 10. Diagrams of its angular dependence are qualitatively the same as for the polarization P in figure 7. The value of χ is given in units of the dimensionless scaling parameter $\chi_{sc} = d_0^2/(3\varepsilon_0\Gamma a^3)$ which is the ratio of the units for P in d_0/a^3 and E in Γ/d_0 (to compare, in $BaTiO_3$ the magnitude of χ_{sc} is about $0.45 \cdot 10^3$ [7]). As the field dependence of $P = f(E)$ is nonlinear, the susceptibility χ is introduced as the derivative $\chi = \varepsilon_0^{-1}(\partial P/\partial E)$ [14]. As expected, χ goes to zero in sufficiently strong electric field. In a very strong electric field, at saturation, where P is constant, we have $\chi = \varepsilon_0^{-1}(P/E)$. At this point only the displacive contribution to permittivity ε is left, whereas the difference in their angular dependence increases with the field (figure 7). These specific properties of permittivity follow directly from the eight-minimum shape of the APES induced by the local PJTE and, as shown above in figure 8, the dynamics of localization of the nuclear motion in the lowest wells that are along the applied electric field.

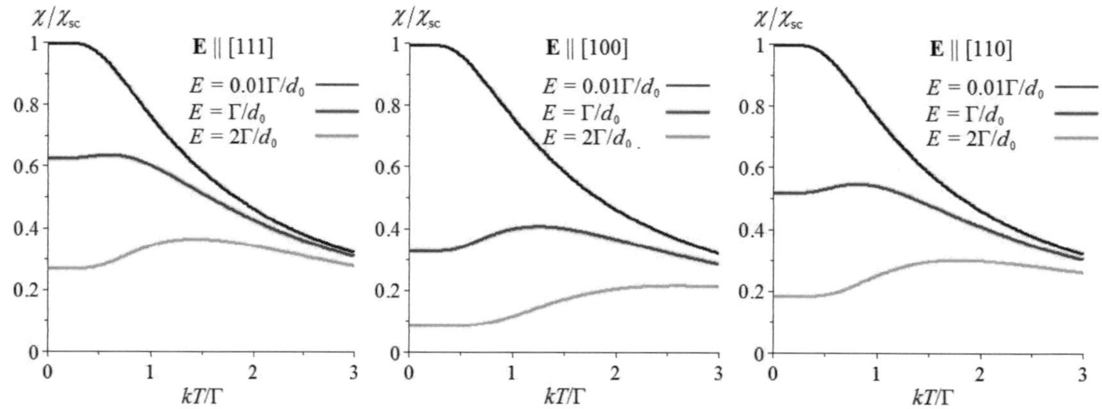

Figure 10. Temperature dependence of the dielectric susceptibility χ [in units of $\chi_{sc} = d_0^2/(3\Gamma\varepsilon_0 a^3)$] of the cluster $[BO_6]$ at three directions of the external electric field, **E** || [111], **E** || [100] and **E** || [110], and three absolute values, $E = 0.01\Gamma/d_0$, Γ/d_0, and $2\Gamma/d_0$. The temperature is given in dimensionless units kT/Γ.

The angular dependence of the dielectric susceptibility can be modeled by the first terms in the expansion in cubic harmonics,

$$\chi(E) \approx \chi_0(E) + \chi_4(E)Y_4(\theta,\varphi) \tag{7}$$

where χ_0 is the isotropic contribution, χ_4 is the anisotropic amplitude, and $Y_4(\theta, \varphi)$ is the fourth order cubic harmonic, $Y_4(\theta, \varphi) = (x^4 + y^4 + z^4 - \frac{3}{5}r^4)/r^4$. By fitting the numerical values of the anisotropy in figure 9 for $kT = \Gamma$ we find:

$$\chi_0 \approx 0.89 - 0.43\frac{Ed}{\Gamma} + 0.058\left(\frac{Ed}{\Gamma}\right)^2, \quad \chi_4 \approx 0.017 - 0.39\frac{Ed}{\Gamma} + 0.102\left(\frac{Ed}{\Gamma}\right)^2 \tag{8}$$

where $E = |\mathbf{E}|$, $d = |\mathbf{d}|$, and Γ is the tunneling parameter, same as above.

The temperature dependence of χ is shown in figure 10. As expected, in accordance with equation (4), it drops with temperature approaching zero at $T \rightarrow \infty$. At high temperatures all the tunneling states are equally populated, the common factor exp ($-E_n/kT$) can be factored out in the equation (4), while the trace of the operator **d** is zero.

4. Conclusion

In this paper ferroelectric cubic perovskites ABO_3 are considered. All approximations used above and most of the numeric results shown in figures 5 - 10 are based upon estimates made in [5] for barium titanate, the most studied member of the large class of cubic perovskites ABO_3. Similar to the case of $BaTiO_3$, the PJTE in the octahedral B centers results in an eight-minimum APES with a dipole moment in the wells. Although the actual magnitudes of the observable effects vary for different ferroelectric perovskite oxides, qualitatively they are expected to be similar to the ones reported above. The important condition for this similarity is polar off-center instability of the octahedral cluster $[BO_6]$. Resulting from the PJTE, it manifests itself in a strongly anharmonic APES of the ground-state electron term with a maximum at the on-center and a wrapped trough with alternating minimum and saddle-point off-center positions of the metal ion B. The local PJTE in octahedral clusters $[BO_6]$ creates local rotating dipoles. Such systems manifest a significant enhancement of permittivity of orientation origin unexpected without the PJTE.

This is a general feature of all ferroelectric perovskites ABO_3. It follows from their cubic crystal symmetry in the high-temperature paraelectric phase. If, distinguished from the d^0case of $BaTiO_3$, for the central atom B there are several d electrons present in the outer shell or, not $3d$ but, say, $4d$ orbitals (as in $KNbO_3$) are involved in vibronic coupling, still the PJTE is possible [2] provided the fraction F^2/K_0 is strong enough compared to the band gap Δ. Depending on a particular d^n configuration, the corresponding inequality relating F^2/K_0 to Δ includes a numeric factor (e.g., in the d^0 case the inequality is $8F^2/K_0 > \Delta$ with the factor of 8). However, for a ferroelectric perovskite ABO_3 under consideration, its ferroelectric feature with the polar instability of on-center position of the ion B is a direct indication that the corresponding inequality is true.

Also, the enhanced permittivity has unexpected angular dependence. Rotation of the dipoles is hindered by either tunneling or over-the-barrier transitions. The crystal is assumed to be in its paraelectric phase. Therefore, in the absence of external perturbations the rotational motion of the local dipoles is assumed to be uncorrelated (not ordered). Besides, depending on the height of the potential barriers between the wells, the rotation of the dipoles may be free or hindered. With no barriers the APES has a two-dimensional polar trough. In this case the dipoles are freely rotating with some angular frequency.

For perovskite crystals ABO_3 with a PJTE-induced eight-minimum APES at the B centers, the eight tunneling states were used to numerically solve the eigenvalue problem for the full Hamiltonian, including the PJT coupling and the effect of the applied electric field. The temperature-averaged response to the external electric field revealed the field dependent orientation polarization, dielectric susceptibility, and permittivity. Coinciding with a previous estimate by an order of magnitude, the orientation contribution to permittivity (which should not be there in the absence of the PJTE) is very large and (what was not shown before) it has a significant anisotropy. The origin of these effects was illustrated by a qualitative picture of what can be expected from the PJTE induced dipolar eight-minimum APES.

Explored in this paper, the PJTE includes vibronic coupling of HOMO to just LUMO, the excited d orbitals of the central ion B. As it follows from the equation (2), the more opposite-parity excited states are coupled to HOMO, the more polar-flexible the ground state becomes and, correspondingly the stronger is the polar instability of the ion B. Distinguished from the simplest model of this paper, including other excited states adds more parameters of vibronic coupling, other Fs and Δs. Their complex interplay may result in a somewhat change of wrapping of the ground-state trough. For example, as a plausible effect, instead of the eight-minimum shape of the trough there can be a six-minimum APES. As the most stable low-temperature ferroelectric phase this would determine the

IOP Conf. Series: Journal of Physics: Conf. Series **833** (2017) 012012 doi:10.1088/1742-6596/833/1/012012

tetragonal crystal structure. However, discussed in this paper orientation contribution in electric susceptibility in the high-temperature phase is the same in the six-minimum case as well. Similar to eight-minimum case it would result in a considerable enhancement of permittivity due to orientation polarizability of local dipoles in the external electric field.

The results above are obtained as functions of Γ or the ratio Γ/d_0, where Γ is introduced as the tunneling splitting parameter, and d_0 is the dipole moment in the minimum of the APES. The main parameter is the dipole moment d_0 that stands for the polar distortion, and hence it characterizes the magnitude of the observable effects. As for Γ, its main function is to stand for the speed of transitions between the minima which may vary significantly depending on the height of the energy barrier between them. With low barriers these transitions become more of over-the-barrier nature, and at a zero barrier they are free rotations along the PJTE-induced two-dimensional trough. All the conclusions of this paper apply to the case when over-the-barrier mechanism dominates over tunneling. The only expected difference is the activation nature of the corresponding temperature dependence of the orientation part of susceptibility in this case. Qualitatively, the effects described above in this paper will be present in all such cases, subject to the polar instability $d_0 > 0$, but the quantitative description may be different. The "rotation frequency" of the dipoles influences mostly the magnitude of the field at which the orientation polarization is saturated.

References

[1] Bersuker I B 1966 On the origin of ferroelectricity in perovskite-type crystals *Phys. Lett.* **20** 589; 1988 On the vibronic origin of instability in structural phase transitions in condensed matter, *Fiz. Tverd. Tela (Sov.Phys. - Solid State*) 30, 1738; 1995 Recent development of the vibronic theory of ferroelectricity *Ferroelectrics* **164** 75; Polinger V 2013 Ferroelectric phase transitions in cubic perovskites *J. Phys*: *Conf. Series* **428** 012026

[2] Bersuker I B 2012 Pseudo Jahn-Teller origin of perovskite multiferroics, magnetic-ferroelectric crossover, and magnetoelectric effects: the d^0-d^{10} problem *Phys. Rev. Lett.* **108** 137202; 2013 A Local Approach to Solid State Problems: Pseudo Jahn-Teller origin of Ferroelectricity and Multiferroicity *J. Phys*: *Conf. Series* **428** 012028; 2013 The Pseudo Jahn-Teller Effect – a Two-State Paradigm in Formation, Deformation, and Transformation of Molecular Systems and Solids *Chem Rev* **113,** 1351

[3] Bersuker I B and Polinger V Z 1989 *Vibronic Interactions in Molecules and Crystals* (Berlin-Heidelberg: Springer-Verlag) Sect 6.2; Bersuker I B 2006 *The Jahn-Teller effect* (Cambridge: Cambridge University Press) Sect 8.3

[4] Ravel B, Stern E A, Vedrinskii R I and Kraizman V 1998 Local structure and the phase transitions of $BaTiO_3$ *Ferroelectrics* **206-207** 407; Stern E A 2004 Character of Order-Disorder and Displacive Components in Barium Titanate *Phys Rev Lett* **93**, 037601; Völkel G and Müller K A 2007 Order-disorder phenomena in the low-temperature phase of $BaTiO_3$ *Phys Rev* B **76** 094105; Blinc R, Zalar B, Laguta V V and Itoh M 2005 Order-Disorder Component in the Phase Transition Mechanism of ^{18}O Enriched Strontium Titanate *Phys Rev Lett* **94** 147601; Jeong I-K, Lee S, Jeong S-Y, Won C J, Hur N, and Llobet A 2011 Structural evolution across the insulator-metal transition in oxygen-deficient $BaTiO_{3-\delta}$ studied using neutron total scattering and Rietveld analysis *Phys Rev* B **84** 064125

[5] Polinger V, Garcia-Fernandez P and Bersuker I B 2015 Pseudo Jahn-Teller origin of ferroelectric instability in $BaTiO_3$ type perovskites: The Green's function approach and beyond *Physica* B **457** 296

[6] Debye P 1912 Eine Resultate einer kinetischen Theorie der Isolatoren *Phys. Zeitschr.* **13** 97

[7] Bersuker I B 2015 Giant permittivity and electrostriction induced by dynamic Jahn-Teller and pseudo Jahn-Teller effects *Appl Phys Lett.* **107** 202904; 2015 Pseudo Jahn-Teller effect in the origin of enhanced flexoelectricity *Appl Phys Lett* **106** 022903

[8] Sugano S, Tanabe Y and Kamimura H 1970 *Multiplets of Transition-Metal Ions in Crystals* (New York and London: Academic Press) Appendix II

[9] Bersuker I B, Gorinchoi N N and Polinger V Z 1984 On the origin of dynamic instability of molecular systems *Theoret Chim Acta* **66** 161

[10] Öpik and Pryce M H L 1957 Studies of the Jahn-Teller effect. I. A survey of the static problem *Proc Roy Soc London* **238** 425

[11] Gomez M, Bowen S P and Krumhansl J A 1967 Physical properties of an off-center impurity in the tunneling approximation. I. Statics *Phys Rev* **153** 1009

[12] The axes for the molecular orbitals in the paper of Gomez et al [9] do not coincide with the x, y, z ones for the vector of dielectric induction **D**. Therefore their matrix (4) is somewhat different from ours. Beside this in their equation (3.1) the minus sign is omitted.

[13] Traditionally for the polar angle θ the reading is from the "north pole" where $\theta = 0$ to the "south pole" with $\theta = \pi$. In this paper the read-off value $\theta = 0$ corresponds to the equatorial direction whereas the direction to the north pole is at $\theta = \pi/2$, and to the south pole at $\theta = -\pi/2$.

[14] Shore H B 1966 Model Hamiltonian for paraelectric impurities in alkali halides *Phys Rev* **151** 570

IOP Conf. Series: Journal of Physics: Conf. Series **833** (2017) 012013 doi:10.1088/1742-6596/833/1/012013

Dynamical pseudo-Jahn-Teller effect: time evolution of quantum states above avoided crossing of electronic levels

V Hizhnyakov, H Kaasik and T Vaikjärv

Institute of Physics, University of Tartu, W. Ostwaldi Str 1, 50411, Tartu, Estonia

E-mail: hizh@ut.ee

Abstract. Quantum states above the avoided crossing of electronic levels of a vibronic system with pseudo-Jahn-Teller effect are considered both analytically and numerically. Equations for energies and wave functions of the states are derived. Expansions of these functions over numerically exact vibronic states are obtained and time evolution of the states is found by applying the Fock-Krylov theorem.

1. Introduction

Avoided crossing of energy levels plays an important role in many phenomena of physics, chemistry and biology. First time this problem was addressed in connection with Landau-Zener effect [1,2] – the transition between levels with energy separation being a linear function of time and with coupling in the diabatic Hamiltonian being independent of time. At present we know a number of phenomena governed by avoided crossing: energy exchange, charge exchange, mixed valence, predissociation, associative recombination (see e.g. [3-8] and references therein). Note also strong entanglement between electronic and nuclear degrees of freedom in systems with avoided levels crossing which could be exploited in the nascent field of quantum computation [9-11].

In molecular systems avoided crossing is usually considered in terms of pseudo-Jahn-Teller effect (PJTE) [12,13]. In the simplest such a system - dimer [14,15] the main motion above the avoided crossing can be considered in terms of vibrations at the upper branch of the adiabatic potential energy surface (APES). See in this connection the 1979 study of Loorits [16] (published in Russian as a preprint of Institute of Physics, Tartu, Estonia; translated to English in [17], this issue). In this work it was shown that the higher-energy part of the optical spectrum of the system with strong PJTE in excited state consists of series of groups of lines. This result was interpreted in favour of existence of vibrations at the upper branch of the APES: every group of lines corresponds to creation of a particular quantum vibrational state at optical transition in this branch of the APES. Non-adiabatic transitions to another branch of the APES show up in this spectrum as components in groups. A year later analogous groups of spectral lines were numerically found in the optical $E \times e$ problem [18-20]. Corresponding levels in the upper branch of the APES in this case are known as Slonczewski resonaces [21].

Although direct evidence of existence of vibrations at the upper branch of the APES of PJTE have been obtained long ago, as far as we know the properties of these vibrations were not clarified until now. Here these properties are considered. We derive analytical equations describing the energies and the wave functions of these states (see also Ref. [22] where analogous description of Slonczewski resonances in $E \times e$ problem is given). We also find the expansion of these states over the exact vibronic states of the problem. This allows us to get time dependence of the states under consideration by applying the Fock-Krylov theorem [23].

IOP Conf. Series: Journal of Physics: Conf. Series **833** (2017) 012013 doi:10.1088/1742-6596/833/1/012013

If calculated in this way, time dependence of the states under consideration holds for short times – shorter than or comparable with the period of the main mode in the diabatic basis. At large time one needs to take into account the interaction of the system with the vibrational continuum of the environment (phonons in crystals) causing the transfer of the energy to the bulk. This can be done by applying the theory [22, 24-26] allowing one to consider interaction of the Jahn-Teller and pseudo-Jahn-Teller systems with phonons. Using this theory one can calculate rates of decay of the basic vibronic states due to emission of phonons to the bulk. In the current study we use results obtained in Ref. [24] for description of time evolution of the states under consideration. This allows one to find evolution of these PJTE systems also for long times.

Note that the short time-scale dynamics of the PJTE can be also calculated by means of numerical solution of time-dependent Schrödinger equation; see in this connection the multi-configurational time-dependent Hartree method developed in Refs. [27-32]. This method allows one to calculate dynamics of vibronic systems with several (usually up to ten) vibrational degrees of freedom. However, this theory does not allow one to consider long time-scale evolution of vibronic systems which is determined by their interaction with the phonon continuum.

2. Pseudo-Jahn-Teller effect. Quantum states above the avoided crossing

In the simplest case the Hamiltonian of the systems describing PJTE and a dimer can be presented in the form [12-17] $H = H_L \cdot I + H_{eL}$, where

$$\begin{aligned} H_L &= \left(d^2/dx^2 + \omega^2 x^2\right)/2, \\ H_{eL} &= \mathcal{E}\sigma_x + kx\sigma_z \end{aligned} \tag{1}$$

Here H_L is the Hamiltonian of one-dimensional vibrations in harmonic approximation, H_{eL} is the vibronic interaction, x is the reduced displacement of the vibrating system from the equilibrium position (dynamical variable), I is the 2×2 unit matrix , σ_x and σ_z are the Pauli matrices, ω , $\mathcal{E}$ and k are the parameters of the model ($\mathcal{E}$ gives the splitting of the levels at $x = 0$, k is the strength of the vibronic coupling; units with $\hbar = 1$ are used). In case of deep minima of the crossing potentials (with energy $-D$, where $D = k^2/2\omega^2$ is the vibronic stabilization energy) the parameters of the model satisfy the conditions $D \gg \mathcal{E} \gg \omega$. For $\omega \lesssim \mathcal{E}$ this Hamiltonian is used for description of the PJTE [13].

The eigenstates $|\nu\rangle$ of the Hamiltonian H (called vibronic states) can be found by numerical solving of the stationary Schrödinger equation using the eigenstates $|n\rangle$ of the Hamiltonian H_0 of the active mode as the basis [14-17]. The latter states $|n\rangle$ have a quantum number - the integer $n = 0,1,2,...$ which determines the energies of the levels of the mode $\omega(n+1/2)$. There are even (+) and odd (-) solutions of the stationary Schrödinger equation $H|\nu\rangle = E|\nu\rangle$ which have the form

$$\begin{aligned} \left|\Psi_{\nu_+}\right\rangle &= \sum_n \left(c^+_{2n,\nu}|2n\rangle|0\rangle_e + c^+_{2n+1,\nu}|2n+1\rangle|1\rangle_e\right), \\ \left|\Psi_{\nu_-}\right\rangle &= \sum_n \left(c^-_{2n,\nu}|2n\rangle|1\rangle_e + c^-_{2n+1,\nu}|2n+1\rangle|0\rangle_e\right) \end{aligned} \tag{2}$$

where $C^{\pm}_{n,\nu}$ are the solutions of two matrix equations with tridiagonal matrices (see equations (15) in Ref. [17]). Solutions of these equations can be easily found numerically; see Ref. [17] where these solutions were found for rather large values of the vibronic stabilization energy D.

The Hamiltonian H in the adiabatic approximation gets the form (see equations. (27) and (28) in Ref. [17])

$$H_{\pm} = \frac{1}{2}\left(-d^2/dx^2 + x^2 + k^2\mathcal{E}^2/4\left(k^2x^2+\mathcal{E}^2\right)^2\right)\cdot I + \sqrt{k^2x^2+\mathcal{E}^2}\cdot\sigma_z \tag{3}$$

Here H_+ describes in this approximation the motion of the system in the upper branch of the potential surfaces, while H_- describes the motion of it in the lower branch of the potential surfaces. We are interested in the eigenstates of Hamiltonian H_+ describing the quantum states above the avoided crossing. Using the dimensionless coordinate $r = x\sqrt{2}D^{1/6}$ the Hamiltonian H_+ gets the form

$$H_+ = -d^2/dr^2 + \sqrt{r^2+\varepsilon^2} + \Omega^2 r^2 + \varepsilon^2/4\left(r^2+\varepsilon^2\right)^2 \tag{4}$$

where $D = k^2/2$, $\Omega = 1/2D^{1/3}$, $\varepsilon = \mathcal{E}/D^{1/3}$. We consider the case when $\Omega \lesssim \varepsilon \ll 1$ and restrict ourselves to the consideration of states with the characteristic values of $r > \varepsilon$. In this case in the zeroth approximation one can neglect the terms $\propto \Omega^2$ and $\propto \varepsilon^2$. Then Hamiltonian H_+ gets the following simple form: $H_+^{(0)} = -d^2/dr^2 + |r|$. The solution of this equation is the Airy function [2]:

$$\psi_l(|r|) = \mu_l \mathrm{Ai}(|r| + \mathrm{E}_l),\quad \psi_l(-|r|) = (-1)^n \psi_l(|r|) \tag{5}$$

where $\mathrm{Ai}(r)$ is the Airy function, $\mathrm{E}_l = E_l/D^{1/3}$, E_l is the energy of the level number l, μ_l are the normalization factors:

$$\mu_0 = 1.31,\ \mu_1 = 1.01,\ \mu_2 = 0.936,\ \mu_3 = 0.880,\ \mu_4 = 0.847,\ \mu_5 = 0.817,\ \mu_6 = 0.796 \tag{6}$$

The energies E_l are determined by the boundary conditions. The first order corrections can be found by applying the equation

$$\delta\mathrm{E}_l = \langle\psi_l|(H_+ - \mathrm{E}_l)|\psi_L\rangle = 2\int_0^\infty dr\psi_l(r)(H_+ - \mathrm{E}_L)\psi_l(r) \tag{7}$$

2.1 Even states

Let us consider first the even states. These states satisfy the boundary condition $d\psi(r)/dr\,|_{r=0} = 0$. Using this condition we get the following energies of the first five even levels above the avoided crossing in zeroth approximation: $\mathrm{E}_0 = 1.02$, $\mathrm{E}_2 = 3.25$, $\mathrm{E}_4 = 4.82$, $\mathrm{E}_6 = 6.16$, $\mathrm{E}_8 = 7.37$. Let us consider now the first-order corrections. The term $\sqrt{r^2+\varepsilon^2} - |r|$ gives rather weakly dependent on n correction $\delta\mathrm{E}_{2l} = \varsigma_{2l,\varepsilon}\varepsilon$, where $\varsigma_{2l,\varepsilon} \sim 1$. The term $\left(\varepsilon^2/8D^{2/3}\right)/\left(r^2+\varepsilon^2/D^{2/3}\right)^2$ gives even smaller corrections. However, the term $\Omega^2 r^2$ gives rapidly growing with l corrections $\delta\mathrm{E}'_{2l} = \alpha_{2l}/D^{3/2}$ with $\alpha_0 = 0.22$, $\alpha_2 = 5.69$, $\alpha_4 = 8.67$, $\alpha_6 = 12.43$, $\alpha_8 = 20.29$. Consequently, one can present energies of the even states in the form

$$E_{e,2l} \approx D^{1/3}\left(\mathrm{E}_{2l} + \varepsilon\varsigma_{2l,\varepsilon} + \zeta_{2l}D^{-2/3}\right). \tag{8}$$

2.2 Odd states

Now we consider the odd states. These states satisfy the condition $\psi(r)|_{r=0} = 0$. This condition gives in zeroth approximation the following energies of the levels in the upper branch of the potential surface: $\mathrm{E}_1 = 2.34$, $\mathrm{E}_3 = 4.09$, $\mathrm{E}_5 = 5.52$, $\mathrm{E}_7 = 6.79$, $\mathrm{E}_9 = 7.95$. The first-order corrections of the energies of these states with respect to $\sqrt{r^2+\varepsilon^2} - |r|$ give $\varepsilon^2\eta_{2n+1}$: the first-order corrections stemming from the term $\Omega^2 r^2/2$ equal $\zeta_{2n+1}/D^{2/3}$, where first 5 values of η_{2n+1} and ζ_{2n+1} are:

$\eta_1 = 0.160$, $\eta_3 = 0.063$, $\eta_5 = 0.026$, $\eta_7 = 0.020$, $\eta_9 = 0.016$, and $\zeta_1 = 1.087$, $\zeta_3 = 2.162$, $\zeta_5 = 3.212$, $\zeta_7 = 4.249$, $\zeta_9 = 5.279$. Consequently, the energies of odd terms approximately given by the equation

$$E_{2l+1} \approx D^{1/3}\left(\mathrm{E}_{2l+1} + \varepsilon^2\eta_{2l+1} + \zeta_{2l+1}D^{-2/3}\right) \tag{9}$$

The wave functions of the first 6 levels (3 even and 3 odd) are depicted in figure 1.

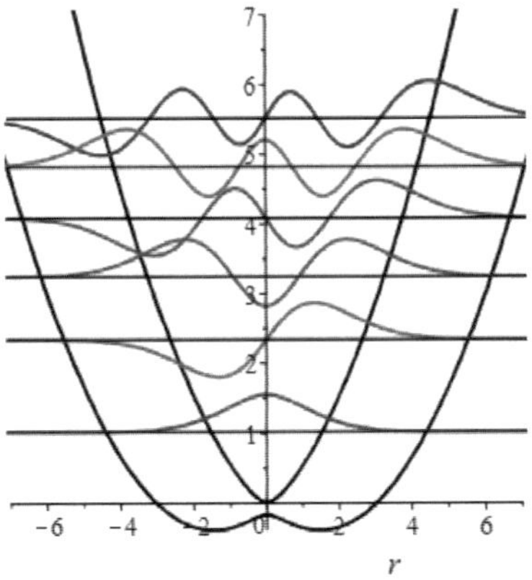

Figure 1. Adiabatic potential energy surfaces in case of PJTE (thick black curves) and first six vibrational states above the avoided crossing (energies E_n – straight black lines, wave functions are shown by curves); $D = 27$, $\varepsilon = 1$
.

3. Time evolution of vibrational levels in the upper states

The wave functions $\psi_n(x)$ describe only dependence of full wave functions on the vibrational coordinates. The electronic part of the wave functions in the adiabatic approximation above the avoided crossing for $\mathcal{E} > 0$ equals $|e_+\rangle = \left(|0\rangle_e + |1\rangle_e\right)/\sqrt{2}$. The full states in this approximation read $|\tilde{\psi}_l\rangle = |\psi_l\rangle|e_+\rangle$. These states are not the stationary ones; they can be presented as the wave-packets of the stationary states:

$$|\tilde{\psi}_l\rangle = \sum\nolimits_\nu \left(S^+_{\nu,l}|\Psi_{\nu+}\rangle + S^-_{\nu,l}|\Psi_{\nu-}\rangle\right) \tag{10}$$

where

$$\begin{aligned} S^{\pm}_{\nu,2l} &= 2^{-1/2}\sum_n c^{\pm}_{2n,\nu}\langle 2n|\psi_{2l}\rangle, \\ S^{\pm}_{\nu,2l+1} &= 2^{-1/2}\sum_n c^{\pm}_{2n+1,\nu}\langle 2n+1|\psi_{2l+1}\rangle \end{aligned} \tag{11}$$

The amplitudes $c^{\pm}_{n,\nu}$ (see equations (2)) are found at the diagonalization of the vibronic matrix; see Ref. [17]. The overlapping integral $\langle n|\psi_l\rangle$ of the wave function of the harmonic oscillator with the wave function of the state under consideration differs from zero if both, l and n are even or odd. It equals

$$\langle n|\psi_l\rangle = 2\sqrt{1/2^n n!\sqrt{2\pi}}\,D^{1/12}\mu_l\int_0^\infty dr H_n(r/\sqrt{2}D^{1/6})e^{-r^2/4D^{1/3}}Airy(r-r_l)$$

where H_n is the Hermite polynomials. This integral can be easily calculated for any given D and r_l. The density function

$$\rho_l(\mathrm{E}) = \sum_\nu\left(\left|S^+_{\nu,l}\right|^2\delta(\mathrm{E}-E^+_\nu) + \left|S^-_{\nu,l}\right|^2\delta(\mathrm{E}-E^-_\nu)\right) \tag{12}$$

IOP Conf. Series: Journal of Physics: Conf. Series **833** (2017) 012013 doi:10.1088/1742-6596/833/1/012013

gives the spectrum of these states. According to the Fock-Krylov theorem, the square root of the absolute value of its Fourier transform

$$L(t) = \left| \sum_{\nu} \left|S_{\nu+,l}\right|^2 e^{iE_{\nu+}t} + \left|S_{\nu-,l}\right|^2 e^{iE_{\nu-}t} \right|^2 . \qquad (13)$$

describes the time evolution (decay) of this state. Thus, to find the time evolution mentioned one needs to calculate the amplitudes $S_{n,\nu\pm}$. This can be done using equations (5) and (11). Equation (13) does not account for the decay of the vibronic levels caused by the interaction with phonons. This decay will be taken into account by adding the imaginary summands $i\gamma_{\nu\pm}$ to the energies $E_{\nu\pm}$ in equation (13). The method of calculations of the decay factors $\gamma_{\nu\pm}$ was developed in [24].

4. Numerical results

Presented here analytical theory of the quantum states above the avoided crossing of APESs allows one to calculate the functions $L_n(t)$ describing the decay of the population probability of these levels. Using given above equations (2), (5), (11) and (13) we performed calculations of $L_n(t)$ for six lowest levels of the system with $D=25$, $\varepsilon=0.5$, $\varepsilon=5$ and $D=50$, $\varepsilon=0.5$, $\varepsilon=5$ (we use units $\hbar=\omega=1$). In our calculations we used 2000 x 2000 and 2500 x 2500 vibronic matrices with no remarkable differences in the results. To calculate amplitudes $S^{\pm}_{\nu,l}$ (see equations (11)), we took into account 61 and 71 lowest states $|n\rangle$ also with no remarkable difference in the results. The calculated decay-functions $L(t)$ are presented in figures 2 – 4. Several characteristic features of the decays should be noted.

First, the fastest stages of the time evolutions under consideration have characteristic time $\lesssim 0.1$. For usual parameters of molecular systems this corresponds to femtosecond times. At this stage, the larger is the vibronic interaction D the faster is the short time-scale dynamics (see figures 2 and 3). These conclusions are in agreement with the result obtained in [27-32] by means of the multi-configurational time-dependent Hartree method.

Another essential characteristic of the decay laws is the presence of the oscillatory features in the $t \lesssim 1$ region for all levels except the lowest ones. This-type oscillatory time-dependence of the probability of population of a quantum state is characteristic for systems with discrete energy spectrum weakly interacting with quasi-continuum of states. Physically such regime of decay corresponds to the situation when the energy transferred from one sub-system to another can return back. The quasi-continuum mentioned corresponds here to the exact vibronic states with almost equidistant energy levels with the mean energy quantum being close to unity. The probability to return back increases with the level number: the higher is the level the more states below exist from which the system can return back The reason of absence of the oscillatory features for the lowest levels is the following: when the system leaves this level it leaves also the upper branch of the APES and cannot return back.

IOP Conf. Series: Journal of Physics: Conf. Series **833** (2017) 012013 doi:10.1088/1742-6596/833/1/012013

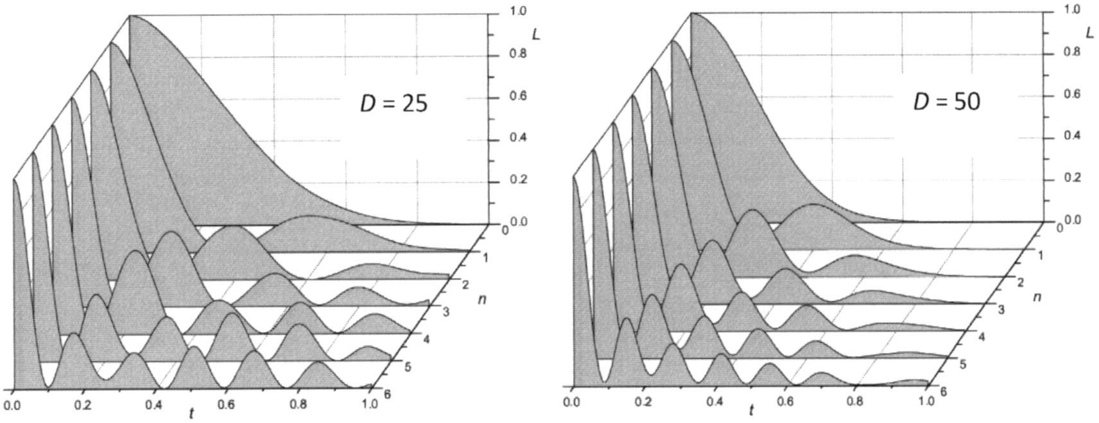

Figure 2. Short-time evolution of the states: $L(t)$ for levels $n = 0 - 6$ at $\varepsilon = 0.5$, $\gamma = 0.05$.

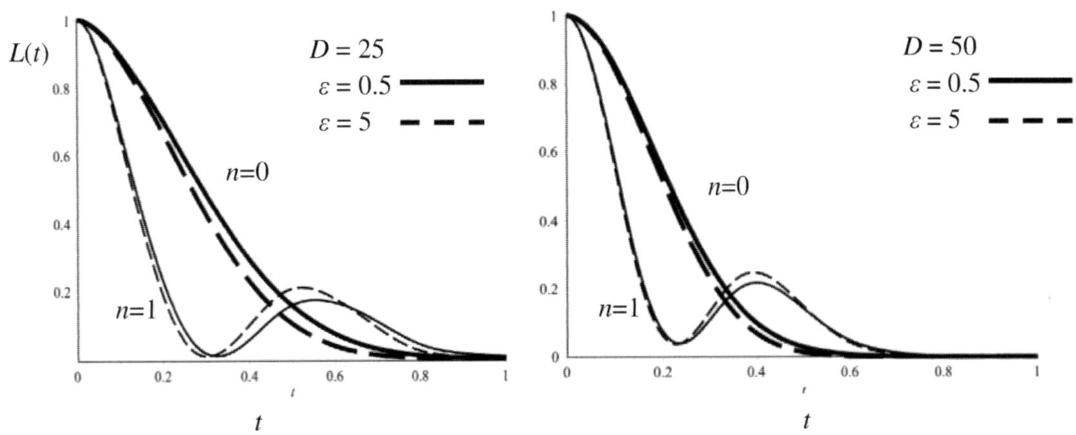

Figure 3. Effect of ε in short time evolution is small: $L(t)$ for levels $n = 0$ and 1, $\gamma = 0.05$.

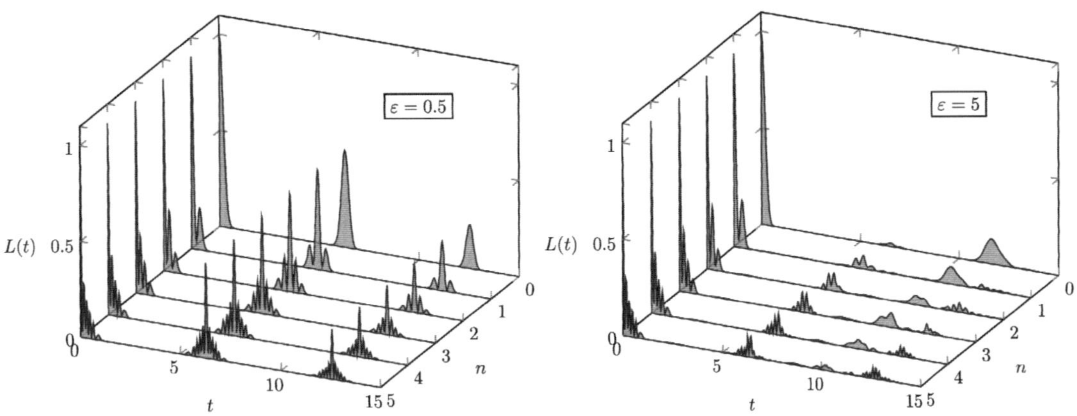

Figure 4. Long time evolution of the states with n=0...5 for $\varepsilon = 0.5$, $D = 50$, $\gamma = 0.05$.

One more characteristic feature of decays of the population probability is the repeating of the probability maxima with the mean period close to 2π corresponding to the period of the main vibration. This period appears due to the broad almost equidistant energy spectrum of the highly excited vibronic system with large D. It is a consequence of the Floquet's theorem which in this case tells us that the periodically time-dependent system has periodic energy spectrum. The decay of the periodicity at large time in our case corresponds to the violation of the equidistant levels. This violation is essential for large ε but it is rather small in case of small ε. Therefore the decay mentioned is quite small in last case and it is remarkable in the first case. Due to the same reason the repeating period for large ε differ from 2π. One more reason of the decay in the long-time-scale is the vibronic interaction with phonons, which in our case is taken into account by the factor $e^{-\gamma t}$, where γ is the decay constant (for calculation of γ see Ref. [24]).

5. Summary

Pseudo-Jahn-Teller effect (PJTE) in a simplest vibronic system with two quasi-degenerate electronic states and a single non-totally symmetric vibrational mode (with the frequency ω) is considered. An analytical theory of the quantum states above the avoided crossing of the electronic levels in this system is proposed. It is found that in case of large PJTE stabilization energy and small or moderate electronic splitting energy the wave functions of these quantum states are approximately given by Airy functions. The knowledge of these functions allows us to calculate the time evolutions of the probability of population of these states. To this end the equations describing decomposition of these functions over numerically exact vibronic states are derived. This allows us to calculate these time evolutions by applying the Fock-Krylov theorem.

A remarkable features of the found time dependencies is existence of three different time scales: 1) the primary dephasing decay with femtosecond ($t \sim 0.1\omega^{-1}$) characteristic duration, 2) the oscillatory stage for $t \lesssim 1/\omega$ corresponding to energy transfer forth and back between the states under consideration and the vibronic states and 3) the repeating of the probability maxima with relatively long mean period $\sim 2\pi$ – a consequence of the Floquet's theorem. At longer times the decrease of the probabilities under consideration is determined by the vibronic interaction with phonons.

Acknowledgements

This work was supported by institutional research funding IUT2-27 of the Estonian Ministry of Education and Research, and by European Union through the European Regional Development Fund (projects EU49695 and 3.2.0101.11-0029).

References

[1] Landau L D 1932 *Phys. Z. Sowietunion* **2** 46
Landau L D and Lifshitz E M 1977 *Quantum Mechanics: Nonrelativistic Theory* 3rd ed (New York: Pergamon) pp 342–51
[2] Zener C 1932 *Proc. R. Soc. London* A **137** 696
[3] Pauling L and Wilson E B Jr 1935 *Introduction to Quantum Mechanics* (New York: McGraw-Hill) Section 41
[4] O'Malley T F 1971 *Advances in Atomic and Molecular Physics* ed D R Bates and I Esterman (New York: Academic) vol 7 p 233
[5] Piepho S.B, Krauzs E R and Schatz P N 1978 *J. Am. Chem. Soc.* **100** 2996
[6] Wong K Y and Schatz P N 1981 *Progr. Inorg. Chem.* **28** 369
[7] Hendrickson D N 1990 *Mixed Valence Systems: Applications in Chemistry, Physics and Biology (NATO ASI Series)* ed K Prassides (Dordrecht: Kluwer) pp 67–90
[8] Tsukerblat B S, Palii A V, Gamurar V Ya and Kishinevsky H V 1991 *Phys. Lett.* A **158** 341–4
[9] Stoneham A M 2009 *Physics 2* p 34
[10] Englman R and Yahalom A 2001 *Vibronic Interactions: Jahn-Teller Effect in Crystals and*

Molecules (NATO Science Series) ed M D Kaplan and G O Zimmerman (Dordrecht: Kluwer Academic) pp 5–14

[11] McKemmish L K, McKenzie R H, Hush N S and Reimers J R 2011 *J. Chem. Phys.* **135** 244110

[12] Bersuker I B 2001 Modern aspects of the Jahn–Teller effect theory and applications to molecular problems *Chem. Rev.* **101** 1067–114

[13] Bersuker I B 2013 Pseudo-Jahn-Teller effect – a two-state paradigm in formation, deformation, and transformation of molecular systems and solids *Chem. Rev.* **113** pp 1351–90

[14] Fulton R L and Gouterman M 1961 *J. Chem. Phys.* **35** 1059

[15] Fulton R L and Gouterman M 1964 *J. Chem. Phys.* **341** 2280

[16] Лооритс В А 1979 Проявление резонанса Ферми в оптических спектрах динамического псевдо-эффекта Яна-Теллера, *Институт физики АН ЭССР, Академия Наук Эстонской ССР, Отделение физико-математических и технических наук, Препринт А-11* (in Russian)

[17] Loorits V 2017 Manifestation of Fermi resonance in optical spectra of dynamical pseudo-Jahn-Teller effect *J. Phys. Conf. Ser.* (this issue)

[18] Loorits V 1980 *Izv. Akad. Nauk. ESSR, Fiz-Mat.* **29** 208

[19] O'Brien M C M and Evangelou S N 1980 *Solid State Commun.* **36** 29

[20] Koppel H, Haller E, Cederbaum L S and Domcke W 1980 *Mol. Phys.* **41** 669

[21] Slonczewski J 1963 *Phys. Rev.* **131** 1596

[22] Pae K and Hizhnyakov V 2016 *J. Chem. Phys.* **145** 064108

[23] Fock V A and Krylov S N 1947 *Zh. Eksp. Teor. Fiz.* **17** 93

[24] Hizhnyakov V, Pae K and Vaikjärv T 2012 *Chem. Phys. Lett.* **525–6** 64

[25] Pae K and Hizhnyakov V 2013 *J. Chem. Phys.* **138** 104103

[26] Pae K and Hizhnyakov V 2014 *J. Chem. Phys.* **141** 234113

[27] Meyer H D, Manthe U and Cederbaum L S 1990 *Chem. Phys. Lett.* **165** 73

[28] Manthe U, Meyer H D, and Cederbaum L S 1992 *J. Chem. Phys.* **97** 199

[29] Beck M 2000 *Phys. Rep.* **324** 1

[30] Köppel H, Döscher M, Bâldea I, Meyer H-D and Szalay P G 2002 *J. Chem. Phys.* **117** 2657

[31] Meyer H D and Worth G A 2003 *Theor. Chem. Acc.* **109** 251

[32] Faraji S, Meyer H D and Köppel H 2008 *J. Chem. Phys.* **129** 074311

IOP Conf. Series: Journal of Physics: Conf. Series **833** (2017) 012014 doi:10.1088/1742-6596/833/1/012014

Diabatic models with transferrable parameters for generalized chemical reactions

Jeffrey R Reimers[1,2], Laura K McKemmish[3], Ross H McKenzie[4] and Noel S Hush[5,6]

[1] International Centre for Quantum and Molecular Structure, College of Sciences, Shanghai University, Shanghai 200444 China
[2] School of Mathematical and Physical Sciences, University of Technology Sydney, NSW 2007 Australia
[3] Department of Physics and Astronomy, University College London, Gower Street, London WC1E 6BT UK
[4] School of Mathematics and Physics, The University of Queensland, QLD 4072 Australia
[5] School of Chemistry, The University of Sydney, Sydney, NSW 2006 Australia
[6] School of Molecular Biosciences, The University of Sydney, NSW, 2006 Australia

E-mail: jeffrey.reimers@uts.edu.au, reimers@shu.edu.cn

Abstract. Diabatic models applied to adiabatic electron-transfer theory yield many equations involving just a few parameters that connect ground-state geometries and vibration frequencies to excited-state transition energies and vibration frequencies to the rate constants for electron-transfer reactions, utilizing properties of the conical-intersection seam linking the ground and excited states through the Pseudo Jahn-Teller effect. We review how such simplicity in basic understanding can also be obtained for *general* chemical reactions. The key feature that must be recognized is that electron-transfer (or hole transfer) processes typically involve one electron (hole) moving between two orbitals, whereas general reactions typically involve two electrons or even four electrons for processes in aromatic molecules. Each additional moving electron leads to new high-energy but interrelated conical-intersection seams that distort the shape of the critical lowest-energy seam. Recognizing this feature shows how conical-intersection descriptors can be transferred between systems, and how general chemical reactions can be compared using the same set of simple parameters. Mathematical relationships are presented depicting how different conical-intersection seams relate to each other, showing that complex problems can be reduced into an effective interaction between the ground-state and a critical excited state to provide the first semi-quantitative implementation of Shaik's "twin state" concept. Applications are made (i) demonstrating why the chemistry of the first-row elements is qualitatively so different to that of the second and later rows, (ii) deducing the bond-length alternation in hypothetical cyclohexatriene from the observed UV spectroscopy of benzene, (iii) demonstrating that commonly used procedures for modelling surface hopping based on inclusion of only the first-derivative correction to the Born-Oppenheimer approximation are valid in *no* region of the chemical parameter space, and (iv), demonstrating the types of chemical reactions that may be suitable for exploitation as a chemical qubit in some quantum information processor.

IOP Conf. Series: Journal of Physics: Conf. Series **833** (2017) 012014 doi:10.1088/1742-6596/833/1/012014

1. Introduction

Though chemical reactions involve complicated restructuring of the electronic and nuclear configuration and dynamics, it can be useful to use the simplification of envisaging reactions as occurring along a single reaction coordinate. This basic concept forms the conceptual framework of Transition-State Theory and much of modern chemistry. Modern electronic-structure calculation methods can predict wide ranges of properties to within the accuracy of experimental measurements, capturing all chemical and spectroscopic features of the system, but such approaches rarely provide insight into why processes occur and have to be repeated for each small system variation. Diabatic models can be used to interpret these calculations, however, revealing the chemical features controlling the process. Basically, simple forms, called diabatic surfaces, are used to describe the reactants and products, and then these surfaces are coupled together in some way to mimic the Born-Oppenheimer surfaces coming from the electronic-structure calculations. Our purpose here is to do this in such a way that the parameters involved can be transferred between systems, and that their values can be used to express why processes happen in terms of simple and intuitive chemical descriptions.

Parallel but equivalent ways of doing this have been developed since the 1950's assuming that the diabatic states are harmonic and that the coupling between them is linear. One of these, whose symbolism and approach we follow herein, is Hush's theory of adiabatic electron transfer [1-7], with many critical results also being established by others working in the electron-transfer field [8-12]. Closely related are the polaron models from solid-state physics [13-14] and the Rashba spin Hamiltonian [15], but a different approach from the field of spectroscopy has been the Pseudo-Jahn-Teller effect (PJTE) developed by Bersuker and others [16-20] that focuses on spectroscopic states as "reactants" and "products" coupled through Herzberg-Teller [21] vibronic coupling [22-25]. At the basic level, these approaches are all equivalent [26], and can remain so even after extension to include multiple nuclear motions [27] (as is required to describe processes involving conical intersections) or multiple electronic states [28], but the convenience with which they may be applied to different problems varies.

The basic model is sketched in figure 1, as it is usually presented in adiabatic electron-transfer theory relevant to general chemical reactivity. In simple cases involving symmetric isomerization reactions, only 3 parameters are often sufficient to describe complex Born-Oppenheimer ground and excited electronic states, increasing to 4 parameters to include any reaction asymmetry. In figure 1, these parameters are represented as the resonance energy J, the reorganization energy λ, the reaction exothermicity $-E_0$, and the geometry difference between the diabatic reactants and products $2Q_m$. Another parameter that is involved but rarely manifested is the reduced mass m for the nuclear motion. These parameters can be represented in many other ways, however, with e.g., in the literature of PJTE it being common to replace λ and Q_m with the force constant k and vibronic coupling constants α. As shown later, transformation between these representations involve no approximation and hence all observable properties are equally described independent of the mathematical representation used. The advantage of the parameter representation shown in the figure is that all parameters are transferrable and can be independently measured directly by simple experiments, and all parameters are easy to deduce from high-level calculations, leading to the widespread adoption of this representation. Also, we will show that these parameters result is a great simplification of the problem compared to the PJTE representation, allowing physical insight to be deduced. However they are represented, the model parameters control a very wide range of observable quantities, from geometries to infrared spectra to electronic spectra to redox potentials to spin distributions to reaction rate constants. The minimum number of parameters required increases when bystander chemical processes and their associated diabatic surfaces interfere with the processes of interest, but here our focus is on how such scenarios can still be reduced to a minimalist description.

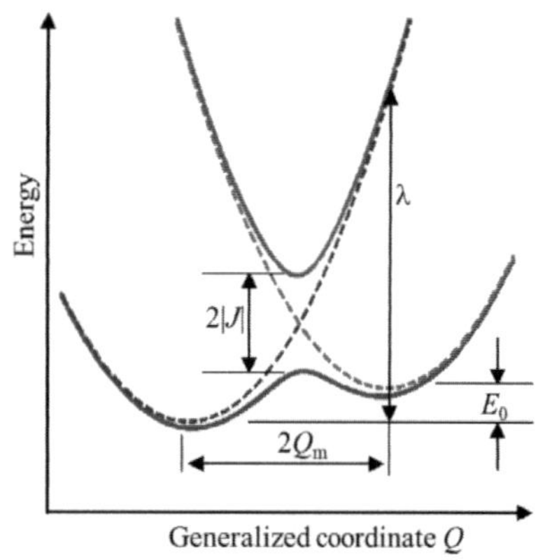

Figure 1. The one-mode two-state diabatic model for chemical reactions [8-10,13-14,16,29], indicating the diabatic surfaces for the reactant (blue dash) and product (red dash) states whose geometries differ by $2Q_m$ in the reaction coordinate Q with energy asymmetry E_0, as well as the effect of the resonance energy J and the reorganization energy λ in determining the Born-Oppenheimer adiabatic ground-state (*g*, purple) and excited-state (*s*, green) potential energy surfaces. Reproduced from McKemmish L, McKenzie R H, Hush N S and Reimers J R 2011 *J. Chem. Phys.* **135**, 244110 with permission of the American Institute of Physics.

Our aim here is to allow general chemical reactivity and spectroscopy to enjoy the well-established benefits coming from adiabatic electron-transfer theory. Hush focused on electron-density changes smoothly throughout the transition from reactants to products, manifesting through manifold properties accessible to high-level calculations and experimental observations [1-7,29-41]. These properties include not just the rate constants for electron-transfer reactions but also ground-state energies and vibration frequencies as well as electronic spectral properties. An important phenomenon identified using this approach is *intervalence spectroscopy* [5] involving intermolecular motion of charge induced by light absorption. This led to the design and synthesis of the Creutz-Taube ion [42], an unprecedented chemical species in which two Ru atoms appear to share a π bond across an intervening pyrazine bridging ligand. It also forms the basis for understanding of primary charge separation in natural and artificial solar-energy harvesting systems used in plants, bacteria, organic photovoltaics and artificial photosynthesis, as well as the reverse process, light emission in organic light-emitting diodes. A critical feature of the approach is that it leads to very many analytical or numerically exactly-solvable relationships linking system properties, as recently summarized in a number of contexts [28,43-46]. Exemplary in Hush's description of intervalence compounds is that he showed how from measurement of the visible absorption spectrum of the dye Prussian blue, it was possible to deduce kinetics parameters critical to the understanding of the charge conductivity of the material [5]. A related modern application was the demonstration of how protein mutagenesis controls the output voltage produced following primary charge separation during bacterial photosynthesis [47] and the development of a formalism for understanding modern Stark Spectroscopy [48] and its application to photosynthetic systems [49]. The critical feature is the universality of the model parameters for the qualitative description of system properties. Simple models can also be extended to universal quantitative ones provided that full quantum solutions [26,50], sufficient interfering processes [43,51-52], and motions orthogonal to the reaction coordinate are also included [27,53-54]. The assignment [55] of the Q-band spectrum of chlorophyll-a, arguably the world's most important chromophore whose properties display a strong pseudo-Jahn-Teller effect, following 50 years of intense debate, provides another example of the power of this approach.

Looking at the wider perspective, since the inception of the quantum theory of matter, the application of diabatic models in this capacity was envisaged. This is clear in the founding works, of London [56-57] as well as von Neumann and Wigner [58] for general adiabatic and non-adiabatic reactions, Eyring and Polanyi in the LEP potential-energy surface for reactions [59-60], Horiuti and Polanyi [61] for proton transfer reactions, Wall and Glockler [62] for isomerization reactions like ammonia inversion, and Hush for hydrogen transfer reactions [29]. Central qualitative ideas became imbedded in chemical culture such as symmetry breaking [6,16-20,63-67], understanding of aromaticity through the concept of resonance energy [68] and Shaik's "twin state" concept [69-73], as well as the *reaction force* description of classic chemical reactions [74-76] that leads naturally to the

IOP Conf. Series: Journal of Physics: Conf. Series **833** (2017) 012014 doi:10.1088/1742-6596/833/1/012014

Hammond-Leffler postulate [77-78], etc. Shaik's twin-state description is particularly provocative as it tells that the properties of the ground-state of a molecule are intricately coupled to the properties of an excited "twin" state.

This throws into dismay claims often made about density-functional theory (DFT) since its inception that it is only a *ground-state theory* as only the *ground-state density* is directly considered. However, the changes to the ground state caused by any perturbation can *always* be expressed in terms of response functions that mix in excited states, so DFT ground-state potential-energy surfaces always depict how the calculation method represents *both* ground and excited states. A simple example of this is that it was once believed that generalized-gradient functionals correctly described polyacetylene and general extended π systems as they predict realistic values for bond-length alternations and other singlet ground-state geometry features. However, because of its asymptotic potential error it grossly misrepresents excited states, leading to very poor values of molecular polarizabilities and eventually to the discovery that it predicts a triplet ground-state for polyacetylene rather than the observed singlet one [79]. So from a qualitative perspective, simple diabatic models and what they imply about chemical reactivity and spectroscopy have indeed had a profound effect on general chemical thinking.

However, the same cannot be said for semi-quantitative analyses. For example, Shaik's qualitative description of twin states effects controlling the properties of aromatic compounds [69-73] was never converted into quantitative modelling, like what flows easily from adiabatic electron-transfer theory. Parameters are never transferred between systems and molecules, e.g., a completely different diabatic model would be developed for chemical and spectroscopic processes of the ammonium cation than that which would be developed for processes of ammonia itself. At one level, different simple parameters describing these two molecules are demanded as electron correlation effects reshape and reorder orbitals as a result of the change in electron occupation. This effect, e.g., provides the mechanism by which molecular ionization energies and electron affinities calculated by Hartree-Fock theory as energy differences between species with different electron numbers differ from those obtained using Koopmans' theorem. However, such changes are generally small and one expects that the chemical and structural properties of the ammonium cation should, to a realistic level of approximation, be specifiable simply in terms of the analogous properties of ammonia itself. At an even more basic level, for different properties of the same molecule like its ground state structure and spectroscopy, its excited state structure and spectroscopy, and its reactivity in each state are usually generated *individually* for each process of concern rather than developing a *single* model for *all* properties. The dreams of the founders of quantum chemical theory have thus been realized in full only for electron transfer reactions.

In this review, we present a fundamental feature only just recognized [28] as a defining element of molecular potential-energy surfaces. Utilizing it, diabatic models can be produced for many types of chemical processes that involve transferrable parameters- parameters whose values depict wide-ranging properties, parameters who values can be transferred between systems. This provides a unification of manifold descriptions of chemical processes taught to students into a single conceptual formalism. Applications of this formalism to understanding why the chemistry of first-row elements is so different to that of later row ones, to understanding aromaticity, to designing chemical quantum computers, and to understanding the basics of Born-Oppenheimer breakdown and non-adiabatic chemical processes. In terms of understanding general Born-Oppenheimer breakdown, the one-dimensional model described herein obviously cannot deal with processes involving direct conical intersections, but analogues are drawn to known fully general mathematical results coming from the field of Catastrophe Theory [80-83] which indicate that the primary conclusions reached provide examples of generally expected results.

2. Semi-quantitative application of transferrable diabatic models to systems other than electron transfer: multiple coupled conical-intersection seams

The reason that parameters previously derived for general chemical reactions cannot be transferred between properties or between systems is that the effect of the number of electrons involved in the chemical processes was not taken into account analytically within the models [28]. This means that the effect of varying electron number gets treated implicitly rather than explicitly, modifying the values of deduced parameters.

Comparison is made between different scenarios in which one, two, or four electrons are critically involved are sketched in figure 2. The one-electron scenario (or one-hole scenario) is typical of electron-transfer reactions: just two molecular orbitals are involved, and two possible diabatic electronic states are produced named *G* (often corresponding closely to the Born-Oppenheimer adiabatic ground-state *g*) and it's singly excited state *S* (often corresponding closely to the Born-Oppenheimer adiabatic excited state *s*). This scenario generates a single conical intersection seam depicting locations at which the two adiabatic states become degenerate, and through the PJTE [17-20], this seam implicitly controls all system properties. Standard relations of adiabatic electron-transfer theory arise from this scenario [26,28,44,84].

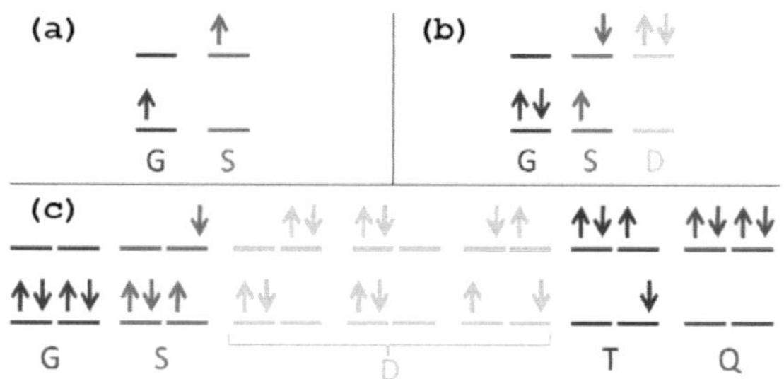

Figure 2. Molecular-orbital occupancies of delocalized diabatic states for scenarios with (a) one electron (or hole) in two orbitals, (b) two electrons in two orbitals, and (c) four electrons in four orbitals. Reproduced with permission from Reimers, McKemmish, McKenzie and Hush 2015 *Phys. Chem. Chem. Phys.* **17** 24598 with permission of the PCCP Owner Societies.

For a chemical reaction involving a molecule with a typical closed-shell electron configuration, two electrons are involved in the occupied orbital of interest whilst the other orbital of interest must be unoccupied. An example of this is say the ammonia inversion reaction for which the critical orbitals are the nitrogen lone-pair orbital *n* and the symmetric σ^*_{NH} orbital. However, as figure 2 shows, the two electrons may be distributed in *three* ways, the ground diabatic state *G* and the single-excited diabatic state *S* as before, but now also a doubly excited diabatic state *D*. If strong vibronic coupling between the *n* and σ^*_{NH} orbitals dominates variations in the structural, chemical, and spectroscopic properties of the system, then this coupling manifests *equally strongly* between *G* and *S* as between *S* and *D*. Three critical conical intersection seems thus connect the adiabatic surfaces, a low energy one between *g* and *s* that is typically critical to understanding ground-state properties, plus higher-energy ones that are more difficult to access between *g* and *d* and between *s* and *d*. Because *D* strongly perturbs *S*, the shape of the *g*-*s* seam is also strongly perturbed and no longer reflects the pure interaction between *S* and *G*.

For the inversion reaction of the ammonia cation, only one electron is present in the lone-pair orbital and hence only two adiabatic states can be generated using these orbitals; this situation is directly analogous to electron-transfer problems. The shape of the conical intersection seam is distorted significantly for ammonia inversion, however, accommodating the addition of the *D* state. Simple diabatic models focusing only on the apparent critical states, the *G* and *S* diabatic states, fail to include this distortion and so the values of the diabatic-model parameters are perceived to be quite different for the neutral and cationic molecules.

Shown also in figure 2 is the scenario for an aromatic molecule like benzene. For it, the HOMO and LUMO orbitals are doubly degenerate, so critical chemical processes such as the isomerization between its Kekulé structures that are dominated by excitations involving these orbitals intrinsically

IOP Conf. Series: Journal of Physics: Conf. Series **833** (2017) 012014 doi:10.1088/1742-6596/833/1/012014

involve 7 coupled diabatic states, the G and S states as before, now three different ways of making a D states, plus in addition a triply excited state T and quadrupally excited state Q. All diabatic states interact with each other to generate a multitude of conical-intersection seams with greatly varying shapes. The properties of the critical lowest-energy seam, that between g and s, are modified to accommodate all other higher-energy seams. So whilst a single calculation based on the PJTE [17-20] manifested by the g-s seam is likely to be very apt for the quantitative modelling of critical observed properties, the deduced parameters will not be global properties of the coupled orbital system but instead effective parameters depicting the modification of the intrinsic properties associated with the presence of the higher-energy seams.

To produce simple semi-quantitative models involving universally apt diabatic parameters, one must allow multiple seams with their varying shapes to be explicitly representable within the mathematical description used [28]. To do this, the number of diabatic electronic states involved must be expanded from two, as used in previous typical calculations exploiting the PJTE, to encompass all strongly coupled states and their generated conical-intersection seams. This is clearly 3 states for typical closed-shell chemical reactions but a useful approximation for aromatic reactions is to represent the 3 D states by single state, making a 5-state model. We start with the traditional PJTE Hamiltonian [16-20] for a two-state linearly coupled system

$$\mathbf{H}^{\mathbf{2D}} = \begin{bmatrix} T + \frac{k}{2}Q^2 & \alpha Q \\ \alpha Q & T + \frac{k}{2}Q^2 + 2J \end{bmatrix} \tag{1}$$

which describes the G and S states produced when one electron (or hole) is distributed in two orbitals, where Q is the reaction coordinate, k is the force constant depicting harmonic vibration in this mode, J is the electronic coupling (or *resonance energy*), α is the vibronic coupling constant, and $T = -\frac{\hbar^2}{2m}\frac{\partial^2}{\partial Q^2}$ is the nuclear kinetic energy operator where m is the effective mass of the oscillator. Equation (1) is sometimes called Holstein Hamiltonian [13-14] and is equivalent in form to the Rabi Hamiltonian for spin dynamics [85-86] (as the full Jahn-Teller Hamiltonian is equivalent to the Rashba spin Hamiltonian [15]). Assuming that electron-correlation effects do not significantly modify the properties of the involved orbitals as orbital occupancy is changed, it can be immediately generalized [27-28,54] to include D and then T and Q as [28]

$$\mathbf{H}^{\mathbf{3D}} = \begin{bmatrix} T + \frac{k}{2}Q^2 & \alpha Q & 0 \\ \alpha Q & T + \frac{k}{2}Q^2 + 2J & \alpha Q \\ 0 & \alpha Q & T + \frac{k}{2}Q^2 + 4J \end{bmatrix} \tag{2}$$

for two electrons in two orbitals, and

$$\mathbf{H}^{\mathbf{5D}} = \begin{bmatrix} T+\frac{k}{2}Q^2 & \alpha Q & 0 & 0 & 0 \\ \alpha Q & T+\frac{k}{2}Q^2+2J & \alpha Q & 0 & 0 \\ 0 & \alpha Q & T+\frac{k}{2}Q^2+4J & \alpha Q & 0 \\ 0 & 0 & \alpha Q & T+\frac{k}{2}Q^2+6J & \alpha Q \\ 0 & 0 & 0 & \alpha Q & T+\frac{k}{2}Q^2+8J \end{bmatrix} \quad (3)$$

for four electrons in four orbitals. The form shown for the two-state system is typical of vibronic coupling Hamiltonians used in spectroscopy but can be transformed by a 45° rotation of the electronic basis set. This makes the Hamiltonian typically used in electron-transfer theory using a basis set representing diabatic reactant and product states having different equilibrium geometries, as depicted in figure 1:

$$\mathbf{H}^{\mathbf{2L}} = \begin{bmatrix} T+\frac{k}{2}(Q+Q_{\mathrm{m}})^2+J-\frac{\lambda}{4} & J \\ J & T+\frac{k}{2}(Q-Q_{\mathrm{m}})^2+J-\frac{\lambda}{4} \end{bmatrix}, \quad (4)$$

where $Q_{\mathrm{m}} = \alpha / k$ and $\lambda = 2kQ_{\mathrm{m}}^2 = \frac{2\alpha^2}{k} = 2\alpha Q_{\mathrm{m}}$. In this representation, the coordinates $\pm Q_{\mathrm{m}}$ specify the geometries of the uncoupled reactant and product species, whilst the role of J as the resonance energy linking the reactants to products is clearly displayed. The representations $\mathbf{H}^{\mathbf{2L}}$ used in adiabatic electron-transfer theory and $\mathbf{H}^{\mathbf{2D}}$ used in PJTE theory are formally equivalent and lead to identical Born-Oppenheimer potential energy surfaces and indeed all observable properties. While Similar transformations yield new equations for description of the other scenarios in terms of localized chemical structures:

$$\mathbf{H}^{\mathbf{3L}} = \begin{bmatrix} T+\frac{k}{2}(Q+\sqrt{2}Q_{\mathrm{m}})^2+2J+\frac{\lambda}{2} & -\sqrt{2}J & 0 \\ -\sqrt{2}J & T+\frac{k}{2}Q^2+2J+\lambda & \sqrt{2}J \\ 0 & \sqrt{2}J & T+\frac{k}{2}(Q-\sqrt{2}Q_{\mathrm{m}})^2+2J+\frac{\lambda}{2} \end{bmatrix} \quad (5)$$

and

$$\mathbf{H}^{\mathbf{5L}} = \begin{bmatrix} T+\frac{k}{2}(Q+\sqrt{3}Q_{\mathrm{m}})^2-\frac{3\lambda}{4} & J+\frac{2J}{\sqrt{3}} & 0 & J-\frac{2J}{\sqrt{3}} & 0 \\ J+\frac{2J}{\sqrt{3}} & T+\frac{k}{2}(Q+Q_{\mathrm{m}})^2-\frac{\lambda}{4} & \frac{-4J}{\sqrt{3}} & 0 & -J+\frac{2J}{\sqrt{3}} \\ 0 & \frac{-4J}{\sqrt{3}} & T+\frac{k}{2}Q^2 & \frac{4J}{\sqrt{3}} & 0 \\ J-\frac{2J}{\sqrt{3}} & 0 & \frac{4J}{\sqrt{3}} & T+\frac{k}{2}(Q+Q_{\mathrm{m}})^2-\frac{\lambda}{4} & -J-\frac{2J}{\sqrt{3}} \\ 0 & -J+\frac{2J}{\sqrt{3}} & 0 & -J-\frac{2J}{\sqrt{3}} & T+\frac{k}{2}(Q-\sqrt{3}Q_{\mathrm{m}})^2-\frac{3\lambda}{4} \end{bmatrix} \quad (6)$$

IOP Conf. Series: Journal of Physics: Conf. Series **833** (2017) 012014 doi:10.1088/1742-6596/833/1/012014

The localized and delocalized representations are formally equivalent but expose critical properties in different ways. The delocalized (spectroscopic extended PJTE) description shows how vibronic coupling acts as a ladder operator to successively couple each diabatic state to the previous one, whilst the localized (kinetics) description readily depicts chemical reactions and shows what happens as conical intersections are approached (i.e., as $J \to 0$). In this limit, the off-diagonal matrix elements disappear in the localized representation, revealing Born-Oppenheimer surfaces depicting molecular isomers with distinctly different equilibrium geometries. However, these equilibrium geometries occur at $Q = \pm Q_{\mathrm{m}}$ for $\mathbf{H}^{\mathbf{2L}}$, $Q = 0, \pm\sqrt{2}Q_{\mathrm{m}}$ for $\mathbf{H}^{\mathbf{3L}}$, and $Q = 0, \pm Q_{\mathrm{m}}, \pm\sqrt{3}Q_{\mathrm{m}}$ for $\mathbf{H}^{\mathbf{5L}}$. Hence the fundamental properties of the lowest-energy conical-intersection seam involved with the ground state change significantly as more and more coupled diabatic excited states are added. The parameters describing individual conical intersections and the perceived two-state PJTE are therefore not transferrable ones; transferrable parameters are only generated when all interacting diabatic states are included in the mathematical description.

Both sets of extended Hamiltonians can be transformed into other equivalent forms that reduce the complex problems to either uncoupled Born-Oppenheimer representations (shown in cyan) or else just coupled pairs of diabatic states. The coupled pairs of states then appear to be simple two-dimensional problems equivalent to problems from adiabatic electron-transfer theory and traditional PJTE theory. For reactions involving a single doubly occupied orbital and single unoccupied orbital, this yields

$$\mathbf{H}^{\mathbf{3D}'} = \begin{bmatrix} T + \frac{k}{2}Q^2 & 0 & \sqrt{2}\alpha Q \\ 0 & T + \frac{k}{2}Q^2 + 2J & 0 \\ \sqrt{2}\alpha Q & 0 & T + \frac{k}{2}Q^2 + 4J \end{bmatrix} \tag{7}$$

$$\mathbf{H}^{\mathbf{3L}'} = \begin{bmatrix} T + \frac{k}{2}(Q + \sqrt{2}Q_{\mathrm{m}})^2 + 2J + \frac{\lambda}{2} & 0 & 2J \\ 0 & T + \frac{k}{2}Q^2 + 2J + \lambda & 0 \\ 2J & 0 & T + \frac{k}{2}(Q - \sqrt{2}Q_{\mathrm{m}})^2 + 2J + \frac{\lambda}{2} \end{bmatrix} \tag{8}$$

and indicates that the single-excitation can be decoupled whilst the ground state and doubly excited state appear to be connected analogously to the states involved in an electron-transfer application *except* for a renormalization of the parameters: the apparent vibronic coupling constant and nuclear displacement are $\sqrt{2}$ times larger than expected, whilst the apparent resonance energy is doubled. For the case of aromatic processes, the situation is not so simple with equation (6) indicating a rescaling of the displacement by a factor of $\sqrt{3}$ in the weak-coupling limit of $2|J|/\lambda \ll 1$, but in the alternate strong-coupling limit appropriate to benzene, the Hamiltonians can be transformed instead into

$$\mathbf{H}^{5\mathrm{D}'} \approx \begin{bmatrix} T+\frac{k}{2}Q^2 & 0 & 0 & 0 & 2\alpha Q \\ 0 & T+\frac{k}{2}Q^2+2J & 0 & 0 & 0 \\ 0 & 0 & T+\frac{k}{2}Q^2+4J & 0 & 0 \\ 0 & 0 & 0 & T+\frac{k}{2}Q^2+6J & 0 \\ 2\alpha Q & 0 & 0 & 0 & T+\frac{k}{2}Q^2+8J \end{bmatrix} \tag{9}$$

and

$$\mathbf{H}^{5\mathrm{L}'} \approx \begin{bmatrix} T+\frac{k}{2}(Q+2Q_\mathrm{m})^2 & 0 & 0 & 0 & 4J \\ 0 & T+\frac{k}{2}Q^2+2J & 0 & 0 & 0 \\ 0 & 0 & T+\frac{k}{2}Q^2+4J & 0 & 0 \\ 0 & 0 & 0 & T+\frac{k}{2}Q^2+6J & 0 \\ 4J & 0 & 0 & 0 & T+\frac{k}{2}(Q-2Q_\mathrm{m})^2 \end{bmatrix} \tag{10}$$

indicating a doubling of the displacement and quadrupling of the resonance energy.

Equations (7)-(10) capture the critical features associated with Shaik's concept that the ground-state has a "twin" excited state with intimately linked properties [69-73]. Previously, however, the twin state was presumed to be the singly excited state, but these equations show that the twin state is always the most excited state possible for the system in hand- the singly excited state for electron transfer, the doubly excited state for most chemical reactions, and the quadrupally excited state for aromatic systems. In this process, general chemical reactions involving many electrons and many interrelated conical-intersection seams are mapped onto a simple two-state model. In aromatic molecules the singly excited state is actually twinned to the triply excited state [28].

Ten chemical processes for which transferrable diabatic-model parameters have been determined are shown in figure 3, with their parameters given in table 1. These span a wide range of chemical phenomena from isomerization to aromaticity to small-molecule spectroscopy to organic conductors to natural and artificial molecules for solar-energy conversion. A critical property considers whether the ground-state has a single well or a double minimum: a double well is found whenever [44]

$$\left(\frac{E_0}{\Delta E}\right)^2 < \left[1-\left(\frac{2|J|}{\lambda}\right)^{2/3}\right]^3\left[1+\left(\frac{2J}{\lambda}\right)^2\right] \text{ or } \left(\frac{E_0}{\lambda}\right)^2 < \left[1-\left(\frac{2|J|}{\lambda}\right)^{2/3}\right]^3\left[1+\left(\frac{2J}{\lambda}\right)^2\right]^2 \tag{11}$$

where $\Delta E = (\lambda^2 + 4J^2)^{1/2}$ is the electronic energy gap. For weak coupling with $2J/\lambda \to 0$, a double well occurs whenever $|E_0|/\lambda < 1$ but for symmetric systems with $E_0 = 0$, symmetry breaking occurs whenever [6,16]

$$\frac{2|J|}{\lambda} < 1 . \tag{12}$$

Hence from table 1, benzene has high symmetry ($2|J|/\lambda$ = 3.3) whilst ammonia has low symmetry ($2|J|/\lambda$ = 0.8).

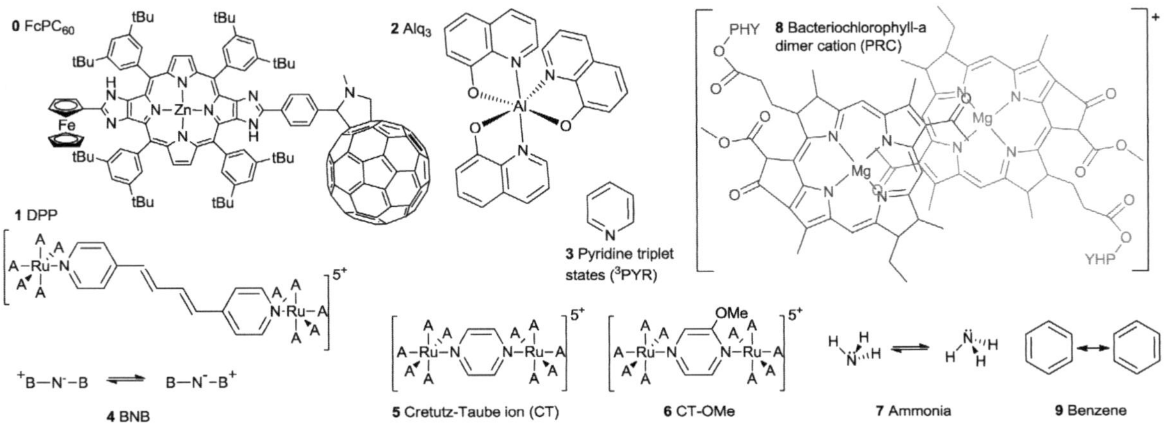

Figure 3. Some sample molecular systems (**0** [87-89], **1** [90-91], **2** [92-93], **3** [46,94], **4** [95], **5** [96], **6** [97], **7** [43,46], **8** [47,50], and **9** [46]) with electronic states that can be described using two coupled diabatic potential-energy surfaces. OMe is methoxy, PHY is phytyl; ^{t}Bu is tertiary butyl; A is ammonia; $FcPC_{60}$ is Zinc, [[5,10,16,21-tetrakis[3,5-bis(1,1-dimethylethyl)phenyl]-13-[4-(1',5'-dihydro-1'-methyl-2'*H*-[5,6]fullereno-C_{60}-I_h-[1,9-*c*]pyrrol-2'-yl)phenyl]-1,12-dihydro-23*H*,25*H*-diimidazo[4,5-*b*:4',5'-*l*]porphin-2-yl-κN^{23},κN^{24},κN^{25},κN^{26}]ferrocenato(2-)]-, (*SP*-4-1); Alq3 is *mer*-tris(8-hydroxyquinolinato)aluminum(III); DPP is ruthenium(5+), decaammine[μ-[4,4'-[(1*E*,3*E*)-1,3-butadiene-1,4-diyl]bis[pyridine-κ*N*]]]di- (9CI). Reproduced from Reimers J R, McKemmish L, McKenzie R H, and Hush N S 2015 *Phys. Chem. Chem. Phys.* **17** 24640 with permission of the PCCP Owner Societies.

Table 1. Diabatic-model parameters describing the chemical and spectroscopic properties of the 10 molecular systems shown in figure 3.

System	$\frac{2\lvert J\rvert}{\lambda}$	$\frac{\hbar\omega}{\Delta E}$	$\frac{E_0}{\hbar\omega}$	Q_c	$\frac{\Delta\Delta E^\dagger}{\hbar\omega}$
0 $FcPC_{60}$	0.029	0.15	-13	0.06	45
1 DPP	0.043	0.08	0	0.11	11
2 Alq3	0.08	0.16	0	0.14	6.3
3 ^{3}PYR	0.3	0.095	1.3	0.67	0.28
4 BNB	0.74	0.18	0	1.11	0.10
5 CT	0.80	0.089	0	1.68	0.045
6 CT-OMe	0.80	0.089	1.5	1.68	0.045
7 NH_3	0.80	0.006	0	6.45	0.003
8 PRC	1.8	0.41	0.6	1.39	0.065
9 Benzene	3.3	0.010	0	12.6	0.001

3. Usefulness of having transferrable parameters

Once the simple two-state one-mode model is established, many analytical results are available relating ground-state and excited-state structural, spin density, electrochemical potential, and kinetic properties (see, e.g., [2,5-6,8-10,13-14,16-20,27-29,34,44,48,98-100]), with other equations also available for multi-mode situations (see e.g., [27,54,99]). Standard relationships from parallel PTJE studies [16-20] can also be applied. Such relationships provide insight into fundamental chemical processes. Here we consider how benzene chemistry can be related to carbon chemistry involving standard single and double bonds, as well as to understanding why first-row chemistry is so different to the rest.

3.1. Deducing the CC bond lengths in ethane and ethylene from properties of benzene

Standard relationships from adiabatic electron-transfer theory or PJTE theory apply directly to the ground state and its twin state, and it is possible to generate many parallel ones relating to properties of other states involved in multi-electron systems [28]. Here we provide a simple example of the power of this approach, determining the bond-length alternation in hypothetical cyclohexatrine (a Kekulé structure of benzene) from only the observed ground-state and singly-excited state spectral properties and the atomic masses [28]. The motion involved is sketched in figure 4 in terms of an internal bending angle ϕ that produces C-C bonds of alternating length. Formally, the resulting "bond-length alternation" is defined as the difference between the bond lengths of the resulting single bonds and the double bonds, and the deduced value of 0.109 Å is in excellent agreement with the value of 0.113 Å estimated based on analysis of observed bond lengths for related conjugated molecules [101].

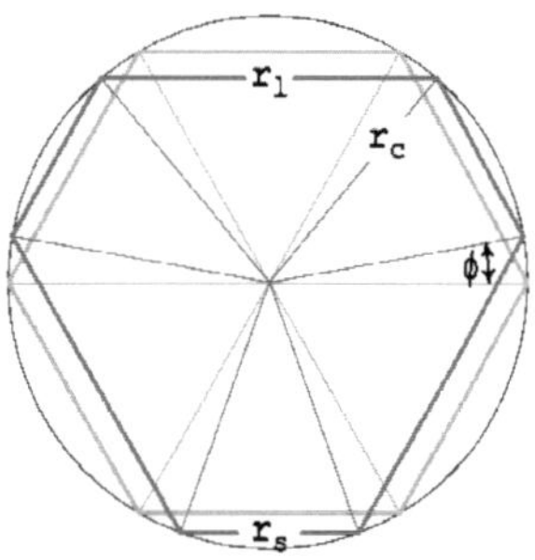

Figure 4. The Kekulé distortion in benzene is produced by motion involving the internal angle ϕ, manifesting long (l) single-bonds and short (s) double-bonds within the structure having bond-length alternation $r_l - r_s \approx 0.042\phi$ Å/°. Reproduced with permission from Reimers, McKemmish, McKenzie, Hush, Phys. Chem. Chem. Phys., 17 (2015) 24598 with permission of the PCCP Owner Societies.

From equation 3, the vertical excitation energy of the twin Born-Oppenheimer state q from the benzene ground state g is $8|J|$, whilst that for the singly excited state s is $2|J|$. The observed excitation energy for s is 4.77 eV and hence $|J|$ = 2.4 eV. The relationships between the Born-Oppenheimer vibration frequencies in the various states are [28]

$$\frac{2|J|}{\lambda} = \frac{\omega_s^2}{\omega_s^2 - \omega_g^2} = \frac{\omega_q^2 + \omega_g^2}{\omega_q^2 - \omega_g^2} . \tag{13}$$

Using the observed values of ω_g = 1309 cm^{-1} and ω_s = 1564 cm^{-1} (n.b., the excited state frequency exceeds the ground state frequency owing to the underlying double-well structure) yields $2|J|/\lambda$ = 3.34. Misinterpretation of s as the twin state q of g would have yielded $2|J|/\lambda$= 5.68 instead; using the correct value then leads to λ= 1.43 eV, otherwise λ= 0.84 eV. The force constant k for the vibration can be determined from the known [102] moment of inertia associated with the rotation by ϕ combined with the diabatic vibration frequency ω. This is given simply by $\omega = \omega_s$ when all states are included and by $\omega = \omega_s(1+\lambda/2|J|)^{1/2}$ if the twin state is incorrectly identified as s instead of q. Using the correct relationship leads to k = 0.43 eV/°^{2}, otherwise 0.37 eV/°^{2}. The force constant is related to the geometry ϕ_{m} corresponding to hypothetical cyclohexatriene without any resonance stabilization by

$$\lambda = 2k\left(\frac{\phi_{\mathrm{m}}}{2}\right)^2 \tag{14}$$

if all states are included (equation 10), leading to ϕ_{m} = 2.58° and hence a bond length alternation of 0.109 Å, close to the value deduced [101] from experimental data of 0.113 Å. Assuming that s is the twin state would have lead instead to (equation 4)

$$\lambda = 2k\phi_{\mathrm{m}}^2 \tag{15}$$

which, when combined with the analogous previous results, leads to a bond-length alternation of just 0.045 Å.

We see that recognition of the presence of higher-energy conical intersection seams has significant repercussions for the actual properties of the lowest-energy seam and hence the ground-state structure. Recognition of the intrinsic relationships involved leads to equations depicting many properties in terms of simple expressions. Shaik's "twin state" concept led to qualitative understanding of various key properties [61,62,63,64,65] but never led to quantitative analysis owing to its misidentification of the actual state that is twin to the ground state.

3.2. Why ammonia is so different to phosphine, stilbene, and the other group XV hydrides

The inversion reaction in the XH_3 series of molecules involves the mixing of the X lone-pair orbital *n* with the symmetric antibonding orbital σ^*_{NH} as a function of the out-of-plane torsional angle τ that distorts the geometry from planar to pyramidal [43]. The HXH bond angle θ also changes with this motion simple according to

$$2\cos\theta = 3\sin^2\tau - 1 \ . \qquad (16)$$

Table 3 indicates that the observed [28] bond angles contract from the hypothetical value of 120° for a planar structure to 108° for NH_3 to 93° for PH_3 for 90° in BiH_3. A critical question is as to why there is such a large difference between ammonia and phosphine, with then only little change as one goes further down the periodic table. Another question is as to what limiting angle would be expected if the periodic table could be extended down to many more rows, whilst a third question concerns what happens when one of these molecules is ionized to make XH_3^+. Traditional chemical explanations such as Valence-Shell Electron-Pair Repulsion Theory (VSEPR) [103-105] either just decree N to be tetrahedral and P to be octahedral or else assign atom sizes to fit the results, but they do not explain *why* N is so different to P. VSEPR provides no insight into what the chemistry might be like on later rows of the periodic table, and whilst it explains the basic result that the bond angles in the radical cations are larger than those in the neutral molecules, it makes no quantitative prediction of the magnitude of the effect. Traditional analyses made using PJTE approaches [106] analyzed in terms of change to the resonance energy, force constant, and vibronic coupling constants accurately describe calculated and experimental data but do not perceive the relationships between the key variables and provide no explanation as to why these quantities change in the way they do between NH_3 and PH_3 and therefore have not led to basic understanding.

Table 3. Estimates of the critical parameters for XH_3 inversion reactions coming from (i) observed or calculated data for the ground-state equilibrium torsional angle τ_e (or equivalently bond angle θ_e) and barrier height $\Delta E^\dagger$, J_u and λ_u, and (ii) from calculated vertical transition energies at the planar and ground-state equilibrium geometries to the twin state *q* [43], J_2 and λ_2.

	Observed						EOM-CCSD/VTZ								
XH_3	τ_e °	θ_e °	$\Delta E^\dagger$ eV	$2J_u/\lambda_u$	$4J_u$ eV	$2\lambda_u$ eV	τ_e °	θ_e °	$\Delta E^\dagger$ eV	$2J_u/\lambda_u$	$4J_u$ eV	$2\lambda_u$ eV	$2J_2/\lambda_2$	$4J_2$ eV	$2\lambda_2$ eV
NH_3	21.4	107.5	0.22	0.82	23[a]	28[a]	22.8		0.27	0.79	20.4[a]	26[a]	0.85	38	45
PH_3	32.9	93.3	1.38[b]	0.48	10[b]	21[b]	32.5		1.5	0.50	12.1	24	0.53	14	27
AsH_3	33.8	92.0	1.38[b]	0.44	8[b]	17[b]	33.5		1.82	0.45	11.0	24	0.48	13	26
SbH_3	34.2	91.5	1.63[b]	0.41	8[b]	19[b]	33.8		1.99	0.44	10.9	25	0.43	10	24
BiH_3	35.1	90.3	1.67[b]	0.36	6[b]	16[b]	34.7		2.65	0.38	10.7	28	0.34	9	25

[a]: large errors arise from valence/Rydberg orbital inversion modifying τ_m combined with the instability of equation (16) as of $2J_u/\lambda_u \rightarrow 1$.

[b]: $\Delta E^\dagger$ extrapolated from observed transitions and/or calculated data only up to 0.5 eV.

While just two orbitals are sufficient to describe the chemistry of the inversion reaction, attempting to apply the appropriate diabatic models leads immediately to difficulties as most observed spectra are dominated by Rydberg transitions rather than the valence states of interest. This creates not only the experimental challenge of isolating key high-energy spectroscopic transitions but also the conceptual one of understanding how the Rydberg states interact with and so distort the valence-state surfaces. As a result, the basic 3-state (*G*, *S*, *D*) diabatic description must be augmented by adding the lowest Rydberg state *R*, its double excitation *DR*, and the combined Rydberg-Valence double excitation *RV* [43]. High-level calculations provide a large amount of data and allow all transitions to be followed as a function of the torsional (or bond) angle changes. However, the interesting regions of the potential-energy surfaces range out to torsional angles of 50° (bond angles down to 68°) and so anharmonic effects also become important. The simplest quantitative diabatic model that could describe the shapes of all key adiabatic potential-energy surfaces out to such angles was found to have 11 parameters: the basic 3 diabatic-model parameters expanded to 5 to allow for (small) changes in orbital properties with occupation, expanded to 8 to include anharmonicity, and finally expanded to 11 to include the Rydberg states. While this complexity seems to destroy the desired simplicity and generality of the diabatic chemical description, at least 24 parameters are required to describe the same data using a (non-physical) Taylor expansion of each of the calculated adiabatic surfaces, whilst all of the 11 diabatic-model parameters have clear physical meaning. The power of the diabatic approach remains apparent as it led to the reassignment of many observed high-energy bands in the absorption spectrum of ammonia, bands previously described as Rydberg transitions that were actually intense valence absorptions instead [43]. What results is a comprehensive analysis of a wide range of experimental and high-level computational data, isolating key excited-state resonances located well above the ionization continua of the XH_3 series.

Such a quantitative analysis provides the basis for qualitative understanding. Assuming that the notion of twin states still holds, the properties of the ground state *g* and its twin-state *d* may be considered in isolation, allowing effective 2-state model parameters to be deduced from the properties of the ground-state surface and the vertical excitation energy of the twin state at the planar geometry. As the diabatic-model parameters depict many different observable properties, this process may be conducted in many different ways and the process only makes clear sense if the deduced parameters are insensitive to the method used. Here we report in table 3 on two possible ways of doing this.

First, we consider the vertical excitation energies to the twin state *d*, labelling the deduced parameters as J_2 and λ_2, using the EOM-CCSD [107-108] method and valence triple-zeta [109-111] (VTZ) basis sets to calculate the transition energies using MOLPRO [112]. The important relationships here are simply [28,43]

$$\begin{aligned} 4|J_2| &= \text{vertical excitation energy at the planar geometry } \tau=0 \\ 2\lambda_2 &= \text{vertical excitation energy at the ground-state adiabatic minimum } \tau=\tau_e \, . \end{aligned} \tag{17}$$

Next, we determine parameters named J_u and λ_u, from just the values of the adiabatic ground-state well depth $\Delta E^\dagger$ and equilibrium angle τ_e using

$$\begin{aligned} \left(\frac{2J_u}{\lambda_u}\right)^2 &= 1-\left(\frac{\tau_e}{\tau_m}\right)^2 \\ \lambda_u &= 4\Delta E^\ddagger \left(1-\frac{2|J_u|}{\lambda_u}\right)^{-2} . \end{aligned} \tag{18}$$

This approach requires knowledge of the third parameter τ_m (equivalent to Q_m in equations (4)-(10)) specifying the torsional angle in the absence of resonance. Many alternative ways of independently estimating all three parameters are known, and it was found that these result in τ_m (and hence the XHX angle θ_m) being identified as universal parameters for the whole XH_3 series [43]. This can be understood by noting that the diabatic orbitals can be identified as *sp* hybrids based on the natures of the interacting *n* and σ^*_{NH} orbitals. Simple geometry assuming that the diabatic angles are produced by maximizing overlap between the hydrogen 1*s* orbitals and the three X orbitals orthogonal to a *sp* hybrid, as shown in figure 5, then leads to

$$\tau_m = \mathrm{atan}\frac{1}{2} = 26.6^\circ, \qquad \theta_m = \mathrm{acos}\frac{-1}{5} = 101.5^\circ \tag{19}$$

and is appropriate for the electron-transfer-type case in which one electron is involved in the reaction. This scenario therefore describes the bond angle in the limit of no resonance energy for the XH_3^+ series. When two electrons are involved, the effect of the *S-D* interaction is to distort the ground state, scaling τ_m out to

$$\tau_m = \sqrt{2}\,\mathrm{atan}\frac{1}{2} = 37.6^\circ, \qquad \theta_m = \mathrm{acos}\left[\frac{3}{2}\sin^2\left(\sqrt{2}\,\mathrm{atan}\frac{1}{2}\right) - \frac{1}{2}\right] = 86.7^\circ \tag{20}$$

for XH_3. These equations provide the answer to the fundamental question as to what would the limiting geometry be for X being on the infinite row of the periodic table. Assuming that the diabatic angles universally take this value, J_u and λ_u can be immediately determined from the adiabatic well depth and equilibrium geometry using equation 16.

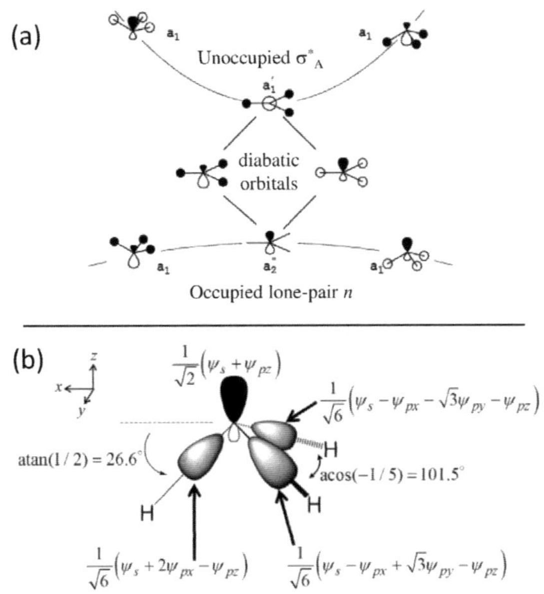

Figure 5. (a) Modified Walsh diagram showing how *sp* hybridized diabatic orbitals on X interfere to produce pure *p* and *s* adiabatic orbitals at the planar geometry, and how the adiabatic orbitals decouple as XH_3 molecules distort. (b) An X *sp* diabatic lone-pair orbital $2^{-1/2}(\psi_s + \psi_{pz})$ and its 3 orthogonal $s^{1/2}p^{5/2}$ bondable hybrid orbitals at a torsional angle of τ_m = 26.6° and HXH bond angle θ_m = 101.5°. Reproduced with permission from Reimers, McKemmish, McKenzie, Hush, Phys. Chem. Chem. Phys., 17 (2015) 24618 with permission of the PCCP Owner Societies.

Values of J_u and λ_u deduced from experimental data and from CCSD/VTZ data are given in table 3. These show the same trends but absolute values differ by up to a factor of two. Such sensitivity arises as the values of $2J_u/\lambda_u$, which are also given in the table, are not far from one, the limit at which Equation (16) becomes unstable. Improved calculations at the CCSD(T)/aug-cc-pwCVQZ level [43] give results much closer to the experimental ones, with the differences between the calculated results serving to highlight how difficult it can be to extract simple characteristic data for real systems displaying manifold bystander processes. Comparing values deduced by two very different approaches utilizing computed data, J_u and λ_u compare well to J_2 and λ_2 except for the case of NH_3. The poor agreement for NH_3 stems from the strong interaction between the valence and Rydberg states manifested in the calculations and so is itself a useful result, pointing to a key physical feature.

Most importantly, all three methods used to determine the critical ratio $2J/\lambda$ reported in table 3 yield very similar results, indicating that this feature can be robustly determined from experimental and (medium quality) calculated data. The subtle effects that dominate excitation energies and transition-state energies especially when $2J/\lambda \approx 1$ do not influence this ratio to the same extent as they influence the perceived resonance and reorganization energies themselves. Equation (16) shows that just their ratio controls the equilibrium geometry and hence it is this robustly determined property that is responsible for the large change in bond angle observed between ammonia and phosphine. Indeed, for ammonia, these ratios are much larger (ca. 0.8) than they are for PH_3 – BiH_3 (ca. 0.5 – 0.4), explaining the parallel abrupt change in geometry observed. The critical question concerning why nitrogen prefers tetrahedral coordination whilst phosphorous and the heavier elements prefer octahedral coordination then can be answered by finding why the value of $2J/\lambda$ changes so abruptly between NH_3 and PH_3. Table 3 shows that there is an abrupt change in *both* J and λ between NH_3 and PH_3 but that the change in J appears as the square of that in λ. So there is a common process involved in modifying both quantities, with the resonance energy being a clearer indicator. Hence we focus on why this quantity changes so much.

Equation (2) indicates that $4|J|$ is essentially the increase in energy when *both* electrons are taken from the lone-pair orbital n to the σ^*_{XH} orbital at the planar geometry. We find that while the lone-pair orbital change smoothly down the periodic table, discontinuous changes occur to σ^*_{XH}. Figure 6 indicates the source of this discontinuity [43], showing the σ^*_{NH} and σ^*_{AsH} orbitals evaluated using minimal (STO-3G) basis sets as well as those augmented by a single Rydberg s function. For NH_3, the antibonding orbital *contracts* significantly when the Rydberg orbital is present. This happens because the valence orbital is higher in energy than the Rydberg orbital, and these two significantly interact leading to "bonding" and "antibonding" pairs. The combination that is most relevant to the twin state of the ground state is the antibonding combination, one which has a *nodal surface* separating its inner (valence dominated) and outer (Rydberg dominated) regions. The presence of this nodal surface *compresses* the valence orbital. However, the orbital ordering reverses between NH_3 and PH_3, so that for heavier elements the most relevant combination is the bonding one. This allows the antibonding orbital to *delocalize* and spread out. The twin state is made through double-occupation of the antibonding orbital. When this orbital is compressed, the intra-orbital electron-electron repulsion energy is dramatically increased, increasing the orbital energy and hence the resonance energy. As the properties of the ground state are intimately linked to the properties of its twin state, the properties of ammonia are distinctly different to those of the rest of the Group-XV hydrides. This effect is likely to be generally descriptive of the properties of first-row elements compared to those of later rows. Virtual-orbital mixing is a long-recognized process affecting *excited-states* known as "Rydbergization" [113-114]. What we see here is that the ground-state is the "twin-state" to an affected excited state, so that whatever happens to the excited state also becomes important for understanding *ground-state* properties.

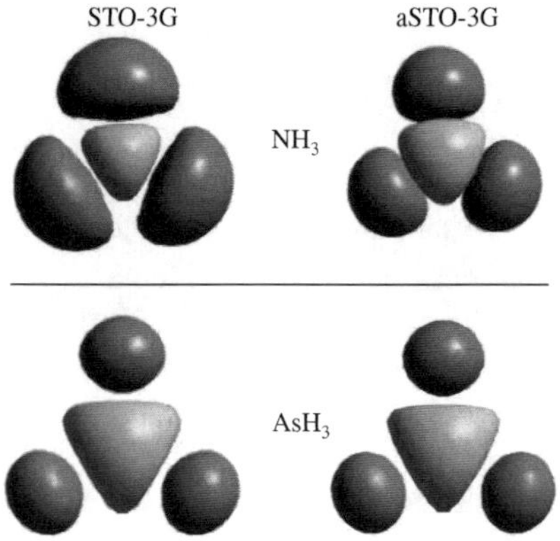

Figure 6. Isodensity surfaces of the unoccupied σ^*_A valence orbital of NH_3 and AsH_3 that controls hybridization, at its planar D_{3h} structure, evaluated using the Hartree-Fock method. Reproduced with permission from Reimers, McKemmish, McKenzie and Hush 2015 Phys. *Chem. Chem. Phys.* **17** 24618 with permission of the PCCP Owner Societies.

4. Usefulness of having a general model for chemical processes

Being able to describe wide-ranging chemical processes using a consistent model enables global questions to be addressed concerning the types of chemistry appropriate for certain needs, and the accuracy of computational approaches used to model chemical processes. We consider general issues associated with failure of the Born-Oppenheimer approximation, and the question as to what type of chemical reaction would be best suited for application in a chemical quantum qubit.

4.1. Fundamental nature of Born-Oppenheimer breakdown

The Born-Oppenheimer approximation may be applied to any of the previous Hamiltonians by parametrically diagonalizing them as a function of the nuclear coordinate. For the two-state systems $\mathbf{H}^{\mathbf{2L}}$ and $\mathbf{H}^{\mathbf{2D}}$, we add an energy asymmetry E_0 to allow for exothermic or endothermic reactions as sketched in figure 1, revising them to

$$\mathbf{H}^{\mathbf{2L}}=\begin{bmatrix} T+\frac{\hbar\omega}{2}(Q+Q_\mathrm{m})^2+J-\frac{\lambda}{4} & J \\ J & T+\frac{\hbar\omega}{2}(Q-Q_\mathrm{m})^2+J-\frac{\lambda}{4}+E_0 \end{bmatrix} \text{ and} \tag{21}$$

$$\mathbf{H}^{\mathbf{2D}}=\begin{bmatrix} T+\frac{\hbar\omega}{2}Q^2+\frac{E_0}{2} & \alpha Q-\frac{E_0}{2} \\ \alpha Q-\frac{E_0}{2} & T+\frac{\hbar\omega}{2}Q^2+2J+\frac{E_0}{2} \end{bmatrix}, \tag{22}$$

where now dimensionless normal coordinates are used for simplicity ($Q \to Q/(m\hbar\omega)^{1/2}$, $k \to \hbar\omega$, $T \to -\hbar\omega/2\times\partial^2/\partial Q^2$) [102]. Diagonalization of either of these equations leads to Born-Oppenheimer [115-117] adiabatic ground-state and excited-state potential-energy surfaces $\varepsilon_\pm(Q)$, as shown in figure 1, where [7]

$$\varepsilon_\pm(Q)=\frac{E_0}{2}+\frac{\lambda}{4}+\frac{\hbar\omega}{2}Q^2\pm\left[\left(\frac{E_0}{2}-\hbar\omega Q_\mathrm{m}Q\right)^2+J^2\right]^{1/2}. \tag{23}$$

Using them, the original Hamiltonians may be *equivalently* represented in the Born-Oppenheimer electronic-state basis as

$$\mathbf{H}^{\mathbf{BO}}=\begin{bmatrix} T+\varepsilon_-(Q)+\Delta H^{\mathrm{DC}}(Q) & \Delta P^{\mathrm{FD}}(Q)\frac{\partial}{\partial Q}+\Delta H^{\mathrm{SD}}(Q) \\ \Delta P^{FD}(Q)\frac{\partial}{\partial Q}+\Delta H^{\mathrm{SD}}(Q) & T+\varepsilon_+(Q)+\Delta H^{\mathrm{DC}}(Q) \end{bmatrix}. \tag{24}$$

Almost all modern calculations are performed neglecting all terms except $T+\varepsilon_\pm(Q)$, known as the Born-Oppenheimer approximation:

$$\mathbf{H}^{\mathbf{BO}}\approx\begin{bmatrix} T+\varepsilon_-(Q) & 0 \\ 0 & T+\varepsilon_+(Q) \end{bmatrix}. \tag{25}$$

This is a type of *adiabatic* approximation as there are no terms linking the different electronic states and hence the problem separates into independent ones for the electronic and nuclear motions. Including the diagonal correction $\Delta H^{\mathrm{DC}}(Q)$ only also constitutes an adiabatic approximation known as the Born-Huang approximation [116-117]:

$$\mathbf{H}^{\mathrm{BO}} \approx \begin{bmatrix} T+\varepsilon_-(Q)+\Delta H^{\mathrm{DC}}(Q) & 0 \\ 0 & T+\varepsilon_+(Q)+\Delta H^{\mathrm{DC}}(Q) \end{bmatrix}, \tag{26}$$

which is often described as embodying a mass-dependent correction to the two Born-Oppenheimer molecular potential-energy surfaces

$$\varepsilon_\pm^{\mathrm{BH}}(Q)=\varepsilon_\pm(Q)+\Delta H^{\mathrm{DC}}(Q)\ . \tag{27}$$

The terms $\Delta P^{\mathrm{FD}}(Q)$ and $\Delta H^{\mathrm{SD}}(Q)$ provide first-derivative (momentum) and second-derivative (kinetic energy) nonadiabatic corrections that couple the two adiabatic basis states.

For the two-state one-harmonic-model, the three correction terms to the Born-Oppenheimer approximation evaluate as [44]

$$\begin{aligned} \Delta P^{\mathrm{FD}}(Q) &= \frac{\hbar\omega}{2}\frac{Q_c}{Q_c^2+(Q-Q_x)^2} \\ \Delta H^{\mathrm{DC}}(Q) &= \frac{1}{2\hbar\omega}\left(\Delta P^{\mathrm{FD}}(Q)\right)^2 \\ \Delta H^{\mathrm{SD}}(Q) &= \frac{2}{\hbar\omega}\frac{Q-Q_x}{Q_c}\left(\Delta P^{\mathrm{FD}}(Q)\right)^2 \end{aligned} \tag{28}$$

where Q_x is the location of the intersection of the localized diabatic surfaces,

$$Q_x=\frac{E_0}{\lambda}Q_m \tag{29}$$

and Q_c is the diameter of the cusp formed at the crossing point,

$$Q_c=\frac{2|J|}{\lambda}Q_m\ . \tag{30}$$

The critical role of the cusp diameter in determining the properties of chemical reactions was first recognized by London [57,118], forming the basis of the adiabatic theory of electron transfer developed by Hush [2,5]. As a conical intersection seam is approached, $Q_c \to 0$ and so all three Born-Oppenheimer corrections diverge to infinity.

Equation (26) presents a hierarchy of terms with the diagonal correction and second-derivative correction scaling as the square of the first-derivative one. Such scaling is in general independent of the number of nuclear degrees of freedom considered [119-122]. Based upon this hierarchy, for non-adiabatic electron-transfer reactions and other processes occurring near conical intersections, it is typical to only include the first-derivative term in calculations. Away from conical intersections say in the reactant or product valleys, the influence of the derivative corrections is greatly reduced owing to the large energy gap that develops between the two adiabatic surfaces, but such a scaling does not apply to the diagonal correction. It is therefore common practice when considering ground-state vibration frequencies and redox potentials to include only it when accurate calculations are desired.

The one-mode two-state model is useful in that it allows numerically exact results to be deduced for all observable properties simply and easily though solution of (any one of) the diabatic Hamiltonians [44]. It can also be extended to multi-mode multi-state problems including modes of different symmetry [27] (e.g., the electrochemical, spin resonance, electronic spectroscopy, vibrational spectroscopy, Stark spectroscopy, etc., of the special-pair radical cation in bacterial photosynthesis can be quantitatively modelled [47,49] in quick time using a 70-mode 4-state model [50], the Q-band spectra of the chlorophyllides can be fitted using a real-time algorithm involving 52 modes and 2 states [55], etc.). However, solution of the problem using Born-Oppenheimer basis functions is intrinsically unstable, requiring vibrational basis sets with dimensions found to scale empirically as conical intersections are approached as $1/6Q_c^2$ [44]!. A simple way to understand this result is to note that the diagonal correction is sharply spiked at the crossing point $Q = Q_X$ of the diabatic surfaces and adds an extra height of

$$\Delta\Delta E^{\ddagger} = \left(\Delta H^{DC}(Q)\right)_{\max} = \frac{\hbar\omega}{8Q_c^2} \qquad (31)$$

to the Born-Oppenheimer transition-state energy (or saddle point energy for the inverted region) $\Delta E^{\ddagger}$. This correction also grows alarmingly as conical intersections are approached, as shown in table 1 for some illustrative examples. Essentially we see that surface-hopping or other non-adiabatic calculations cannot be performed accurately unless all vibrational levels of energy less than $\Delta\Delta E^{\ddagger}$ are included in the basis set.

This general requirement for the convergence of any numerical calculation of full quantum dynamics performed using the Born-Oppenheimer representation immediately tells that the diagonal correction is globally important. Across the full parameter space of the model, we found *no* region in which including only the first-derivative correction in calculations led to results similar to the exact ones [44]. The widely applied practice of using just this term in non-adiabatic calculations is therefore invalid. On the contrary, we found that not only for processes occurring at potential-energy minima but also for processes near conical intersections, the diagonal correction *alone* may give useful results. Many processes labelled "non-adiabatic" and treated by surface-hopping mechanisms may indeed be much more accurately modelled as an adiabatic process on the Born-Huang ground-state surface, with crossing from reactants to products blocked simply by the high transition-state energy induced by $\Delta\Delta E^{\ddagger}$ [44]. For example, in the inverted region in which the coupling is small (see [44] for the general case) and $\lambda > -E_0$ so that reactions become slower as they become more exothermic [8,10,123], no adiabatic transition-state is present in the Born-Oppenheimer surface and hence chemical reactions are interpreted usually in terms of surface hopping, but the associated Born-Huang surface almost always *does* contain an adiabatic transition state whose properties are a primary controlling element. However, in general, inclusion of *all three* Born-Oppenheimer corrections is required in order to obtain accurate results for processes near conical intersections.

Herein, all results are presented for a one-dimensional model descriptive of many processes that occur on slices formed near a multi-dimensional conical intersection. As these slices approach the conical intersection seam (i.e., as $J \to 0$), the cusp diameter approaches zero, dramatically enhancing the *height* of the barrier $\Delta\Delta E^{\ddagger}$ as it narrows its *width*. That all three terms must be considered in full quantum dynamics is not an accident but a *fundamental property* of conical intersections.

Conical intersections form an example of a pitchfork bifurcation cusp catastrophe [80-83]. They thus form a subset of a well-studied area of mathematics. One of the general properties of such systems is that it is not possible to expand dynamics near the cusp using approximate hierarchical methods: dynamics becomes intrinsically chaotic, and only extensive complete solutions yield accurate results. This property is dimension independent, with the results found by comparing approximate and numerically exact quantum dynamics in the one-dimensional case providing a typical example. The one-dimensional model thus provides a critical test case for more general methods.

4.2. Choice of chemical reaction for use as a chemical quantum qubit

Chemical qubits have been suggested as possibly providing the basis for some quantum information processing technology [124], and have even been postulated as having a role in the phenomenon of consciousness [125-126]. Chemical qubits are based on the entanglement of nuclear and electronic motions involved in isomerization reactions. Armed with a model that allows all such processes to be consistently treated, it is possible to consider the whole chemical parameter space available and hence to recommend what types of reactions may prove suitable. Concerning a possible role in consciousness, we showed that the originally proposed model [125-126] involving tubulin-dimer isomerization was fundamentally flawed in that the proposed vibration was in fact an irreversible chemical reaction [127] and also that the proposed mechanism for coherence was in general untenable

IOP Conf. Series: Journal of Physics: Conf. Series **833** (2017) 012014 doi:10.1088/1742-6596/833/1/012014

[128]. As no specific replacement qubit has been suggested, we have concluded that at the moment their proposal is without any current basis for scientific discussion [129].

More generally, the issue of coherence is critical when it comes to considering chemical quantum computers: qubit operation requires very long coherence times and hence great isolation of the qubit from its environment, yet function in some quantum information device requires the limiting classical states of the qubit to be easily read and written, requiring strong interaction [130]. Entanglement is the critical element of a quantum qubit and while any amount of entanglement is in principle useful [130], realistic devices require a large amount of entanglement of order $S = 1$, where S is the von Neumann entropy of the system. This entropy ranges from zero for a classical system to unity for a completely entangled qubit. Evaluated across the parameter space of the single-mode two-state chemical model [46], the developed entanglement is shown in figure 7. Different plots show the entanglement as a function of $2J/\lambda$ and $\hbar\omega/\Delta E$ for different values of the energy asymmetry $E_0/\hbar\omega$, with the Born-Oppenheimer surfaces sketched at a characteristic point on each plot. For symmetric isomerization reactions like ammonia inversion, the entanglement is near unity whenever $2J/\lambda < 1$ (localized double-well potential) and $\hbar\omega/\Delta E < 1$ (away from the region where the dynamic Jahn-Teller effect becomes important with clear vibrational energy levels forming in each well). This result documents the initial intuition leading to the postulate that chemical qubits could be useful for quantum information processing [124].

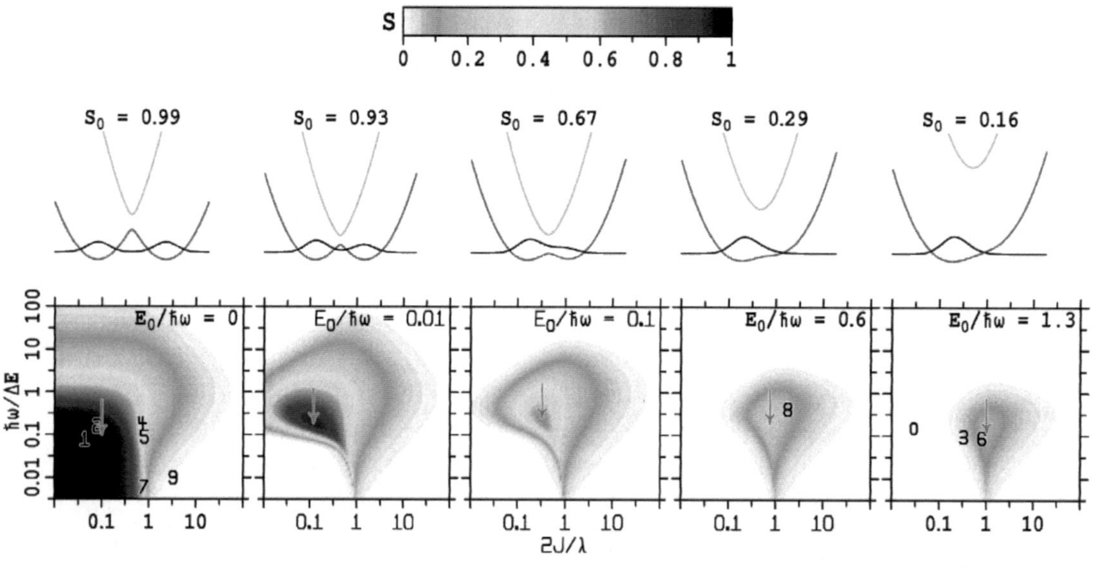

Figure 7. The entanglement (von Neumann entropy) S_0 of the ground vibronic level is shown as a function of $2J/\lambda$ and $\hbar\omega/\Delta E$ for various values of the energy asymmetry $E_0/\hbar\omega$ taken as the effect of local environment on an otherwise symmetric isomerization reaction [46]. The Born-Oppenheimer potential-energy surfaces and associated entanglement are indicated for the key situation for each $E_0/\hbar\omega$ value depicted by the red arrows. Values appropriate to the 10 sample molecular systems from figure 1 and table 1 are indicated, with molecules having large $|E_0/\hbar\omega|$ grouped together on the right-most plot.

The simplest way to model the effects of environment on coherence and hence entanglement is to treat the environment as simply generating random fluctuations in the energy asymmetry E_0. Figure 7 shows what happens to the highly desired scenario for symmetric reactions when a small amount of asymmetry modelled as $E_0 / \hbar\omega = 0.01$ is added. Entanglement is lost over much of the parameter space, remaining robust only when the coupling is strong and only a few vibrational levels fill the double-well region. Any practical qubit would need to be constructed using isomerization reactions having this very restrictive property. Increasing the asymmetry to the large but not unprecedented value of $E_0 / \hbar\omega = 0.1$, the maximum achievable entanglement becomes $S_0 = 0.67$ in a very small region of the parameter space, with higher levels of asymmetry seemingly preventing any practical device. Of the molecular systems considered, the dipyridyl polyene **1** and the Alq3 **2** sit in the region of greatest entanglement and greatest environmental immunity, whilst BNB **4** and the Creutz-Taube ion **5** sit in a region where the entanglement never gets much more than one half but it is quite stable to environmental perturbation. Whether or not the ground-state vibrational density is bimodal (i.e., zero-point vibration is supported in each well) is a critical indicator of qubit performance [46].

5. Conclusions

The success in modern times in accurate modelling of complex data involving the PJTE and related effects has not been matched by parallel increases in conceptual understanding. Individual molecular properties are always complex, and understanding this complexity is always essential for accurate modelling. However, key aspects of the chemical process can usually be described in terms of some basic phenomenon, providing fundamental understanding. We show how such simplicity of understanding can be revealed amidst the complexity of results coming from high-level calculations and detailed experimental data, highlighting the roles played by high-energy states in controlling the properties of the low-energy states of direct interest. Simplicity has rarely been found in standard applications of PJTE theory as these are usually cast in terms of the resonance energy, the force constant, and the vibronic coupling constant as fundamental variables, yet by convolving the force constant and the vibronic-coupling constant to produce the reorganization energy means that many critical properties including the presence/absence of symmetry breaking is controlled simply by the dimensionless ratio of the resonance and reorganization energies. Application of this transformation is always possible and will always aid PJTE studies. More significantly, however, PJTE is usually applied only between the states of direct interest, producing numerically accurate methods involving effective two-state parameters rather than global transferrable parameters involved when all states sharing the same vibronic coupling are included in the analysis. Without understanding the overall structure of many interacting conical-intersection seams, basic simplicity cannot be obtained.

Simplicity delivers models with transferrable parameters usable in a wide range of related scenarios. Shaik's concept that the ground-state is twinned to an excited state whose properties are directly linked is converted into a semi-quantitative analysis method. Conceptually, this means that the properties of the ground state can be understood through understanding the properties of its twin state and the relationships between them. This reveals that Rydbergization of the excited states of the first-row elements is the property that makes them so different to the elements in the later rows of the periodic table. It also shows how isomerization and symmetry breaking can be understood in general terms for all chemical processes, allowing a consistent platform for teaching chemistry to be established that is fully compatible with teaching in condensed-matter physics. Other general questions can also be addressed like what chemical reactions are likely to be useful as the basis of quantum qubits, and, through exact quantum solution of simple model problems, what is required in high-level calculations taking on modern challenges in understanding non-adiabatic chemistry. Gaining an intrinsic understanding of chemistry and how it functions then leads back to the development of better methods for rigorous modelling.

References

[1] Hush N S 1952 *J Chem Phys* **20** 1660
[2] Hush N S 1958 *J Chem Phys* **28** 962
[3] Hush N S 1961 *Soviet electrochemistry: Proceedings of the Fourth Conference on Electrochemistry 1956*, ed Frumkin A N (New York: Consultants Bureau) p 99

[4] Hush N S 1961 *Trans Farad Soc* **57** 577
[5] Hush N S 1967 *Prog Inorg Chem* **8** 391
[6] Hush N S 1975 *Chem Phys* **10** 361
[7] Hush N S 1980 *NATO Adv Study Inst Ser, Ser C* **58** 151
[8] Kubo R and Toyozawa Y 1955 *Prog Theor Phys* **13** 160
[9] Marcus R A 1956 *J Chem Phys* **24** 966
[10] Levich V G and Dogonadze R R 1959 *Dokl Akad Nauk SSSR Ser Fiz Khim* **124** 123
[11] Jortner J 1976 *J Chem Phys* **64** 4860
[12] Warshel A 1982 *J Phys Chem* **86** 2218
[13] Holstein T 1959 *Annals of Physics* **8** 325
[14] Holstein T 1959 *Annals of Physics* **8** 343
[15] Stroppa A, Barone P, Di Sante D, Cuoco M, Picozzi S and Whangbo M-H 2016 *Int J Quantum Chem* **116** 1442
[16] Öpik U and Pryce M H L 1957 *Proc R Soc London, A* **238** 425
[17] Jahn H A and Teller E 1937 *Proc Roy Soc A* **161** 220
[18] Bersuker I B 2013 *Chem Rev* **113** 1351
[19] García-Fernández P, Aramburu J A, Moreno M, Zlatar M and Gruden-Pavlović M 2014 *J Chem Theory Comput* **10** 1824
[20] Pearson R G 1983 *J Molec Str THEOCHEM* **103** 25
[21] Herzberg G and Teller E 1933 *Z Phys Chem* **21** 410
[22] Craig D P 1950 *J Chem Soc* 59
[23] Fulton R L and Gouterman M 1961 *J Chem Phys* **35** 1059
[24] Ross I G 1975 *Isr J Chem* **14** 118
[25] Fischer G 1984 *Vibronic Coupling* London: Academic Press)
[26] Piepho S B, Krausz E R and Schatz P N 1978 *J Am Chem Soc* **100** 2996
[27] Reimers J R and Hush N S 2004 *Chem Phys* **299** 79
[28] Reimers J R, McKemmish L, McKenzie R H and Hush N S 2015 *Phys Chem Chem Phys* **17** 24598
[29] Hush N S 1953 *J Polymer Sci* **11** 289
[30] Hush N S and Pople J A 1955 *Trans Faraday Soc* **51** 600
[31] Hush N S 1957 *Z Elektrochem Angewandte Physik Chem* **61** 734
[32] Hush N S 1957 *J Chem Phys* **27** 612
[33] Hush N S and Pryce M H L 1957 *J Chem Phys* **26** 143
[34] Hush N S 1958 *Discuss Faraday Soc* **No. 26** 145
[35] Hush N S and Pryce M H L 1958 *J Chem Phys* **28** 244
[36] Hush N S 1960 *Disc Farad Soc* **29** 113
[37] Blackledge J and Hush N S 1963 *J Electroanal Chem (1959-1966)* **5** 435
[38] Hush N S and Blackledge J 1963 *J Electroanal Chem (1959-1966)* **5** 420
[39] Case B, Hush N S, Parsons R and Peover M E 1965 *J Electroanal Chem (1959-1966)* **10** 360
[40] Hush N S 1968 *Electrochim Acta* **13** 1005
[41] Hush N S and Segal G A 1968 *Discuss Faraday Soc* **No. 45** 23
[42] Creutz C and Taube H 1969 *J Am Chem Soc* **91** 3988
[43] Reimers J R, McKemmish L, McKenzie R H and Hush N S 2015 *Phys Chem Chem Phys* **17** 24618
[44] Reimers J R, McKemmish L, McKenzie R H and Hush N S 2015 *Phys Chem Chem Phys* **17** 24640
[45] McKemmish L, McKenzie R H, Hush N S and Reimers J R 2015 *Phys Chem Chem Phys* **17** 24666
[46] McKemmish L K, McKenzie R H, Hush N S and Reimers J R 2011 *J Chem Phys* **135** 244110
[47] Reimers J R and Hush N S 2004 *J Am Chem Soc* **126** 4132
[48] Reimers J R and Hush N S 1991 *J Phys Chem* **95** 9773

[49] Kanchanawong O, Dahlbom M G, Treynor T P, Reimers J R, Hush N S and Boxer S G 2006 *J Phys Chem B* **110** 18688
[50] Reimers J R and Hush N S 2003 *J Chem Phys* **119** 3262
[51] Reimers J R, Shapley W A and Hush N S 2003 *J Chem Phys* **119** 3240
[52] Reimers J R, Shapley W A, Rendell A P and Hush N S 2003 *J Chem Phys* **119** 3249
[53] Reimers J R 2015 *Aust J Chem* **68** 1202
[54] Reimers J R and Hush N S 1996 *Chem Phys* **208** 177
[55] Reimers J R, Cai Z-L, Kobayashi R, Rätsep M, Freiberg A and Krausz E 2013 *Sci Rep* **3** 2761
[56] London F 1928 *Z Phys* **46** 455
[57] London F 1932 *Z Phys* **74** 143
[58] von Neumann J and Wigner E 1929 *Phys Z* **30** 467; translated in 2000 Quantum Chemistry 8 25
[59] Eyring H and Polanyi M 1931 *Z Phys Chem Abt B* **12** 279
[60] Evans M G and Polanyi M 1938 *Trans Faraday Soc* **34** 11
[61] Horiuti J and Polanyi M 2003 *J Molec Catalysis A* **199** 185
[62] Wall F T and Glockler G 1937 *J Chem Phys* **5** 314
[63] Bader R F W 1962 *Can J Chem* **40** 1164
[64] Salem L and Wright J S 1969 *J Amer Chem Soc* **91** 5947
[65] Pearson R G 1969 *J Amer Chem Soc* **91** 4947
[66] Pearson R G 1971 *Accounts Chem Res* **4** 152
[67] Levin C C 1975 *J Am Chem Soc* **97** 5649
[68] Pauling L 1960 *The Nature of the Chemical Bond*: Oxford University Press)
[69] Zilberg S, Haas Y, Danovich D and Shaik S 1998 *Angew Chem, Int Ed* **37** 1394
[70] Shaik S, Zilberg S and Haas Y 1996 *Acc Chem Res* **29** 211
[71] Shaik S, Shurki A, Danovich D and Hiberty P C 1996 *J Am Chem Soc* **118** 666
[72] Shaik S, Shurki A, Danovich D and Hiberty P C 2001 *Chem Rev (Washington, D C)* **101** 1501
[73] Zilberg S and Haas Y 2011 *J Phys Chem A* **115** 10650
[74] Toro-Labbe A 1999 *J Phys Chem A* **103** 4398
[75] Toro-Labbe A, Gutierrez-Oliva S, Murray J S and Politzer P 2007 *Mol Phys* **105** 2619
[76] Toro-Labbe A, Gutierrez-Oliva S, Murray J S and Politzer P 2009 *J Mol Model* **15** 707
[77] Hammond G S 1955 *J Am Chem Soc* **77** 334
[78] Leffler J E 1953 *Science* **117** 340
[79] Cai Z-L, Sendt K and Reimers J R 2002 *J Chem Phys* **117** 5543
[80] Saunders P T 1980 *An introduction to catastrophe theory* Cambridge: Cambridge University Press)
[81] Xu F 1990 *Z Phys Chem* **166** 79
[82] Krokidis X, Silvi B, Dezarnaud-Dandine C and Sevin A 1998 *New J Chem* **22** 1341
[83] Wales D J 2001 *Science* **293** 2067
[84] Piepho S B 1990 *J Am Chem Soc* **112** 4197
[85] Rabi I I 1936 *Physical Review* **49** 324
[86] Rabi I I 1937 *Physical Review* **51** 652
[87] Lee S-H, Larsen A G, Ohkubo K, Cai Z-L, Reimers J R, Fukuzumi S and Crossley M J 2012 *Chem Sci* **3** 257
[88] Curiel D, Ohkubo K, Reimers J R, Fukuzumi S and Crossley M J 2007 *Phys Chem Chem Phys* **9** 5260
[89] Lee S-H, Blake I M, Larsen A G, McDonald J A, Ohkubo K, Fukuzumi S, Reimers J R and Crossley M J 2016 *Chem Sci* **7** 6534
[90] Woitellier S, Launay J P and Spangler C W 1989 *Inorg Chem* **28** 758
[91] Reimers J R and Hush N S 1990 *Inorg Chem* **29** 3686
[92] Brinkmann M, Gadret G, Muccini M, Taliani C, Masciocchi N and Sironi A 2000 *J Am Chem Soc* **122** 5147
[93] Yin S, Li L, Yang Y and Reimers J R 2012 *J Phys Chem C* **116** 14826

[94] Cai Z-L and Reimers J R 2000 *J Phys Chem A* **104** 8389
[95] Stanton J F 2010 *J Chem Phys* **133** 174309
[96] Reimers J R, Wallace B B and Hush N S 2008 *Phil Trans Roy Soc A* **366** 15
[97] Hush N S 1982 *Mechanistic Aspects of Inorganic Reactions*, ed Rorabacher D B and Endicott. J F (Washington DC: American Chemical Society) p 301
[98] Reimers J R and Hush N S 1991 *Mixed Valence Systems: Applications in Chemistry, Physics, and Biology*, ed Prassides K (Dordrecht: Kluwer Acad. Publishers) p 29
[99] Reimers J R, Hughes J M and Hush N S 2000 *Biochemistry* **39** 16185
[100] Cave R J and Newton M D 1996 *Chem Phys Lett* **249** 15
[101] Berry R 1961 *J Chem Phys* **35** 2253
[102] Wilson E B, Decius J C and Cross P C 1955 *Molecular Vibrations: The Theory of Infrared and Raman Vibrational Spectra* New York: McGraw-Hill Book Company)
[103] Gillespie R J and Nyholm R S 1957 *Quart Rev Chem Soc* **11** 339
[104] Gillespie R J and Robinson E A 1996 *Angew Chem Internat Ed* **35** 495
[105] Gillespie R J 2008 *Coord Chem Rev* **252** 1315
[106] Bersuker I B, Gorinchoi N N and Polinger V Z 1984 *Theor Chim Acta* **66** 161
[107] Stanton J F and Bartlett R J 1993 *J Chem Phys* **98** 7029
[108] Korona T and Werner H-J 2003 *J Chem Phys* **118** 3006
[109] Kendall R A, Dunning T H, Jr. and Harrison R J 1992 *J Chem Phys* **96** 6796
[110] T.H. Dunning J, Peterson K A and Wilson A K 2001 *J Chem Phys* **114** 2944
[111] Peterson K A, Figgen D, Goll E, Stoll H and Dolg M 2003 *J Chem Phys* **119** 11113
[112] Knowles P J, Lindh R, Manby F R, Schütz M et al. 2010 *MOLPRO, version 2010.1, a package of ab initio programs* Birmingham: University of Birmingham)
[113] Mulliken R S 1976 *Acc Chem Res* **9** 7
[114] Mulliken R S 1977 *Chem Phys Lett* **46** 197
[115] Born M and Oppenheimer R 1927 *Ann Phys* **84** 457
[116] Born M and Huang K 1954 *Dynamical Theory of Crystal Lattices* Oxford: Clarendon)
[117] Ballhausen C J and Hansen A E 1972 *Annu Rev Phys Chem* **23** 15
[118] London F 2000 *Quantum Chemistry* **8** 32
[119] Thompson T C, Truhlar D G and Mead C A 1985 *J Chem Phys* **82** 2392
[120] Kendrick B K, Mead C A and Truhlar D G 2002 *Chem Phys* **277** 31
[121] Zhu C, Jasper A W and Truhlar D G 2004 *J Chem Phys* **120** 5543
[122] Jasper A W and Truhlar D G 2011 *Adv Ser Phys Chem* **17** 375
[123] Levich V G and Dogonadze R R 1959 *Proc Akad Nauk SSSR* **29** 9
[124] Ferguson A J, Cain P A, Williams D A and Briggs G A D 2002 *Phys Rev A* **65** 034303
[125] Penrose R and Hameroff S R 1995 *J Conscious Stud* **2** 99
[126] Hameroff S and Penrose R 1996 *Math Comp Simul* **40** 453
[127] McKemmish L K, Reimers J R, McKenzie R H, Mark A E and Hush N S 2009 *Phys Rev E* **80** 021912
[128] Reimers J R, McKemmish L K, McKenzie R H, Mark A E and Hush N S 2009 *Proc Nat Acad Sci* **106** 4219
[129] Reimers J R, McKemmish L K, McKenzie R H, Mark A E and Hush N S 2014 *Phys Life Rev* **11** 101
[130] Bremner M J, Dawson C M, Dodd J L, Gilchrist A, Harrow A W, Mortimer D, Nielsen M A and Osborne T J 2002 *Phys Rev Lett* **89** 247902

IOP Conf. Series: Journal of Physics: Conf. Series **833** (2017) 012015 doi:10.1088/1742-6596/833/1/012015

Manifestation of Fermi resonance in optical spectra of dynamical pseudo-Jahn-Teller effect

V Loorits

Forecast Models Department, Estonian Environment Agency, Mustamäe tee 33, Tallinn, Estonia, 10616

E-mail: Vello.Loorits@envir.ee

Abstract. Shape of optical spectrum of vibronic transition in one-dimensional model taking into account dynamical pseudo-Jahn-Teller effect was studied. Numerical calculations at large values of interaction parameter revealed periodic character of the envelope. This was interpreted as manifestation of Fermi resonance.

1. Introduction

Let's consider a vibronic system with non-degenerate ground state and two nearby excited states, where excited states interact through low-symmetry vibrations. In harmonic and linear approximations in respect to interaction the vibrational movement of such system in ground state and excited state is described by the following Hamiltonians, respectively:

$$H_{\mathrm{g}} = E_{\mathrm{g}} + H_{\mathrm{L}}$$
$$H_{\mathrm{e}} = E_{\mathrm{e}} + H_{\mathrm{L}} + H_{\mathrm{eL}} \tag{1}$$

where E_g and E_e are average energies of electrons in these states, and

$$H_{\mathrm{L}} = \sum_s \frac{\hbar\omega_s}{2}\left(-\frac{\partial^2}{\partial\xi_s^2} + \xi_s^2\right),$$
$$H_{\mathrm{eL}} = \begin{pmatrix} + & - \\ \Gamma & V \\ V & -\Gamma \end{pmatrix}, \tag{2}$$
$$V = \sum_s \hbar\omega_s k_s \xi_s.$$

Shape of the optical spectrum corresponding to vibronic transitions from ground state to excited electronic states is determined by the function

$$I(E) = \langle g|M^+|a\rangle F_{ab}\left(E - E_e + E_g\right)\langle b|M|g\rangle \tag{3}$$

where

$$F(E) = \frac{1}{2\pi}\int_{-\infty}^{\infty} \mathrm{d}\tau\, e^{\mathrm{i}E\tau}\left\langle e^{\mathrm{i}(H_{\mathrm{L}}+H_{\mathrm{eL}})\tau} e^{\mathrm{i}H_{\mathrm{L}}\tau}\right\rangle. \tag{4}$$

Here M is operator of interaction of optical electrons with light, and brackets < > denote statistical averaging over vibrational states of H_{L}.

Due to averaging odd powers of ξ vanish, so in our diabatic representation [1] the density distribution function *F*(*E*) is diagonal:

$$F(E)=\begin{pmatrix} F_+(E) & 0 \\ 0 & F_-(E) \end{pmatrix}. \tag{5}$$

Consequently,

$$I(E)=\left|\langle +|M|g\rangle\right|^2 F_+(E)+\left|\langle -|M|g\rangle\right|^2 F_-(E). \tag{6}$$

Calculating *F*(*E*) according to the equation (4) we meet difficulties well-known in theory of dynamical Jahn-Teller effect which do not permit to find explicit expression. Therefore we have to use various approximations. However, moments of $F_\pm(E)$ are calculated exactly. First 7 of them are:

$$\begin{aligned}
&\mu_0=1,\\
&\mu_1=\pm\Gamma,\\
&\mu_2=S_2+\Gamma^2,\\
&\mu_3=S_3\pm\Gamma S_2\pm\Gamma^3,\\
&\mu_4=S_4+3S_2^2+2\Gamma^2S_2+\Gamma^4,\\
&\mu_5=S_5+10S_3S_2\pm\Gamma\left(3S_2^2-S_4\right)+2\Gamma^2S_3\pm2\Gamma^3S_2\pm\Gamma^5,\\
&\mu_6=S_6+15S_4S_2+10S_3^2+15S_2^3\pm2\Gamma\left(4S_3S_2-S_5\right)+\\
&+3\Gamma^2\left(3S_2^2+S_4\right)S_3+3\Gamma^4S_2+\Gamma^6.
\end{aligned} \tag{7}$$

Here

$$\begin{aligned}
&S_n=\int_{-\infty}^{\infty}\mathrm{d}E\;E^n\varphi(E),\\
&\varphi(E)=\sum_s\frac{k_s^2}{2}\left[(\bar{n}_s+1)\delta(E-\hbar\omega_s)+\bar{n}_s\delta(E+\hbar\omega_s)\right],\\
&\bar{n}_s=\mathrm{e}^{-\hbar\omega_s/k_\mathrm{B}T}\Big/\sum_{s'}\mathrm{e}^{-\hbar\omega_{s'}/k_\mathrm{B}T}.
\end{aligned} \tag{8}$$

Unfortunately it is practically impossible to restore the form of *F*(*E*) from its first moments. This becomes evident when we calculate *F*(*E*) in semi-classical approximation that holds for strong vibronic interaction:

$$\begin{aligned}
F^0(E)&=\frac{1}{2\pi}\int_{-\infty}^{\infty}\mathrm{d}\tau\;\mathrm{e}^{\mathrm{i}E\tau}\left\langle e^{-\mathrm{i}H_{\mathrm{el}}\tau}\right\rangle=\frac{1}{\pi}\mathrm{Im}\left\langle\frac{1}{E-\mathrm{i}0-V}\right\rangle=\\
&=\frac{1}{\sqrt{2\pi S_2}}\mathrm{e}^{-\frac{E^2-\Gamma^2}{2S_2}}\begin{pmatrix}\sqrt{\left|\frac{E+\Gamma}{E-\Gamma}\right|} & 0\\ 0 & \sqrt{\left|\frac{E+\Gamma}{E-\Gamma}\right|}\end{pmatrix}.
\end{aligned} \tag{9}$$

Equation (9) implies that both polarizations of the semi-classical spectrum have a gap in the range $E\in[\Gamma,-\Gamma]$ and a singular point of type $(E\mp\Gamma)^{-1/2}$ at one of its boundaries.

We also note that at $\Gamma=0$ the problem is equivalent to the one of singlet-singlet transition: equation (4) turns into well-known Lax equation:

$$F_0(E)=\frac{1}{2\pi}\int_{-\infty}^{\infty}\mathrm{d}\tau\ \mathrm{e}^{\mathrm{i}E\tau}\mathrm{e}^{\tilde{\varphi}(r)-\tilde{\varphi}(0)}, \tag{10}$$

where

$$\tilde{\varphi}(\tau)=\int_{-\infty}^{\infty}\mathrm{d}E\ e^{-iE\tau}\varphi(E), \tag{11}$$

and the semi-classical spectrum is given by Gaussian:

$$F_0^0(E)=\frac{1}{\sqrt{2\pi S_2}}\mathrm{e}^{-\frac{E^2}{2S_2}} \tag{12}$$

2. Interaction with one vibration

Now we return to quantum-mechanical calculations. Let's find $F(E)$ in case of interaction with only one vibration. In this case the calculation is much simpler and quantum effects are more clearly manifested. The problem is to find solutions for the following vibrational equation for excited electronic states:

$$\left(H_{\mathrm{L}}+H_{\mathrm{eL}}\right)\begin{pmatrix}|\nu+\rangle\\|\nu-\rangle\end{pmatrix}=E_\nu\begin{pmatrix}|\nu+\rangle\\|\nu-\rangle\end{pmatrix}. \tag{13}$$

Here

$$\begin{aligned}H_{\mathrm{L}}&=\frac{\hbar\omega}{2}\left(-\frac{\mathrm{d}^2}{\mathrm{d}\xi^2}+\xi^2\right),\\ H_E&=\hbar\omega\begin{pmatrix}G & k\xi\\ k\xi & -G\end{pmatrix}.\end{aligned} \tag{14}$$

In the basis of eigenstates of vibrational Hamiltonian H_L the equation (13) splits into two equations with tridiagonal matrices:

$$\begin{aligned}&\begin{pmatrix}G & \sqrt{D} & 0 & 0 & \dots\\ \sqrt{D} & 1-G & \sqrt{2D} & 0 & \dots\\ 0 & \sqrt{2D} & 2+G & \sqrt{3D} & \dots\\ & & \sqrt{3D} & 3-G & \dots\\ \dots & \dots & \dots & \dots & \dots\end{pmatrix}\begin{pmatrix}c_{0\nu}^+\\ c_{1\nu}^-\\ c_{2\nu}^+\\ c_{3\nu}^-\\ \dots\end{pmatrix}=\nu\begin{pmatrix}c_{0\nu}^+\\ c_{1\nu}^-\\ c_{2\nu}^+\\ c_{3\nu}^-\\ \dots\end{pmatrix},\\ &\begin{pmatrix}-G & \sqrt{D} & 0 & 0 & \dots\\ \sqrt{D} & 1+G & \sqrt{2D} & 0 & \dots\\ 0 & \sqrt{2D} & 2-G & \sqrt{3D} & \dots\\ & & \sqrt{3D} & 3+G & \dots\\ \dots & \dots & \dots & \dots & \dots\end{pmatrix}\begin{pmatrix}c_{0\nu}^-\\ c_{1\nu}^+\\ c_{2\nu}^-\\ c_{3\nu}^+\\ \dots\end{pmatrix}=\nu\begin{pmatrix}c_{0\nu}^-\\ c_{1\nu}^+\\ c_{2\nu}^-\\ c_{3\nu}^+\\ \dots\end{pmatrix},\end{aligned} \tag{15}$$

Here $G=\Gamma/\hbar\omega$ and $D=k^2/2=S_2/(\hbar\omega)^2$ are dimensionless parameters of splitting and vibronic interaction, and ν is quantum number of eigenstate of $H_{\mathrm{L}}+H_{\mathrm{eL}}$:

$$E_\nu=\hbar\omega\left(\nu+\frac{1}{2}\right). \tag{16}$$

Depending on whether v is the eigenvalue of the first or of the second equation, the eigenvector is expressed as

$$\begin{pmatrix} |\nu+\rangle \\ |\nu-\rangle \end{pmatrix} = \begin{pmatrix} c_{0\nu}^{+}|0\rangle + c_{2\nu}^{+}|2\rangle + c_{4\nu}^{+}|4\rangle + \dots \\ c_{1\nu}^{-}|1\rangle + c_{3\nu}^{-}|3\rangle + c_{5\nu}^{-}|5\rangle + \dots \end{pmatrix}, \tag{17}$$

or

$$\begin{pmatrix} |\nu+\rangle \\ |\nu-\rangle \end{pmatrix} = \begin{pmatrix} c_{1\nu}^{+}|1\rangle + c_{3\nu}^{+}|3\rangle + c_{5\nu}^{+}|5\rangle + \dots \\ c_{0\nu}^{-}|0\rangle + c_{2\nu}^{-}|2\rangle + c_{4\nu}^{-}|4\rangle + \dots \end{pmatrix}, \tag{18}$$

where $|0\rangle, |1\rangle, |2\rangle, \dots$ are eigenvectors of the Hamiltonian H_L:

$$\begin{aligned} &H_L|n\rangle = E_n|n\rangle \\ &E_n = \hbar\omega\left(n + \frac{1}{2}\right), \\ &n = 0, 1, 2, \dots \end{aligned} \tag{19}$$

From (15) it is evident that stationary states of the Hamiltonian $H_L + H_{eL}$ are divided into two classes. Consequently, there exists an operator with two eigenvalues, which commutes with the Hamiltonian. In the coordinate representation, its effect can be determined by the following expression:

$$P\begin{pmatrix} \psi_+(\xi) \\ \psi_-(\xi) \end{pmatrix} = \begin{pmatrix} \psi_+(-\xi) \\ -\psi_-(-\xi) \end{pmatrix}. \tag{20}$$

The eigenvalues of this operator are 1 and -1, and the eigenvectors are vectors, in which one of the components is even and the other odd function of ξ:

$$\begin{aligned} &P\begin{pmatrix} g(\xi) \\ u(\xi) \end{pmatrix} = \begin{pmatrix} g(\xi) \\ u(\xi) \end{pmatrix}, \\ &P\begin{pmatrix} u(\xi) \\ g(\xi) \end{pmatrix} = -\begin{pmatrix} u(\xi) \\ g(\xi) \end{pmatrix}, \end{aligned} \tag{21}$$

where $g(-\xi) = g(\xi)$ and $u(-\xi) = -u(\xi)$.

Due to tridiagonality of energy matrices in (15) the amount of computations and necessary storage are not excessive, allowing computations at large values of D and G, i.e. in the range of transition to semi-classical spectrum. Further economy of storage and computation time appears at T=0, then transition occurs only from the vibrational ground state. In this case

$$F_{\pm}(E) = \sum_{\nu} |\langle \nu \pm |0\rangle|^2 \delta(E - E_\nu + E_0), \tag{22}$$

and we may compute only one component of eigenvectors $c_{0\nu}^{\pm} = \langle \nu \pm |0\rangle$.

As a result, the required computer resources remain within reasonable limits while taking into account several thousands of vibrational states of the Hamiltonian H_L, which in turn allows working with the values of D and G up to several hundred.

Numerical calculations were carried out on a computer M 4030 using the QL-method with shift [2]. The number of elementary rotations appeared to be slightly below N^2, where N is the order of the matrix, and time to perform one rotation - about 2 ms. Therefore, in an hour we can calculate the spectrum, taking into account around thousand vibrational states of H_L.

3. Computed spectra

In the literature already exist results of numerical solution of the equation (13) for small D and G [1]. So we stopped at the large values of D and G, examining the way of reaching the semi-classical limit. It is well known that the envelope of the line spectrum corresponding to the vibronic transition to non-degenerate electronic state (the case of $\Gamma=0$ in our model), even when taking into account only one vibration, already at $D = 5$ is well described by the semi-classical expression (Gaussian curve). But as can be seen from figures 1 and 2, in this model, despite the high value of D, the calculated spectrum is clearly different from the semi-classical spectrum. In the region $E < -\Gamma$ semi-classical approximation works quite well. However, in the region $E > \Gamma$ clear fluctuations of the intensities of the absorption lines are noticeable. Nature of mentioned fluctuations may be well understood in terms of adiabatic potentials, if we take into account mixing of vibrational states of close energies that belong to different electronic states, i.e. Fermi resonance. Indeed, in adiabatic representation that diagonalizes H_{eL}:

$$H'_{eL} = OH'_{eL}O^T = \hbar\omega\sqrt{G^2 + k^2\xi^2}\begin{pmatrix} 1 & 0 \\ 0 & 1 \end{pmatrix}, \tag{23}$$

where

$$O = \begin{pmatrix} \cos\varphi & \sin\varphi \\ -\sin\varphi & \cos\varphi \end{pmatrix}, \quad \varphi = \frac{1}{2}\arctan\frac{k\xi}{G}. \tag{24}$$

The Hamiltonian H_L consists of two parts - kinetic and potential:

$$H_{\mathrm{L}} = T + U = \frac{\hbar\omega}{2}\left\{-\frac{\mathrm{d}^2}{\mathrm{d}\xi^2}\right\} + \frac{\hbar\omega}{2}\xi^2. \tag{25}$$

The latter does not change with the transition to the adiabatic representation, and the first takes the form

$$T' = OTO^T = \frac{\hbar\omega}{2}\left\{-\frac{\mathrm{d}^2}{\mathrm{d}\xi^2} + \left(\frac{\mathrm{d}}{\mathrm{d}\xi}\frac{\mathrm{d}\varphi}{\mathrm{d}\xi} + \frac{\mathrm{d}\varphi}{\mathrm{d}\xi}\frac{\mathrm{d}}{\mathrm{d}\xi}\right)\begin{pmatrix} 0 & 1 \\ -1 & 0 \end{pmatrix} + \left(\frac{\mathrm{d}\varphi}{\mathrm{d}\xi}\right)^2\right\}, \tag{26}$$

where $\dfrac{\mathrm{d}\varphi}{\mathrm{d}\xi} = \dfrac{kG}{2\left(G^2 + k^2\xi^2\right)}$.

To summarize, in adiabatic representation the vibrational Hamiltonian of excited electronic states can be represented as the sum of three parts - the kinetic, potential, and mixing:

$$H'_{\mathrm{L}} + H'_{\mathrm{eL}} = T' + U' + H'_{\mathrm{eL}} = T + W + K. \tag{27}$$

Here

$$\begin{gathered} T = \hbar\omega\left\{-\frac{1}{2}\frac{\mathrm{d}^2}{\mathrm{d}\xi^2}\right\}, \\ \mathrm{W} = \hbar\omega\left\{\frac{1}{2}\xi^2 + \sqrt{\left(G^2 + k^2\xi^2\right)}\begin{pmatrix} 1 & 0 \\ 0 & -1 \end{pmatrix} + \frac{k^2G^2}{8\left(G^2 + k^2\xi^2\right)^2}\right\}, \\ K = \hbar\omega\left\{\left[\frac{\mathrm{d}}{\mathrm{d}\xi}\frac{kG}{2\left(G^2 + k^2\xi^2\right)} + \frac{kG}{2\left(G^2 + k^2\xi^2\right)}\frac{\mathrm{d}}{\mathrm{d}\xi}\right]\begin{pmatrix} 1 & 0 \\ 0 & -1 \end{pmatrix}\right\}. \end{gathered} \tag{28}$$

Usually when calculating the adiabatic potentials the third member of W is neglected. Also kinetic and mixing parts of the Hamiltonian are omitted. All this corresponds to semi-classical approximation.

IOP Conf. Series: Journal of Physics: Conf. Series **833** (2017) 012015 doi:10.1088/1742-6596/833/1/012015

We will, however, proceed differently: we consider $T + W$ as the unperturbed Hamiltonian, and K as a perturbation. Note that both of these operators, as well as the transformation matrix O commute with the operator P. Consequently,

$$P' = OPO^T = P, \tag{29}$$

i.e. similarly to diabatic representation in the adiabatic representation stationary states are also described by vectors with components of different parity. As the Hamiltonian $T + W$ is diagonal and K contributes only in the second order of perturbation theory, in the adiabatic representation one of the two components predominates over the other. The second component is significant only when energies of stationary states of upper and lower potentials are close.

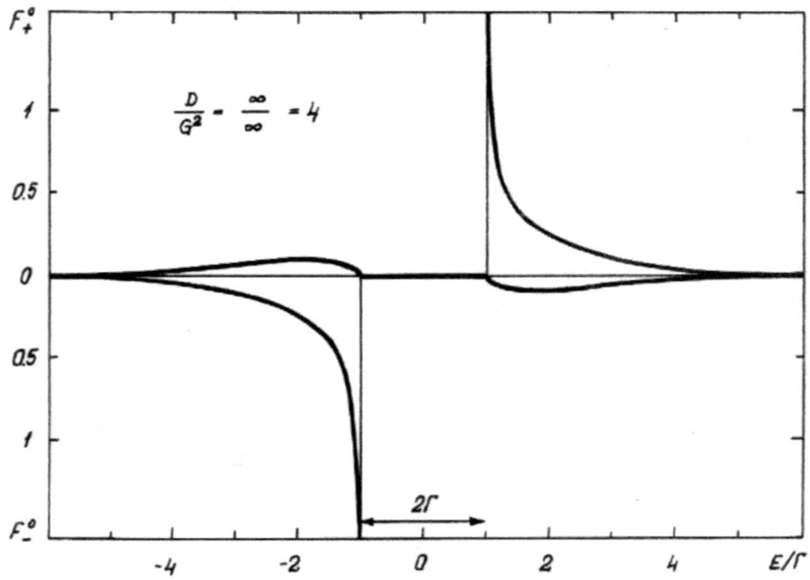

Figure 1. Shapes of functions $F_+^0(E)$ and $F_-^0(E)$ of the absorption spectrum of the two polarizations. $S_2/\Gamma^2 = 4$, semiclassical approximation.

From the shapes of the potentials it is clear that most of the spectrum is located in two areas, respectively, adjacent to $E = -\Gamma$ from the left, and $E = \Gamma$ from the right. In the first of them, the transition occurs to high levels of the lower potential. Therefore, the spectrum there is well described by the semiclassical approximation. The only exception is a neighborhood of the upper barrier point $E \approx -\Gamma$, where it is essential to take into account the kinetic energy operator T.

In the second region $E > \Gamma$ along with levels of lower potential exist lower levels of the upper potential, into which the probability of transition from the ground vibrational state of H_L is high. The distance between these levels is several times greater than the distance between the energy levels of the lower potential that is approximately equal to $\hbar\omega/2$ (see the upper right side of figure 3). If we take into account that the transition can only occur into vibrational states that are even in respect to ξ, strong absorption lines should appear in $F_+(E)$ at frequencies close to even stationary states of the upper potential, and in $F_-(E)$ respectively, close to odd ones. Due to the mixing caused by the operator K these lines must be accompanied by satellites, corresponding to transitions into nearby states of opposite parity of the lower potential. According to Fermi resonance the intensity of these lines should increase with diminishing distance between the respective levels and the levels of the upper potential. All this is indeed observed in our calculated spectrum.

In order to obtain quantitative results in the adiabatic representation one should first solve the Schrodinger equation with the Hamiltonian $T + W$ for high vibrational levels of the lower potential, which can be practically realized only by the WKB method. Such calculations we did not perform, because in the given model they are more complex than the calculations described above and are only of auxiliary importance. Nevertheless, using the simplest version of this method, we calculated positions of the energy levels of the unperturbed Hamiltonian $T + W$ in the area of the upper potential. Results of these calculations are shown in figure 3.

4. Optical spectra of dimers

A problem similar to that considered in this paper arises in the theory of optical spectra of dimers, but there one must also take into account the interaction with another, totally symmetric vibration. This vibration interacts equally with both excited electronic states and does not mix them. Therefore, its effect is reduced to the convolution of the density distribution function $F_{\pm}(E)$ with another density distribution function $F_0(E)$ calculated at the same D, but at G=0 (equation (10)). In our model (one totally symmetric vibration, T=0) $F_0(E)$ is given by the Poisson distribution:

$$F_0(E) = \sum_{n=0}^{\infty} \frac{D^n}{n!} \delta(E - n\hbar\nu). \tag{30}$$

Calculation and analysis of such spectra at moderate D and G was performed in the works [1]. It is easy to obtain the corresponding spectra also at the values of D and G we used, however, even without calculations it is clear that the resulting spectra are strongly smoothed. Taking into account many vibration frequencies and the finite temperature changes the spectra towards the same direction. Therefore, the conditions for observing manifestations of Fermi resonance in real spectra of dimers are not particularly favorable.

This work was first published in Russian as [3].

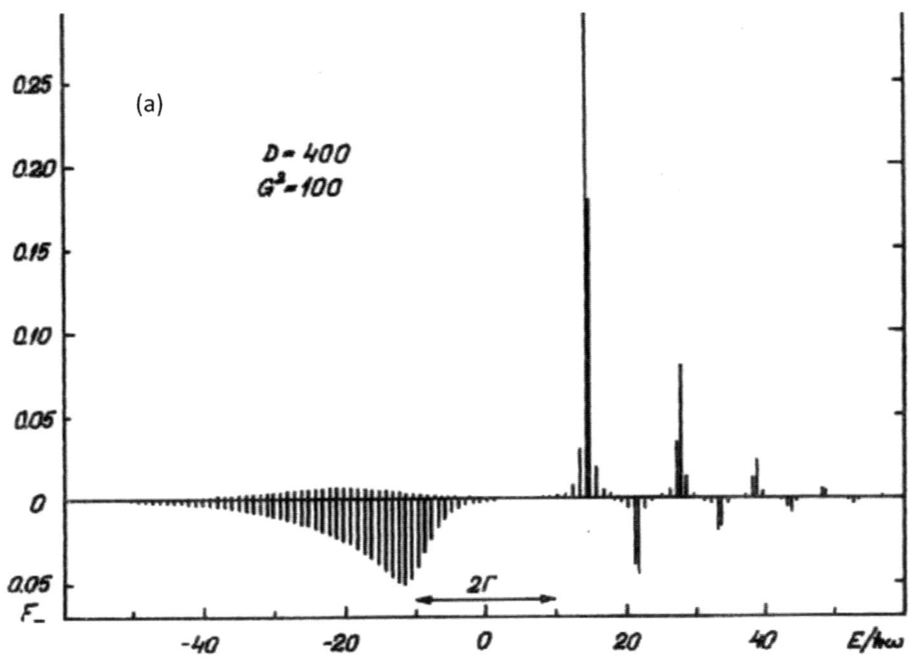

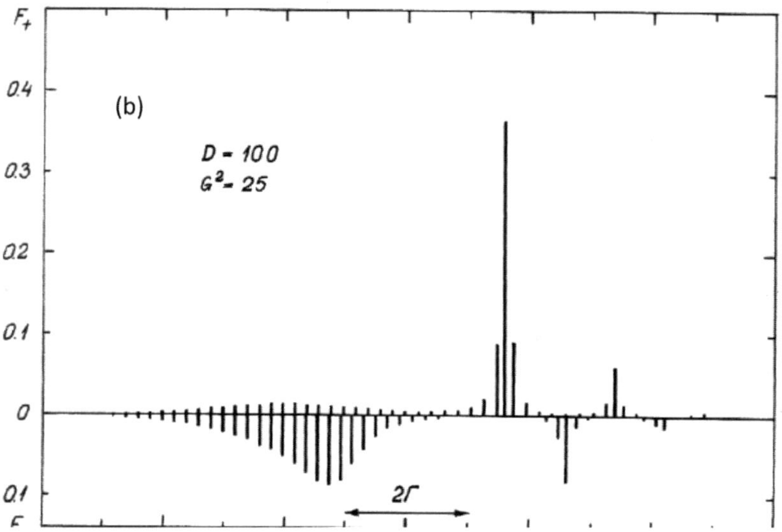

Figure 2. Shapes of functions $F_+^0(E)$ and $F_-^0(E)$ of the absorption spectrum of the two polarizations. $S_2/\Gamma^2 = 4$, quantum calculation. $S_2 = 400(\hbar\omega)^2$, $\Gamma = 10\hbar\omega$ (a); $S_2 = 100(\hbar\omega)^2$, $\Gamma = 5\hbar\omega$ (b).

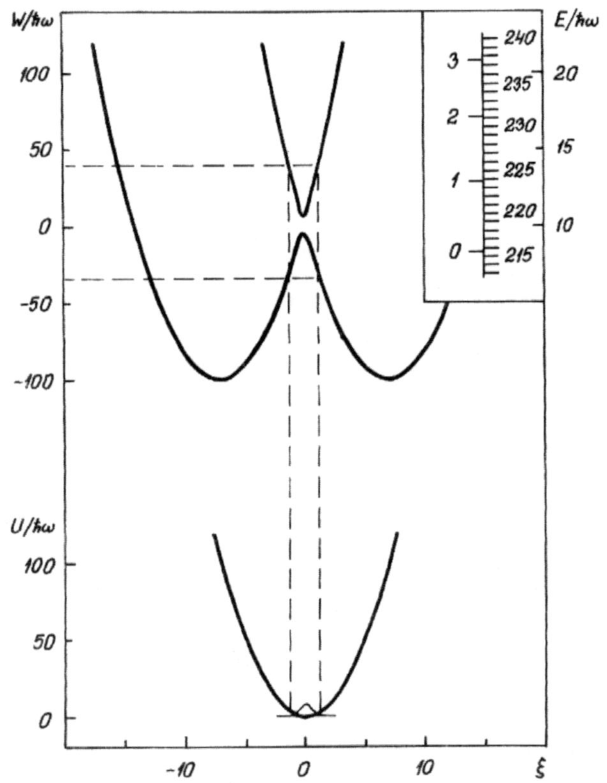

Figure 3. Adiabatic potentials in the ground (U) and excited (W) electronic states. The upper right corner shows the positions of energy levels of potentials of the excited electronic state, calculated by the WKB method.

$S_2/\Gamma^2 = 4$, $S_2 = 100(\hbar\omega)^2$, $\Gamma = 5\hbar\omega$.

IOP Conf. Series: Journal of Physics: Conf. Series **833** (2017) 012015 doi:10.1088/1742-6596/833/1/012015

Acknowledgments
The author is grateful to V.V. Hizhnyakov for valuable discussions.

References
[1] Fulton R L and Gouterman M 1961 *J. Chem. Phy.* **35** 1059; Fulton R L and Gouterman M 1964 *J. Chem. Phys.* **41** 2280; Gouterman M 1965 *J. Chem. Phys.* **42** 352
[2] Wilkinson J H and Reinsch C 1971 *Handbook for Automatic Computation: Volume II: Linear Algebra.* (Berlin, Heidelberg, New York: Springer-Verlag) p 227
[3] Лоорятс В А 1979 Проявление резонанса Ферми в оптических спектрах динамического псевдо-эффекта Яна-Теллера, *Институт физики АН ЭССР, Академия Наук Эстонской ССР, Отделение физико-математических и технических наук, Препринт А-11* (in Russian)

IOP Conf. Series: Journal of Physics: Conf. Series **833** (2017) 012016 doi:10.1088/1742-6596/833/1/012016

Origin of the $U(1)$ field mass in superconductors

Hiroyasu Koizumi

Center for Computational Sciences, University of Tsukuba, Tsukuba Ibaraki 305-8577, Japan

E-mail: koizumi.hiroyasu.fn@u.tsukuba.ac.jp

Abstract. Recently, a new theory for superconductivity has been put forward, in which the persistent current generation is attributed to the emergent singularities of the electronic wave function that are created by the spin-twisting itinerant circular motion of electrons. The persistent current generated by this mechanism behaves in every respect like supercurrent in superconductors, yielding the flux quantum $h/2e$ and the Josephson frequency $2eV/h$, where h is Planck's constant, $-e$ is the electron charge, and V is the voltage across the Josephson junction. The mass generation of the $U(1)$ gauge field (or the Meissner effect) in the new theory is due to the emergence of topological objects, 'instantons' generated by the single-valued requirement of the wave function in the presence of the emergent singularities.

The current standard theory of superconductivity is based on the BCS theory, and explains the emergence of superconductivity as due to the global $U(1)$ gauge symmetry breaking realized by the Cooper pair formation. The $U(1)$ field mass generation is believed to be due to this global $U(1)$ gauge symmetry breaking. However, the feasibility of this mechanism has been questioned since no known interaction can prepare the global $U(1)$ symmetry broken state from the normal state. We argue here that the $U(1)$ mass generation in the BCS superconductor can be attributed to the one by the instanton mentioned above if the Rashba spin-orbit interaction is added. Then, the occurrence of persistent current generation becomes due to the instanton formation, and the role of the Cooper pair formation is to stabilize the instanton by providing an energy gap for perturbative excitations. Upon forming the Cooper pair, the instanton is stabilized and persistent current generation becomes possible. Thus, the superconducting transition temperature coincides with the Cooper pair formation temperature.

1. Introduction

The superconducting state is currently regarded as the global $U(1)$ gauge symmetry broken one brought about by the Cooper pair formation [1]. This point is most clearly seen in the Ginzburg-Landau theory [2], where the superconducting order parameter is given by

$$\langle \psi_\uparrow \psi_\downarrow \rangle \tag{1}$$

where ψ_σ is the field operator for electrons with spin σ, and $\langle \hat{O} \rangle$ means the expectation value of the operator $\hat{O}$ in the superconducting state.

The non-zeroness of the above order parameter violates the global $U(1)$ gauge invariance since it is not invariant with respect to the global $U(1)$ gauge change

$$\psi_\sigma \to e^{ic} \psi_\sigma, \tag{2}$$

where c is a constant.

IOP Conf. Series: Journal of Physics: Conf. Series **833** (2017) 012016 doi:10.1088/1742-6596/833/1/012016

In elementary particle theory, a relativistic extension of the Ginzburg-Landau model, the *abelian Higgs model*, is often used to explain the mass generation in the gauge field [3]. The appearance of the mass here corresponds to the occurrence of the Meissner effect in superconductors.

The current understanding of superconductivity is based on the Cooper pair formation in the BCS theory [4], and the $U(1)$ field mass generation (thus, the occurrence of the Meissner effect) is attributed to the global $U(1)$ gauge symmetry breaking [1]. In this understanding, the Cooper pair formation is the main cause of the occurrence of superconductivity. However, the experimental facts for the cuprate superconductivity indicate that the Cooper pair formation may not be the main cause of the occurrence of superconductivity [5, 6, 7, 8, 9]. Also the validity of the assumed global $U(1)$ gauge symmetry breaking in the current standard theory has been questioned; we list three reasons against the global $U(1)$ gauge symmetry breaking mechanism below.

- The symmetry given in Eq. (2) is actually a redundancy in the quantum theory. Namely, ψ_σ and $e^{ic}\psi_\sigma$ describes the same physics, and the phase change is not a symmetry transformation [10].
- No known interaction can mix states with different charged states, thus, it is impossible to prepare a state in which the phase of the operator in Eq. (1) is physically meaningful [11, 12, 13]. In other words, the assumed superconducting state cannot be reached from a high temperature non-superconducting state by time propagation of the quantum state (which contains the superconductor and its environment) by the Hamiltonian with known interactions.
- If a superconducting state is reachable from a fixed electron number normal state by reducing the temperature through the emission of radiation without contact with anything, the reached superconducting state has the fixed number of electrons, thus, Eq. (1) is zero.

If we insist that the theory of superconductivity must preserve the global $U(1)$ invariance, we need to consider a different $U(1)$ field mass generation mechanism for superconductivity. One candidate is the one put forward by 't Hooft for QCD (quantum chromodynamics) that utilizes the topological objects, 'instantons' [14]. Actually, similar topological objects have been already considered in the $E \otimes e$ dynamical Jahn-Teller problem [15] (we will call them also 'instantons'), and they are known to generate persistent motion even in the ground state. Thus, the $U(1)$ mass generation in superconductivity may be actually due to the appearance of 'instanton', and the persistent current generation is also explained by it.

Let us give a short description for the persistent current generation in the dynamical Jahn-Teller problem. For the $E \otimes e$ Jahn-Teller system, the exact solution for the whole system is obtained; low energy states correspond to those with a half-integral pseudorotational angular momentum. The same problem can be solved using the Born-Oppenheimer approximation. In this approximation, the high-energy electronic problem is solved first and the effective Hamiltonian for the low energy vibrational motion is obtained. The potential energy surface for the vibrational motion has a degenerate point and the real-valued electronic wave functions show sign-change when they are transported around the degenerate point. Since we know that the total wave function is a single-valued function of the coordinates, the vibrational wave function has to have the compensating sign-change. This problem can be handled by introducing a gauge potential, a pure gauge potential with a singularity at the degenerate point [16]. By calculating the energy levels with including the gauge potential in the low energy effective Hamiltonian, the low energy spectrum that agrees with the exact one is obtained. This example indicates that there arise situations where a gauge potential (or a topological term) must be added to the low energy effective Hamiltonian. We call this pure gauge with singularities 'instanton' in the following.

IOP Conf. Series: Journal of Physics: Conf. Series **833** (2017) 012016 doi:10.1088/1742-6596/833/1/012016

The above problem has received a renewed attention when the Berry phase was formulated by Berry [17]. The Berry connection for the $E \otimes e$ is actually the 'instanton' mentioned above. It is noteworthy that an analogous effect has been investigated in the context of topological insulators and persistent current generation in them is one of the hotest topics of materials science[18].

Actually, a new theory for superconductivity that uses this 'instanton' has been already put forward. In this theory, the doubly degenerate E electronic state of the Jahn-Teller problem is replaced by the electron spin degree of freedom. The sign change of wave functions found in the real electronic functions in $E \otimes e$ problem corresponds to the sign change of wave functions for electrons moving circularly with twisting their spin directions. The center of the circular motion is a singularity of the wave function, and the single-valued requirement of the wave function around it gives rise to a pure gauge potential with singularities. The persistent current generated by this gauge potential behaves in every respect like supercurrent in superconductors, yielding the flux quantum $h/2e$ and the Josephson frequency $2eV/h$, where h is Planck's constant, $-e$ is the electron charge, and V is the voltage across the Josephson junction [19, 8].

A notable point about the new superconductivity theory is that it is in accordance with the experimental fact that to observe the ac Josephson effect a dc current feeding is necessary [5, 20] (see Fig. 1). Josephson's prediction is the appearance of ac current by applying a dc voltage across the superconductor-insulator-superconductor (or Josephson) junction [21]; however, a simple application of a dc voltage is impossible since the voltage across the junction becomes zero due to the dc Josephson effect. A clear-cut observation of the ac Josephson effect in reality is the observation of plateaus in the I-V plot, where a dc current is fed and a microwave radiation is applied (see, https://www.youtube.com/watch?v=Z9vBPQrG0U4&nohtml5=False, for such an experiment). In some experiments, radiation that contained a frequency component of the Josephson frequency was detected under the condition that a dc current was fed [22, 23, 24]. In any case, the dc current feeding to the junction is an essential ingredient for the observation of the effect (Fig. 1 (b)); however, this dc current feeding is lacking in the situation considered in the Josephson's derivation and textbook derivations (Fig. 1 (a)).

Recent re-derivations of the ac Josephson effect under the real experimental situation by maintaining the gauge invariance indicate there are two physically different contributions to the Josephson frequency (see Eq. (46) in [5] and Eq. (26) in [20]); one is from the chemical potential difference between the two leads connected to the junction which arises due to the fact that the electric current enters the junction from one end and exists to the other end, and these two ends are contacted with the leads that have different chemical potentials; the other contribution comes from the electric field existing in the region between two superconductors of the junction. These two contributions are the same in the magnitude since the voltage across the junction caused by the electric field balances the chemical potential difference of the leads in the stationary condition. It is important to note that these two contributions are physically different since the chemical potential arises from the many body effect with including the Pauli exclusion principle, while the contribution from the electric field arises from the electromagnetic interaction between the electron and the electric field. Actually, what is included in the Josephson's prediction is the contribution from the electric field, only. He expressed it with the chemical potential difference; here, the fact is implicitly used that the magnitude of the the contribution from the chemical potential difference and that from the electric field are the same. If the dc current feeding boundary condition is adopted for his $2e$ carrier tunneling formalism, the Josephson frequency becomes $4eV/h$, which disagrees with the experiment. This indicates that the actual "ac Josephson effect" is not really due to the Cooper pair tunneling. On the other hand, a satisfactory explanation is given by the new theory [5, 20].

In this paper, we take up the $U(1)$ field mass generation problem in superconductors. We argue that it is attributed to the emergence of 'instanton', which we denote as $\mathbf{A}^{\rm fic} = \frac{c\hbar}{2q}\nabla\chi$

IOP Conf. Series: Journal of Physics: Conf. Series **833** (2017) 012016 doi:10.1088/1742-6596/833/1/012016

Figure 1. Schematic set-up for the Josephson effect experiment. (a) Situation considered by Josephson and found in textbooks [25, 26]. Only current flow between the two superconductors S_L and S_R is taken into account. (b) Actual experimental situation. A dc current flow through the leads connected to the junction exists. Experiments indicate the dc current is an essential ingredient for the observation of the Josephson effects. If the dc current flow exists the contribution described by the dotted arrows arise in addition to the contribution given by the solid arrow.

with $q = -e$. Note that the present paper is not the report of new results; most of the results presented here are already published. Our intention is to provide a seed for further discussion in fundamental aspects of superconductivity, and also to notify the importance of the results and ideas originally come from the Jahn-Teller problem are relevant to the understanding on the phenomenon of superconductivity.

2. The appearance of effective gauge potential: the abelian Higgs model and a model with an $E \otimes e$ Jahn-Teller like coupling from spin-vortex formation

First, we summarize the well-known Higgs mechanism for the $U(1)$ (or abelian) gauge field model [3]. The Lagrangian for this model is given by

$$\mathcal{L} = -\frac{1}{4}F_{\mu\nu}F^{\mu\nu} + [(\partial_\mu - iqA_\mu)\varphi]^\dagger[(\partial_\mu - iqA_\mu)\varphi] + \mu^2\varphi^\dagger\varphi - \lambda(\varphi^\dagger\varphi)^2, \tag{3}$$

where q is the charge on the particle for φ.

Using the polar coordinates $\varphi = \rho e^{i\theta}$, the Lagrangian becomes

$$\mathcal{L} = -\frac{1}{4}F_{\mu\nu}F^{\mu\nu} + \rho^2(\partial_\mu\theta - qA_\mu)^2 + (\partial\rho)^2 + \mu^2\rho^2 - \lambda\rho^4 \tag{4}$$

The effective vector potential B_μ is given by

$$B_\mu = A_\mu - q^{-1}\partial_\mu\theta \tag{5}$$

This effective potential is gauge invariant under the gauge transformation

$$\varphi \to e^{i\alpha}\varphi; \quad qA_\mu \to qA_\mu + \partial_\mu\alpha \tag{6}$$

If we write

$$F_{\mu\nu} = \partial_\mu A_\nu - \partial_\nu A_\mu \approx \partial_\mu B_\nu - \partial_\nu B_\mu \tag{7}$$

by neglecting the contribution

$$q^{-1}(\partial_\mu\partial_\nu - \partial_\nu\partial_\mu)\theta, \tag{8}$$

IOP Conf. Series: Journal of Physics: Conf. Series **833** (2017) 012016 doi:10.1088/1742-6596/833/1/012016

we obtain the following Lagrangian for the $U(1)$ gauge field,

$$\mathcal{L}_{\text{gauge}} = -\frac{1}{4}F_{\mu\nu}F^{\mu\nu} + q^2\rho^2 B_\mu^2. \tag{9}$$

The neglect of the term in Eq. (8) means that the range of the variation of θ is limited so that it can be considered as a single-valued function. Actually, θ is an angular variable with period 2π; thus, the term in Eq. (8) can give rise to a delta function type contribution. However, we neglect it in the following.

The potential energy term in the Lagrangian is given by

$$\mathcal{E}_{\text{pot}} = -\mu^2\rho^2 + \lambda\rho^4. \tag{10}$$

This gives rise to a Mexican hat like shape for the potential energy function with a minimum energy trough at

$$\rho_{\text{min}} = \sqrt{\frac{\mu^2}{2\lambda}} \tag{11}$$

A symmetry breaking is realized by the non-zero expectation value for φ

$$|\langle\varphi\rangle| = \rho_{\text{min}} \tag{12}$$

Then, the gauge field Lagrangian is approximated as

$$\mathcal{L}_{\text{gauge}} \approx -\frac{1}{4}F_{\mu\nu}F^{\mu\nu} + \frac{1}{2}M^2 B_\mu^2; \quad M^2 = 2q^2\rho_{\text{min}}^2, \tag{13}$$

where M is considered as the mass of the gauge field.

From the above Lagrangian, the "Maxwell equations"

$$\partial_\mu F^{\mu\nu} = -j^\nu \tag{14}$$

are obtained, where the electric current j^ν is given by

$$j_\nu = -M^2 B_\nu = -M^2(A_\nu - q^{-1}\partial_\nu\theta) \tag{15}$$

This can be identified as the London equation, leading to the Meissner effect. The above electric current is gauge invariant because B_ν is gauge invariant thanks to the angular variable θ.

In superconductivity φ is taken as

$$\varphi = \psi_\uparrow\psi_\downarrow \tag{16}$$

and q is set to $q = -2e$, where $-e$ is the electron charge. The non-zero expectation value for φ corresponds to Eq. (1).

The above mentioned Higgs mechanism for the $U(1)$ field mass generation in superconductors is one of the essential ingredients of the current standard theory of superconductivity. However, as is explained in Introduction, there are reasons to seek for a different mechanism. In the following, we consider the mass generation by introducing an $E \otimes e$ like coupling term arising from spin vortex formation [8, 19].

First, we identify φ to a two-component electron field given by

$$\varphi = \begin{pmatrix} \psi_\uparrow \\ \psi_\downarrow \end{pmatrix} \tag{17}$$

Next add the following $E \otimes e$ like coupling term between the spin vortex (S_x, S_y, S_z) and the electron spin $\frac{1}{2}\varphi^\dagger \boldsymbol{\sigma}\varphi$,

$$JS_x\varphi^\dagger\sigma_x\varphi + JS_y\varphi^\dagger\sigma_y\varphi = (\psi_\uparrow^\dagger, \psi_\downarrow^\dagger)\begin{pmatrix} 0 & ke^{-i\xi} \\ ke^{i\xi} & 0 \end{pmatrix}\begin{pmatrix} \psi_\uparrow \\ \psi_\downarrow \end{pmatrix} \tag{18}$$

where $JS_x = k\cos\xi$ and $JS_y = k\sin\xi$. Here, we consider the case where the spin-vortex is lying in the xy plane, $(S_x, S_y, 0)$, and the system is uniform in the z direction. k and ξ are functions of x and y, where ξ describes the rotation angle of the spin in the xy plane. Actually, $(S_x, S_y, 0)$ also arises from the spin of electrons, thus, it is obtained by a self-consistent calculation.

The spin-vortex is characterized by the winding number for ξ. Let us define the winding number for ξ along loop C,

$$w_C[\xi] = \frac{1}{2\pi}\oint_C \nabla\xi \cdot d\mathbf{r}. \tag{19}$$

When it is not zero, spin-vortices exist. Centers of spin-vortices are singularities where ξ is not defined. Actually, those points are zeros of the field operator. In order to obtain a new field operator that has zeros at the centers of spin-vortices, we perform the following unitary transformation on φ

$$U\varphi = U\begin{pmatrix} \psi_\uparrow \\ \psi_\downarrow \end{pmatrix}; \quad U = \frac{1}{\sqrt{2}}\begin{pmatrix} e^{i\frac{\xi}{2}} & e^{-i\frac{\xi}{2}} \\ e^{i\frac{\xi}{2}} & -e^{-i\frac{\xi}{2}} \end{pmatrix}, \tag{20}$$

where U diagonalizes the matrix in Eq. (18) as

$$U\begin{pmatrix} 0 & ke^{-i\xi} \\ ke^{i\xi} & 0 \end{pmatrix}U^{-1} = \begin{pmatrix} -k & 0 \\ 0 & k \end{pmatrix} \tag{21}$$

The corresponding unitary transformation on the kinetic energy gives rise to a centrifugal term, which makes the centers of spin-vortices zeros of the new field operator $U\varphi$.

The matrix U is multi-valued if the winding number $w_C[\xi]$ is odd for some loop C. If we insist that the new field operator must be single-valued, this U is not allowed. To make the new field operator single-valued, we add a phase factor $e^{i\frac{\chi}{2}}$ on U, and modify the unitary matrix as

$$U' = e^{i\frac{\chi}{2}}U \tag{22}$$

Thus, the resulting new field operator φ' is given by

$$\varphi' = U'\varphi \tag{23}$$

To make the transformation matrix U' single-valued, χ must satisfy the constraint

$$w_C[\xi] + w_C[\chi] = \text{even number for any } C \tag{24}$$

The added phase in Eq. (22) modifies the gauge potential in the following manner

$$A_\mu^{\text{eff}} = A_\mu + \frac{1}{2q}\partial_\mu\chi \tag{25}$$

since a part of the unitary transformation involves the following transformation

$$e^{i\frac{\chi}{2}}(\partial_\mu - iqA_\mu)e^{-i\frac{\chi}{2}} = \partial_\mu - iq(A_\mu + \frac{1}{2q}\partial_\mu \chi). \tag{26}$$

Note that q is $q = -e$, the electron charge; thus, the phase $\frac{1}{2}\chi$ (period of χ is 2π) is associated for each electron. In contrast to this, the phase θ (period of θ is 2π) is associated for Cooper pair in B_μ in Eq. (5).

The multi-valuedness χ plays an essential role for the single-valuedness of U'. It needs to satisfy the constrain in Eq. (24). Such χ is given in the following form

$$\chi = 2\pi \sum_{a=1}^{N_{\rm center}} w_a \tan^{-1} \frac{y - y_a}{x - x_a} + f(x, y) \tag{27}$$

where $N_{\rm center}$ is the number of the centers of the spin-twisting, (x_a, y_a) are zeros of the field operator, w_a are integers that satisfy the constraint in Eq. (24), and f is single-valued function.

The gradient of function f, ∇f, is determined from the minimal energy condition under the constraint in Eq. (24). Let us consider the chase where χ is independent of time, and the electromagnetic gauge potential is given by $A_\mu = (0, \mathbf{A})$. The energy minimization under the constraint is achieved from the stationary condition for the following functional

$$F[\mathbf{A} + \frac{1}{2q}\nabla\chi] = E[\mathbf{A} + \frac{1}{2q}\nabla\chi] + \sum_{\ell}^{N_{\rm center}} \lambda_\ell \left(\oint_{C_\ell} \nabla\chi \cdot d\mathbf{r} - 2\pi w_{C_\ell} \right) \tag{28}$$

where λ_ℓ are Lagrange multipliers; $N_{\rm loop}$ is the number of independent loops [8, 19].

The stationary condition

$$\frac{\delta F}{\delta \nabla\chi} = 0 \tag{29}$$

yields

$$\mathbf{j} = -\frac{\delta E}{\delta \mathbf{A}} = -2q\frac{\delta E}{\delta\nabla\chi} = 2q\sum_\ell \lambda_\ell \frac{\delta}{\delta\nabla\chi}\oint_{C_\ell} \nabla\chi \cdot d\mathbf{r} = 2q\sum_\ell \lambda_\ell \boldsymbol{\delta}(\mathbf{x}; C_\ell), \tag{30}$$

where $\boldsymbol{\delta}(\mathbf{x}; C_\ell)$ is one of the delta functions introduced by Kleinert [27] given by

$$\boldsymbol{\delta}(\mathbf{x}; C_\ell) = \oint_{C_\ell} \delta^3(\mathbf{x} - \mathbf{y}) d\mathbf{y} \tag{31}$$

Since $\nabla\chi$ is obtained by optimizing the combination $\mathbf{A} + \frac{1}{2q}\nabla\chi$, the change of the gauge in $\mathbf{A}$ is absorbed by the change in $\nabla\chi$ (actually, the change in ∇f). Therefore, $\mathbf{A}^{\rm eff}$ is gauge invariant like B_μ.

Let us calculate the Meissner current in the wave function formalism since it is more transparent. In the wave function formalism [19, 8], the wave function for the ground state is given in the following form,

$$\Psi(\mathbf{r}^{(1)}, \cdots, \mathbf{r}^{(N)}) = \Psi_0(\mathbf{r}^{(1)}, \cdots, \mathbf{r}^{(N)}) e^{-\frac{i}{2}\sum_{\alpha=1}^{N} \chi(\mathbf{r}^{(\alpha)})} \tag{32}$$

where $\mathbf{r}^{(j)}$ is the coordinate of the jth electron and N is the number of electrons. Ψ_0 is a currentless multi-valued wave function, where the multi-valuedness arises from the spin-twisting

IOP Conf. Series: Journal of Physics: Conf. Series **833** (2017) 012016 doi:10.1088/1742-6596/833/1/012016

of the itinerant electrons. The phase factor $e^{-\frac{i}{2}\sum_{\alpha=1}^{N}\chi(\mathbf{r}^{(\alpha)})}$ arises to make the wave function single-valued by compensating the sign change brought about by the spin-twisting.

Using the above wave function, the current density is obtained as

$$\mathbf{j}(\mathbf{x}) = -\frac{q^2\rho(\mathbf{x})}{m}\left(\mathbf{A} + \frac{1}{2q}\nabla\chi\right) = -\frac{q^2\rho}{m}\mathbf{A}^{\text{eff}} \tag{33}$$

where

$$\rho(\mathbf{x}) = \int d^3r^{(2)}\cdots d^3r^{(N)}\left|\Psi_0(\mathbf{x}, \mathbf{r}^{(2)}, \cdots, \mathbf{r}^{(N)})\right|^2 \tag{34}$$

The equation (33) should be compared with Eq. (15); $A^{\text{eff}}_\mu = (0, \mathbf{A}^{\text{eff}})$ replaces the gauge invariant potential B_μ in Eq. (5). It is shown that the electromagnetic field acquired a mass due the appearance of $\nabla\chi$ with fixing the number of electrons.

3. Instanton and $U(1)$ mass generation

There are several different definitions for instanton [28], thus, first, we clarify what we mean by instanton.

Our instanton is

(i) It is a gauge potential given as a pure gauge $iq^{-1}U\partial_\mu U^{-1}$ in the asymptotic region, where U is a unitary matrix.

(ii) It is characterized by a topological integer.

(iii) It is a classical solution to equations derived from the stationary condition for some action.

From the χ given in Eq. (27), we can construct an instanton or a collection of instantons,

$$A^{\text{fic}}_\mu = \frac{1}{2q}\partial_\mu\chi = iq^{-1}e^{i\frac{1}{2}\chi}\partial_\mu e^{-i\frac{1}{2}\chi}, \tag{35}$$

where $U = e^{i\frac{1}{2}\chi}$, and it is a classical solution to equations derived from the stationary condition for Eq. (28). The equation (33) indicates that this instanton produces a mass for the electromagnetic field.

This instanton gives rise to a collection of flux tubes in the z-direction

$$F^{\text{fic}}_{xy} = \partial_x A^{\text{fic}}_y - \partial_y A^{\text{fic}}_x = \frac{\pi}{q}\sum_{a=1}^{N_{\text{inst}}} w_a\delta(x - x_a)\delta(y - y_a) \tag{36}$$

We attribute an instanton for each flux tube; thus, the above collection of flux tubes is attributed to arise from a collection of the same number ($N_{\text{inst}} = N_{\text{center}}$) of instantons.

This field strength can be generalized using the formula by Nambu developed for the dual string [29],

$$F^{\text{fic}}_{\mu\nu} = -\sum_a \frac{\pi w_a}{2q}\epsilon^{\mu\nu\lambda\rho}\int\int d\tau^a d\sigma^a \frac{\partial(y^a_\lambda, y^a_\rho)}{\partial(\tau^a, \sigma^a)}\delta^4(x - y^a) \tag{37}$$

where $\epsilon^{\mu\nu\lambda\rho}$ is the Levi-Civita antisymmetric symbol, $y^a_\mu(\tau^a, \sigma^a)$ represents the position of a point on the world sheet swept out by the ath flux tube (or string).

In a more general form, Eq. (33) is given by

$$j_\nu = -m_V^2 A^{\text{eff}}_\nu \tag{38}$$

It is interesting to note that Nambu derived the solution for $F^{\rm eff}_{\mu\nu} = \partial_\mu A^{\rm eff}_\nu - \partial_\nu A^{\rm eff}_\mu$ by assuming the string field $F^{\rm fic}_{\mu\nu}$. The equation for $A^{\rm eff}_\nu$ is obtained from the conservation of electric charge

$$\partial_\nu j^\nu = -m_V^2 \partial_\nu A^{\rm eff}_\nu = 0. \tag{39}$$

In the present case, $A^{\rm fic}_\nu$, thus also $A^{\rm eff}_\nu$ for A_ν with an arbitrary gauge, is specified from the condition in Eq. (29); further, the conservation of electric charge is automatically satisfied.

4. An $E \otimes e$ like coupling arising from the Rashba spin-orbit coupling

A connection between the new theory based on the instanton and the standard theory will be made if the Cooper pair formation stabilizes the formation of the instanton. If this is achieved, the global $U(1)$ gauge symmetry breaking can be removed from the superconductivity theory.

In order to realize the instanton formation in the BCS superconductor, actually, an additional interaction is needed. We found the Rashba spin-orbit interaction [30] can produce the required instanton [31].

Taking the direction of the internal electric field in the z direction, the Rashba interaction is expressed as

$$(\psi^\dagger_\uparrow, \psi^\dagger_\downarrow) \begin{pmatrix} 0 & k(\partial_x - i\partial_y) \\ k(\partial_x + i\partial_y) & 0 \end{pmatrix} \begin{pmatrix} \psi_\uparrow \\ \psi_\downarrow \end{pmatrix} \tag{40}$$

In a magnetic field, electrons perform cyclotron motion. Taking the center of the cyclotron motion as the origin, we introduce the polar coordinates ρ, ϕ through $x = \rho\cos\phi, y = \rho\sin\phi$. Then, the above term becomes

$$(\psi^\dagger_\uparrow, \psi^\dagger_\downarrow) \begin{pmatrix} 0 & ke^{-i\phi}\left(\partial_\rho - \frac{i}{\rho}\partial_\phi\right) \\ ke^{i\phi}\left(\partial_\rho + \frac{i}{\rho}\partial_\phi\right) & 0 \end{pmatrix} \begin{pmatrix} \psi_\uparrow \\ \psi_\downarrow \end{pmatrix} \tag{41}$$

This term plays the role of Eq. (18).

By setting $\xi = \phi$ in U of Eq. (20), the above term is transformed as

$$U \begin{pmatrix} 0 & ke^{-i\phi}\left(\partial_\rho - \frac{i}{\rho}\partial_\phi\right) \\ ke^{i\phi}\left(\partial_\rho + \frac{i}{\rho}\partial_\phi\right) & 0 \end{pmatrix} U^{-1} = k \begin{pmatrix} \partial_\rho + \frac{1}{2\rho} & \frac{i}{\rho}\partial_\phi \\ -\frac{i}{\rho}\partial_\phi & -\partial_\rho - \frac{1}{2\rho} \end{pmatrix} \tag{42}$$

This indicates that $k(\partial_\rho + \frac{1}{2\rho})$ in the above plays the role of k in Eq. (2). Since the winding number of ξ along this cyclotron orbit is odd, the phase $e^{i\frac{\chi}{2}}$ must be added to U. Thus, the transformation matrix becomes U' in Eq. (22).

A model calculation for a simple cylindrical-symmetry system was performed, previously [31]. It indicates that if the applied magnetic field is sufficiently weak compared with the Rashba interaction, spin-twisting itinerant motion is stabilized; thus, instantons appear.

5. Concluding remarks

In the abelian Higgs model, it is assumed that $\langle\varphi\rangle \neq 0$ is realized in the ground state. However, if we apply it to superconductivity by taking $\varphi = \psi_\uparrow\psi_\downarrow$, the assumption will not be fulfilled since such a ground state cannot be reached from the high temperature normal state by known interactions. Using the state with $\langle\varphi\rangle \neq 0$ is a convenient tool to incorporate electron pairing correlation [32]. However, using the phase of $\langle\varphi\rangle$ as a physically meaningful variable is another assumption that violates the global $U(1)$ gauge invariance. If the global $U(1)$ gauge invariance is not violated in reality, it will yield artifacts. Actually, it leads to a wrong value for the ac

IOP Conf. Series: Journal of Physics: Conf. Series **833** (2017) 012016 doi:10.1088/1742-6596/833/1/012016

Josephson frequency. This point was not apparent since the correct Josephson frequency formula was obtained by Josephson by assuming the boundary condition that is different from the real experimental one. The re-derivations using the real experimental boundary condition indicate that there is an additional contribution which is lacking in the Josephson's derivation. If this contribution is included, the Josephson frequency with $q = -2e$ yields $4eV/h$; this indicates that the observed ac Josephson effect cannot be attributed to the Cooper pair tunneling [5, 20]. We have argued that the origin of the ac Josephson effect is the appearance of χ in Eq. (32) which gives rise to the instanton $\mathbf{A}^{\text{fic}}$.

It is now a common knowledge that the effective low energy Hamiltonian may contain a topological term (or instanton), and we have argued that the superconductivity is a phenomenon of the emergence of instanton. In this theory, the global $U(1)$ gauge symmetry breaking is removed from the theory of superconductivity. It is noteworthy that the new theory can be related to the Ginzburg-Landau theory; the low energy physics described by the phase χ and its conjugate variable, superfluid density ρ, yields the equation for the 'wave function' $\Psi = \rho^{1/2}e^{-i\chi}$, which can be identified as the Ginzburg-Landau equation [8].

References

[1] Anderson P W 1984 *Basic Notions of Condensed Matter Physics* (USA: The Benjamin/Cummings Publishing Company, Inc.)
[2] Ginzburg V L and Landau L D 1950 *Zh. Exsp. Teor. Fiz.* **20** 1064
[3] Higgs P W 1964 *Phys. Rev. Lett.* **13** 508
[4] Bardeen J, Cooper L N and Schrieffer J R 1957 *Phys. Rev.* **108** 1175
[5] Koizumi H 2011 *J. Supercond. Nov. Magn.* **24** 1997
[6] Hidekata R and Koizumi H 2011 *J. Supercond. Nov. Magn.* **24** 2253
[7] Abdullev B, Park C H and Musakhanov M M 2011 *Physica C* **471** 486
[8] Koizumi H, Okazaki A, Abou Ghantous M and Tachiki M 2014 *J. Supercond. Nov. Magn.* **27** 2435–2446
[9] Okazaki A, Wakaura H, Koizumi H, Ghantous M A and Tachiki M 2015 *J. Supercond. Nov. Magn.* **28** 3221–3233
[10] Hamermesh M 1989 *Group Theory and Its Application to Physical Problems* (USA: Dover Publications)
[11] Wick G C, Wightman A S and Wigner E P 1970 *Phys. Rev. D* **1** 3267
[12] Peierls R 1992 *Contemporary Phys.* **33** 221
[13] Leggett A J 2006 *Quantum Liquids: Bose Condensation And Cooper Pairing in Condensed-matter Systems* (UK: Oxford Univ. Press)
[14] 't Hooft G 1986 *Phys. Rep.* **142** 357–387
[15] Bersuker I and Polinger V 1989 *Vibronic Interactions in Molecules and Crystals* (Germany: Springer)
[16] Mead C A and Truhlar D 1979 *J. Chem. Phys.* **70** 2284
[17] Berry M V 1984 *Proc. Roy. Soc. London Ser. A* **391** 45
[18] Stroppa A, Barone P, Picozzi S and Whangbo M H 2016 *Int. J. Quanta. Chem.* **116** 1442
[19] Koizumi H, Hidekata R, Okazaki A and Tachiki M 2014 *J Supercond Nov Magn* **27** 121–132
[20] Koizumi H and Tachiki M 2015 *J. Supercond. Nov. Magn.* **28** 61–69
[21] Josephson B D 1962 *Phys. Lett.* **1** 251
[22] Yanson I K, Svistunov V M and Dmitrenko I M 1965 *J. Exptl. Theoret. Phys. (U.S.S.R.)* **48** 976
[23] Langenberg D N, Scalapino D J, Taylor B N and eck R E 1965 *Phys. Rev. Lett.* **15** 294–297
[24] Giavever I 1965 *Phys. Rev. Lett.* **14** 904
[25] Feynman R P, Leighton R and Sands M 1965 *Lectures on physics* vol 3 (USA: Addison-wesley)
[26] Kittel C 1986 *Introductio to Solid State Physics* 6th ed (USA: Wiley)
[27] Kleinert H 2008 *Multivalued Fields: In Condensed Matter, Electromagnetism, and Gravitation* (Singapore: World Scientific)
[28] Mariño M 2015 *Instantons and Large N: An Introduction to Non-Perturbative Methods in Quantum Field Theory* (UK: Cambridge University Press)
[29] Nambu Y 1974 *Phys. Rev. D* **10** 4261
[30] Rashba E I 1960 *Sov. Phys. Solid State* **2** 1109
[31] Koizumi H and Tachiki M 2015 *J Supercond Nov Magn* **28** 2267–2273
[32] de Gennes P G 1966 *Superconductivity of Metals and Alloys* (W. A. Benjamin, Inc.)

IOP Conf. Series: Journal of Physics: Conf. Series **833** (2017) 012017 doi:10.1088/1742-6596/833/1/012017

Cooper pairs in a two-orbital superconductor: bands filling effect on pair sizes

Grzegorz Litak[1,2], Teet Örd[3], Küllike Rägo[3], Artjom Vargunin[3,4]
[1]Lublin University of Technology, Faculty of Mechanical Engineering, Nadbystrzycka 36, PL-20618 Lublin, Poland
[2]AGH University of Science and Technology, Department of Process Control, Mickiewicza 30, PL-30-059 Krakow, Poland
[3]Institute of Physics, University of Tartu, W. Ostwaldi Str. 1, 50411 Tartu, Estonia
[4]Department of Theoretical Physics, The Royal Institute of Technology, Stockholm, SE-10691 Sweden

E-mail: g.litak@pollub.pl, teet.ord@ut.ee

Abstract. The two-orbital superconducting state is modeled by on-site intra-orbital negative-U Hubbard correlations together with inter-orbital pair-transfer interactions. The influence of bands filling on the temperature dependencies of the sizes of Cooper pairs in different orbitals is analyzed. It is found that the sizes exhibit unconventionally strong variation in the temperature scale for occupations favouring interband proximity effect.

1. Introduction

Multi-component superconductivity models, developed more than fifty years starting from the papers [1, 2, 3], have been an object of growing interest. The research activity in this direction has been especially stimulated by the acceptance of the multi-gap superconductivity in MgB_2 [4], cuprates [5] and iron-arsenic compounds [6].

Due to the interaction between electron pairing amplitudes caused by interband pair-transfer processes the properties of two-band systems may be rather different from the corresponding characteristics in one-band superconductors. In particular, such quantities as coherence lengths cannot be attributed to the definite bands and must be treated as the characteristics of whole two-component condensate [7, 8, 9, 10, 11]. The characteristic lengths of coherency as well as magnetic field penetration depth in a two-orbital superconductor described by the negative-U Hubbard model [12] were examined in [11, 13, 14] as the functions of bands filling and various orbital parameters.

As opposed to coherence lengths which are characteristics of joint condensate, the Cooper pairs remain attached with the certain bands (orbitals).

In what follows we study the behaviour of the sizes of Cooper pairs *vs* temperature for various bands fillings in the two-orbital negative-U Hubbard model of superconductivity. Note that other limits are also possible. Namely, for iron-based systems, magnesium diboride, and other novel superconducting materials transition to BEC and BCS-BEC crossover scenario have been proposed [15]. Such limits open new possibilities of phenomenon descriptions including the reformulation of the Cooper pairs role in superconductivity [16] but they are associated with stronger coupling which is not considered in the present report.

2. Two-orbital superconductivity

We start with the Hamiltonian of the two-orbital model[11, 17] of the following form:

$$\begin{aligned} H &= \sum_{\alpha}\sum_{i,j}\sum_{\sigma}\left[t_{ij}^{\alpha\alpha}+\left(\varepsilon_{\alpha}^{0}-\mu\right)\delta_{ij}\right]a_{i\alpha\sigma}^{+}a_{j\alpha\sigma} \\ &+ \frac{1}{2}\sum_{\alpha}\sum_{i}\sum_{\sigma}U^{\alpha\alpha}n_{i\alpha\sigma}n_{i\alpha-\sigma} \\ &+ \frac{1}{2}\sum_{\alpha,\alpha'}{}^{'}\sum_{i}\sum_{\sigma}U^{\alpha\alpha'}a_{i\alpha\sigma}^{+}a_{i\alpha'\sigma}a_{i\alpha-\sigma}^{+}a_{i\alpha'-\sigma}\,, \end{aligned} \tag{1}$$

where $a_{i\alpha\sigma}^{+}$ ($a_{i\alpha\sigma}$) is the electron creation (destruction) operator in the orbital $\alpha = 1, 2$ localized at the site i; σ is the spin index; $t_{ij}^{\alpha\alpha}$ is the hopping integral; ε_{α}^{0} is the orbital energy; μ is the chemical potential; $U^{\alpha\alpha} < 0$ is the intra-orbital attraction energy; $n_{i\alpha\sigma} = a_{i\alpha\sigma}^{+}a_{i\alpha\sigma}$ is the particle number operator; $U^{\alpha\alpha'}$ with $\alpha \neq \alpha'$ is the inter-orbital interaction energy.

In the reciprocal space one obtains in the mean-field approximation the Hartree-Fock-Gorkov self-consistent equations

$$\sum_{\alpha} n_{\alpha} = N^{-1}\sum_{\alpha}\sum_{\mathbf{k}}\sum_{\sigma}\left\langle a_{\alpha\mathbf{k}\sigma}^{+}a_{\alpha\mathbf{k}\sigma}\right\rangle\,, \tag{2}$$

$$\Delta_{\alpha} = N^{-1}\sum_{\alpha'}U^{\alpha\alpha'}\sum_{\mathbf{k}}\left\langle a_{\alpha'-\mathbf{k}\downarrow}a_{\alpha'\mathbf{k}\uparrow}\right\rangle \tag{3}$$

with

$$\left\langle a_{\alpha\mathbf{k}\sigma}^{+}a_{\alpha\mathbf{k}\sigma}\right\rangle = \frac{1}{2}\left[1-\frac{\tilde{\varepsilon}_{\alpha}(\mathbf{k})}{E_{\alpha}(\mathbf{k})}\tanh\frac{E_{\alpha}(\mathbf{k})}{2k_{B}T}\right]\,, \tag{4}$$

$$\left\langle a_{\alpha-\mathbf{k}\downarrow}a_{\alpha\mathbf{k}\uparrow}\right\rangle = \frac{-\Delta_{\alpha}}{2E_{\alpha}(\mathbf{k})}\tanh\frac{E_{\alpha}(\mathbf{k})}{2k_{B}T}\,. \tag{5}$$

Here n_{α} is the average number of electrons per site and Δ_{α} is the superconductivity gap in the orbital α; N is the number of lattice sites;

$$\tilde{\varepsilon}_{\alpha}(\mathbf{k}) = \varepsilon_{\alpha}(\mathbf{k}) + \frac{1}{2}U^{\alpha\alpha}n_{\alpha} - \mu\,; \tag{6}$$

$\varepsilon_{\alpha}(\mathbf{k})$ is the electron band energy associated with the orbital α and

$$E_{\alpha}(\mathbf{k}) = \sqrt{\tilde{\varepsilon}_{\alpha}^{2}(\mathbf{k}) + |\Delta_{\alpha}|^{2}}\,. \tag{7}$$

The superconducting phase transition temperature T_c is determined by the condition for the existence of non-trivial solutions of the system of equations (3) in the asymptotic limit $\Delta_{1,2} \to 0$.

We carry out the numerical calculations for two-dimensional square lattice with hopping integrals between nearest neighbours $t_{ij}^{\alpha\alpha} = t$ and electron band energies associated with s-orbitals $\varepsilon_{\alpha}(\mathbf{k}) = \varepsilon_{\alpha}^{0} - 2t\left[\cos(ak_x) + \cos(ak_y)\right]$, a is the lattice constant, and $-\frac{\pi}{a} \leq k_{x,y} \leq \frac{\pi}{a}$. The intra- and interorbital interactions are chosen as $U^{11} = -1.5t$, $U^{22} = -2.5t$, $|U^{12}| = |U^{21}| = 0.04t$, and the orbital energy $\varepsilon_{2}^{0} = 0$. In all calculations $k_B = 1$.

The calculated dependency of superconducting phase transition temperature T_c on $n = n_1 + n_2$ is depicted in figure 1 for different values of ε_{1}^{0}. As expected, Van Hove singularity present in the middle of the bands [11, 17, 18] is reflected in the case $\varepsilon_{1}^{0} = \varepsilon_{2}^{0}$ as a maximum in $T_c(n)$ for half filled system ($n = 1$) where the chemical potential μ passes this singular point. If $\varepsilon_{1}^{0} \neq \varepsilon_{2}^{0}$ the function $T_c(n)$ passes two maxima. In the latter case the temperature dependencies of superconducting gaps for the values of n corresponding to the extremum points of $T_c(n)$ are presented in the left column of figure 2. One can observe the substantial interband proximity effect if $n = 1.10$ and the moderate one if $n = 2.90$. For better clarity, the quasiparticles densities of states corresponding to the set of parameters in figure 2 are shown in figure 3.

IOP Conf. Series: Journal of Physics: Conf. Series **833** (2017) 012017 doi:10.1088/1742-6596/833/1/012017

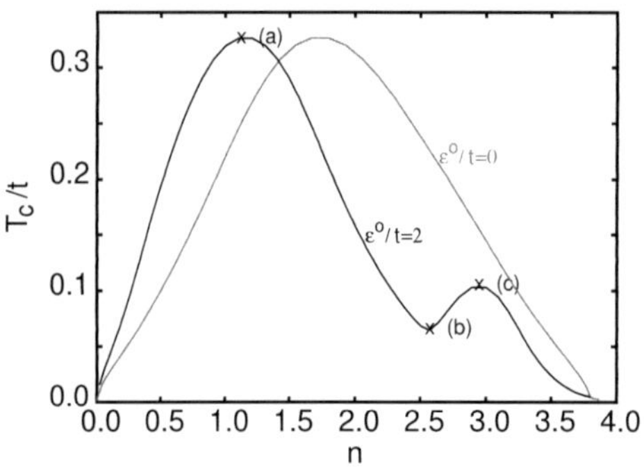

Figure 1. The superconducting phase transition temperature T_c *vs* $n = n_1 + n_2$ for two different $\varepsilon_1^0/t = 0$ and 2, respectively. (a)-(c) denotes characteristic extremum points of the function $T_c(n)$: $n = 1.10$, 2.57, 2.90.

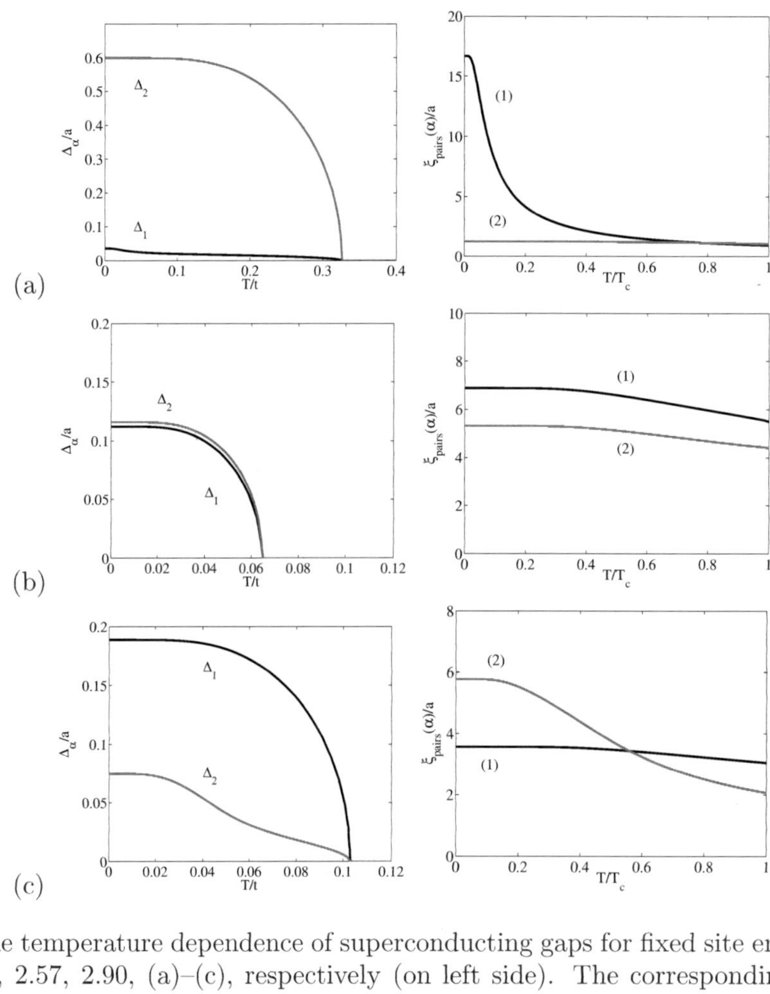

Figure 2. The temperature dependence of superconducting gaps for fixed site energies $\varepsilon_1^0/t = 2$ and $n = 1.10$, 2.57, 2.90, (a)–(c), respectively (on left side). The corresponding temperature dependencies of the Cooper pairs sizes ξ_{pairs} (on right side)

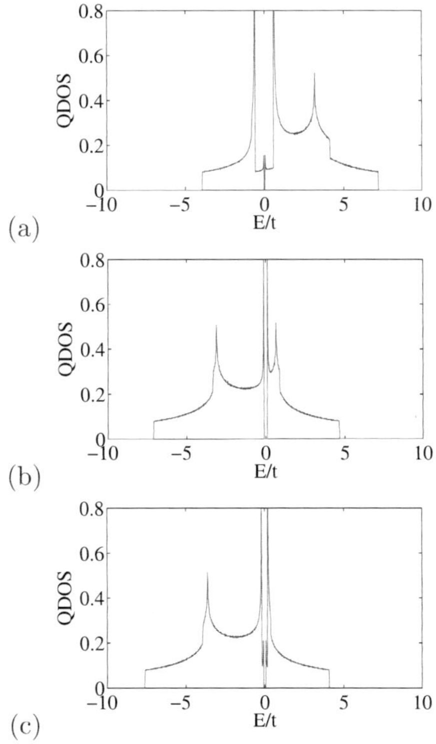

Figure 3. Quasiparticles density of states at $T = 0$ for three cases as in figures 2a-c, respectively. Note, the composite structure of the superconducting gap in figures 3a and c.

3. Sizes of Cooper pairs

The sizes of Cooper pairs ξ_{pair} are estimated from the wave functions of the Cooper pairs $\psi_\alpha(\mathbf{r})$:

$$\xi_{pair}(\alpha) = \sqrt{\frac{\int \mathrm{d}\mathbf{r}|\psi_\alpha(\mathbf{r})|^2\mathbf{r}^2}{\int \mathrm{d}\mathbf{r}|\psi_\alpha(\mathbf{r})|^2}} = \sqrt{\frac{\int \mathrm{d}\mathbf{k}|\nabla_k\psi_\alpha(\mathbf{k})|^2}{\int \mathrm{d}\mathbf{k}|\psi_\alpha(\mathbf{k})|^2}}; \quad \alpha = 1, 2 \tag{8}$$

with

$$\psi_\alpha(\mathbf{k}) = \frac{\Delta_\alpha \tanh \frac{E_\alpha(\mathbf{k})}{2k_BT}}{2E_\alpha(\mathbf{k})}. \tag{9}$$

The right side of figure 2 shows the evolution of the sizes $\xi_{pair}(\alpha)$ with temperature for the same three characteristic band fillings n. For the comparison the temperature dependencies of the superconductivity gap and Cooper pair size in the case of single-orbital negative-U Hubbard model is depicted in figure 4.

As a general observation, we notice that if superconductivity in one band is prevailing, the size of the Cooper pair in the band with weaker superconductivity exhibits strong decrease with the rise of temperature from zero to T_c. Simultaneously, the decrease of the size of the Cooper pair in the band with stronger superconductivity is small. The most prominent example of such behaviour, in the two-orbital model, is depicted in figure 2a. Here the total number of electrons per lattice site $n = 1.10$ corresponds to the left maximum of T_c in figure 1. The latter is caused by the vicinity of the chemical potential to the van Hove singularity (see figure 3a) of the band $\alpha = 2$ which strongly supports the intraband superconducting ordering there. As a result the behaviour of the size of the Cooper pair in the band $\alpha = 2$ is similar to the single-band situation

IOP Conf. Series: Journal of Physics: Conf. Series **833** (2017) 012017 doi:10.1088/1742-6596/833/1/012017

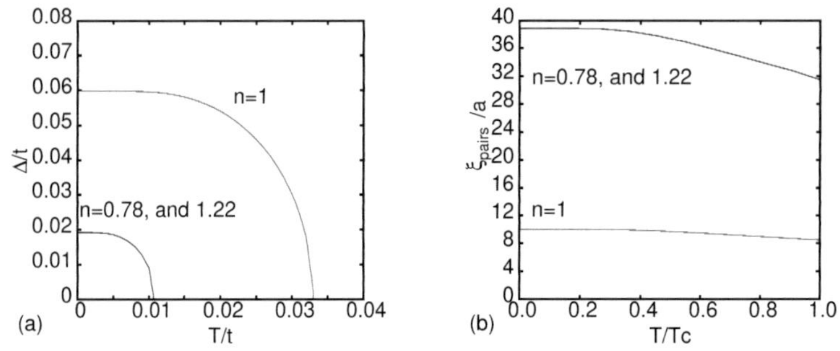

Figure 4. The temperature dependence of the Cooper pair size in the single-orbital negative-U Hubbard model with $U = -1t$ for band filling $n = 0.78$, 1.0 and 1.22.

with weak temperature dependence, c.f. figure 4. For the superconductivity in the band $\alpha = 1$, the intraband contribution is small and the superconducting ordering is mainly caused by the interband proximity effect leading to the strong temperature dependence of the Cooper pair size in this band. The number of electrons $n = 2.90$ corresponds to the right maximum of T_c in figure 1. In this case the chemical potential is close to the van Hove singularity of the band $\alpha = 1$ (see figure 3c) resulting in the stronger decrease of the Cooper pair size in the band $\alpha = 2$ as one can see in figure 2c. For the value $n = 2.57$ the chemical potential position is far enough from the van Hove singularities of the both bands (see figure 3b). In this case, depicted in figure 2b, the gaps $\Delta_{1,2}$ are close and the Cooper pair sizes depend on temperature weakly.

4. Conclusion

We have established that in a two-orbital superconductor the sizes of the Cooper pairs can decrease with the temperature substantially stronger compared to the behavior in single-orbital system. The effect takes place in the band with weaker superconductivity due to the presence of interband proximity effect.

Acknowledgements

The research was supported by the Estonian Research Council through the Institutional Research Funding IUT2-27 and through the grant PUTJD141, and by the Estonian-Polish Joint Research Project under the Agreement on Scientific Cooperation between the Estonian Academy of Sciences and Polish Academy of Sciences 2016-2018.

References

[1] Suhl H , Matthias B T and Walker L P 1959 *Phys. Rev.* **3** 552
[2] Moskalenko V A 1959 *Fiz. Met. Metalloved.* **8** 503
[3] Kondo J 1963 *Progr. Theor. Phys.* **29** 1
[4] Tsuda S, Yokoya T, Takano Y, Kito H, Matsushita A, Yin F, Itoh J, Harima H and Shin S, 2003 *Phys. Rev. Lett.* **91** 127001
[5] Khasanov K, Shengelaya A, Maisuradze A, La Mattina F, Bussmann-Holder A, Keller H and Müller K A 2007 *Phys. Rev. Lett.* **98** 057007
[6] Ding H, Richard P, Nakayama K, Sugawara K, Arakane T, Sekiba Y, Takayama A, Souma S, Sato T, Takahashi T, Wang Z, Dai X, Fang Z, Chen G F, Luo J L and Wang N L 2008 *EPL* **83** 47001
[7] Carlström J, Babaev E and Speight M 2011 *Phys. Rev. B* **83** 174509
[8] Silaev M and Babaev E 2011 *Phys. Rev. B* **84** 094515
[9] Silaev M and Babaev E 2012 *Phys. Rev. B* **85** 134514
[10] Örd T, Rägo K and Vargunin A 2012 *J. Supercond. Novel Magn.* **25** 1351

[11] Litak G, Örd T, Rägo K and Vargunin A 2012 *Acta Phys. Pol. A* **121** 747
[12] Micnas R, Ranninger J and Robaszkiewicz 1990 *Rev. Mod. Phys.* **62** 113
[13] Litak G, Örd T, Rägo K and Vargunin A 2012 *Physica C* **483** 30
[14] Litak G, Örd T, Rägo K and Vargunin A 2014 *Phys. Status Solidi B* **251** 697
[15] Guidini A, Perali A 2014 *Supercond. Sci. Technol.* **27** 124002
[16] Koizumi H 2011 *J. Supercond. Nov. Magn.* **24** 1997
[17] Ciechan A and Wysokiński K I 2009 *Phys. Rev. B* **80** 224523
[18] Markiewicz R S 1997 *J. Phys. Chem. Solids* **58** 1179

IOP Conf. Series: Journal of Physics: Conf. Series **833** (2017) 012018 doi:10.1088/1742-6596/833/1/012018

Superconductivity-induced shift of phonon frequency in the system with itinerant and strongly correlated electrons

Teet Örd, Kadri Veende, Küllike Rägo

Institute of Physics, University of Tartu, W. Ostwaldi Str. 1, 50411 Tartu, Estonia

E-mail: teet.ord@ut.ee

Abstract. We investigate a system consisting of strongly correlated localized and itinerant electron states mixed by optical vibration. The linear vibronic interaction causes the softening of active optical mode and it may induce a structural instability. The modification of renormalized phonon frequency by the gap in the energy of itinerant electrons in the superconducting phase has been established. The disposition of the electron spectrum and chemical potential influences substantially the corresponding effect. It is demonstrated that the presence of van Hove singularity in the center of itinerant electron band introduces qualitative changes into the superconductivity-induced shifts of phonon frequency.

1. Introduction

The mutual influence between superconductivity and phonon dynamics, including structural instability, has been an object of research for many years, see review [1] for earlier publications in this field. A number of effects related to the phonon dynamics, structural instabilities and symmetry breaking present in high temperature superconductors [2, 3] have been investigated intensively both experimentally and theoretically. In particular, various aspects of phonon self-energy effects caused by superconductivity were analyzed in Refs. [4, 5, 6, 7].

In the present contribution we study the changes in phonon frequency, renormalized by the vibronic mixing of itinerant and strongly correlated localized electron states, due to the electron pairing gap in the spectrum of itinerant electrons in the superconducting phase. In the absence of superconductivity, the model was examined in Ref. [8] and the special case of such a scheme was considered as a possible reason for the tetragonal-orthorhombic transition in $La_{2-x}M_xCuO_4$ [9, 10]. The vibronic hybridization of electron states has been accepted as quite general mechanism for structural (ferroelectric) phase transitions [11, 12].

2. Basic equations

We use the following model Hamiltonian for the electron-phonon system:

$$H = H_0 + H_{\mathrm{el-ph}}\,, \quad H_0 = H_{\mathrm{el}} + H_{\mathrm{ph}}\,, \quad H_{\mathrm{el}} = H_{\mathrm{l}} + H_{\mathrm{i}}\,, \tag{1}$$

$$H_{\mathrm{l}} = \sum_j \left(\varepsilon_1 \sum_s X_j^{ss} + \varepsilon_2 X_j^{22} \right) , \tag{2}$$

$$H_{\mathrm{i}} = \sum_{\mathbf{k}} \sum_{s} \tilde{\varepsilon}(\mathbf{k}) a_{\mathbf{k}s}^{+} a_{\mathbf{k}s} + \Delta_{\mathrm{sc}} \sum_{\mathbf{k}} a_{\mathbf{k}\uparrow}^{+} a_{-\mathbf{k}\downarrow}^{+} + \Delta_{\mathrm{sc}}^{*} \sum_{\mathbf{k}} a_{-\mathbf{k}\downarrow} a_{\mathbf{k}\uparrow} , \tag{3}$$

$$H_{\mathrm{ph}} = \sum_{\mathbf{q}} \hbar\omega_{\mathbf{q}} \left(b_{\mathbf{q}}^{+} b_{\mathbf{q}} + \frac{1}{2} \right) , \tag{4}$$

$$H_{\mathrm{el-ph}} = N_0^{-1} \sum_{\mathbf{k}} \sum_{j} \sum_{s} \sum_{\mathbf{q}} g(\mathbf{q}) \mathrm{e}^{i(\mathbf{k}-\mathbf{q})\mathbf{r}_j} a_{\mathbf{k}s}^{+} \left[X_j^{0s} + \eta(s) X_j^{-s2} \right] \left(b_{-\mathbf{q}}^{+} + b_{\mathbf{q}} \right) + \mathrm{h.c.} . \tag{5}$$

Here H_{el} is the Hamiltonian of electron subsystem containing localized strongly correlated electrons (Hamiltonian H_{l}) and itinerant superconducting band electrons (Hamiltonian H_{i}), H_{ph} is the Hamiltonian of phonon subsystem and $H_{\mathrm{el-ph}}$ describes the vibronic mixing of localized and itinerant electron states. In the case of localized electrons the one-electron annihilation and creation operators have been replaced by the Hubbard operators: $d_{js} = X_j^{0s} + \eta(s) X_j^{-s2}$ and $d_{js}^{+} = X_j^{s0} + \eta(s) X_j^{2-s}$ where j is the index of lattice site, s is the spin index and $\eta(\uparrow, \downarrow) = \pm 1$. In the Hamiltonian (2), $\varepsilon_1 = \varepsilon_d - \mu$ and $\varepsilon_2 = 2\varepsilon_d + U - 2\mu$ where ε_d is the energy of a localized electron without Coulomb correlation, $U/2$ is the energy of Coulomb interaction between two electrons per one spin direction and μ is the chemical potential. In the Hamiltonian (3), $\tilde{\varepsilon}(\mathbf{k}) = \varepsilon(\mathbf{k}) - \mu$, $\varepsilon(\mathbf{k})$ is the energy of an itinerant electron in the normal state and Δ_{sc} is the s-wave superconductivity gap. The Hamiltonian (3) can be diagonalized by means of the Bogoliubov-Valatin transformation $a_{\mathbf{k}s} = u_{\mathbf{k}} A_{\mathbf{k}s} + v_{\mathbf{k}}^{*} A_{-\mathbf{k}-s}^{+}$.

For the energy of normal-state itinerant electrons we use the two-dimensional spectrum $\varepsilon(k_x, k_y) = -2t[\cos(k_x a) + \cos(k_y a)]$, where a is the constant of square lattice. The bottom of the energy band $\varepsilon_{\min} = -4t$ and the top of the band $\varepsilon_{\max} = 4t$. The corresponding density of electron states $\rho(\varepsilon)$ reflects the presence of van Hove singularity in the center of the band, see figure 1.

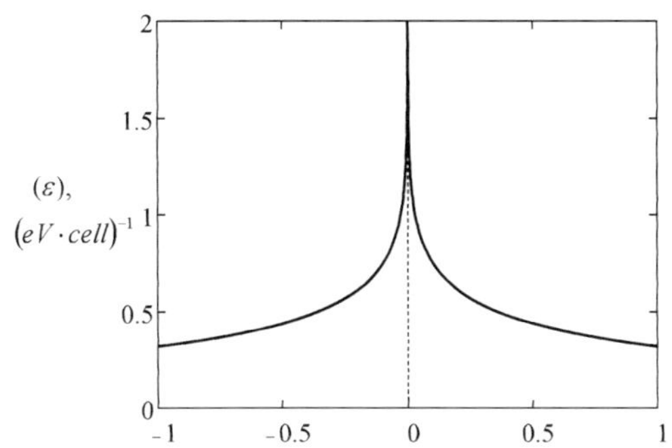

Figure 1. The density of states for the energy band of itinerant electrons with $t = 0.25\,eV$.

3. Renormalized phonon frequency

One can find the renormalized squared phonon frequency by applying the Shrieffer-Wolff transformation [13], generalized for the case of phonon-mediated dynamic hybridization between itinerant and localized electron states, to the Hamiltonian (1).

We will consider the disposition of electron states where the band of itinerant electrons is located between the lower Hubbard (LH) and upper Hubbard (UH) levels, i.e. $\varepsilon_d < \varepsilon_{\min}$ and $\varepsilon_d + U > \varepsilon_{\max}$, see figure 2. Note also that such a set of electron states is roughly similar to the effective electron spectrum in copper-oxide systems [14, 15] stemming from the Emery model [16, 17]. Chemical potential intersects the conduction band of itinerant electrons. In what follows, we neglect the contributions containing AA, A^+A^+, X^{02}, X^{20}, $A_s^+A_{s'}$ and $X^{ss'}$ if $s \neq s'$ as well as $X_jX_{j'}$ if $j \neq j'$. Then, by introducing the average numbers of elementary excitations and electrons, one obtains for the squared phonon frequency renormalized by electron-phonon interaction

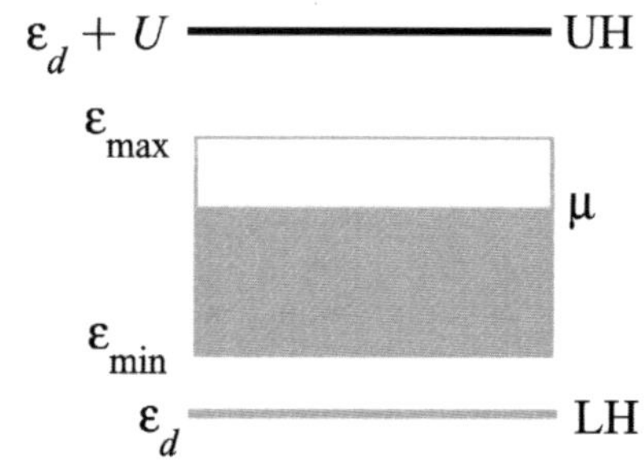

Figure 2. The configuration of electron states used.

$$\Omega_{\mathbf{q}}^2 = \omega_{\mathbf{q}}^2 + \frac{2\,|V(\mathbf{q})|^2}{N_0}$$
$$\times \sum_{\mathbf{k}} \left\{ \frac{1}{2}\left[1 + \frac{\tilde{\varepsilon}(\mathbf{k})}{E(\mathbf{k})}\right] \left[\frac{\nu(\mathbf{k})\,(2n_0 + n_1) - n_1}{E(\mathbf{k}) - \varepsilon_1} + \frac{2n_2 - \nu(\mathbf{k})\,(n_1 + 2n_2)}{\varepsilon_2 - \varepsilon_1 - E(\mathbf{k})}\right]\right.$$
$$\left. + \frac{1}{2}\left[1 - \frac{\tilde{\varepsilon}(\mathbf{k})}{E(\mathbf{k})}\right] \left[\frac{\nu(\mathbf{k})\,(2n_0 + n_1) - 2n_0}{E(\mathbf{k}) + \varepsilon_1} + \frac{\nu(\mathbf{k})\,(n_1 + 2n_2) - n_1}{\varepsilon_2 - \varepsilon_1 + E(\mathbf{k})}\right]\right\}, \tag{6}$$

where

$$V(\mathbf{q}) = g(\mathbf{q})\sqrt{2\omega_{\mathbf{q}}/\hbar}, \quad E(\mathbf{k}) = \sqrt{\tilde{\varepsilon}^2(\mathbf{k}) + |\Delta_{\rm sc}|^2},$$
$$\nu(\mathbf{k}) = \left\langle A_{\mathbf{k}s}^+ A_{\mathbf{k}s} \right\rangle_{H_0}, \quad n_0 = N_0^{-1} \sum_j \left\langle X_j^{00} \right\rangle_{H_0},$$
$$n_1 = N_0^{-1} \sum_j \sum_s \left\langle X_j^{ss} \right\rangle_{H_0}, \quad n_2 = N_0^{-1} \sum_j \left\langle X_j^{22} \right\rangle_{H_0}. \tag{7}$$

It has been assumed that $\varepsilon_{\rm min} - \varepsilon_d \gg \hbar\omega_{\mathbf{q}}$ and $\varepsilon_d + U - \varepsilon_{\rm max} \gg \hbar\omega_{\mathbf{q}}$ in Eq. (6).

In the case $T = 0$ the average numbers $\nu(\mathbf{k}) = n_0 = n_2 = 0$, $n_1 = 1$ and we have for the squared frequency (6) the expression

$$\Omega_{\mathbf{q}}^2 = \omega_{\mathbf{q}}^2 - \frac{2\,|V(\mathbf{q})|^2}{N_0} \sum_{\mathbf{k}} \left\{ \frac{1}{2}\left[1 + \frac{\tilde{\varepsilon}(\mathbf{k})}{E(\mathbf{k})}\right] \frac{1}{E(\mathbf{k}) - \varepsilon_1} + \frac{1}{2}\left[1 - \frac{\tilde{\varepsilon}(\mathbf{k})}{E(\mathbf{k})}\right] \frac{1}{\varepsilon_2 - \varepsilon_1 + E(\mathbf{k})}\right\}, \tag{8}$$

Further, we will analyze the superconductivity-induced shift of squared phonon frequency at zero temperature

$$\frac{\Delta\Omega^2}{\omega^2} = \frac{\Omega^2 - \Omega_{\rm n}^2}{\omega^2}, \tag{9}$$

where $\omega \equiv \omega_{\mathbf{q}}$, $\Omega \equiv \Omega_{\mathbf{q}}$ and $\Omega_{\rm n}$ is the value of Ω if $\Delta_{\rm sc} = 0$.

4. Numerical results and discussion

4.1. Density of electron states with van Hove singularity

By introducing the integration over the energy of itinerant electrons we can represent the squared phonon frequency (8) in the form

$$\frac{\Omega^2}{\omega^2} = 1 - \chi \int_{-4t}^{4t} \mathcal{G}(\varepsilon) \left\{ \frac{1}{2}\left[1 + \frac{\tilde{\varepsilon}(\varepsilon)}{E(\varepsilon)}\right] \frac{1}{E(\varepsilon) - \varepsilon_1} + \frac{1}{2}\left[1 - \frac{\tilde{\varepsilon}(\varepsilon)}{E(\varepsilon)}\right] \frac{1}{\varepsilon_2 - \varepsilon_1 + E(\varepsilon)} \right\} \mathrm{d}\varepsilon\,, \tag{10}$$

where

$$\chi = \frac{2|V|^2}{\omega^2 \pi^2 t}\,, \quad \mathcal{G}(\varepsilon) = h_+^{-1}(\varepsilon) K\left(\frac{h_-(\varepsilon)}{h_+(\varepsilon)}\right)\,, \quad h_\pm(\varepsilon) = 1 \pm \left|\frac{\varepsilon}{4t}\right|\,, \tag{11}$$

and $K(m)$ is the complete elliptic integral of the first kind.

The superconductivity-induced shifts $\Delta\Omega^2/\omega^2$ as the functions of chemical potential position are shown in figures 3, 4 for various dispositions of Hubbard levels. In the figures the energy of lower Hubbard leval ε_d is fixed and the energy of the higher Hubbard level $\varepsilon_d + U$ is tuned by the variation of U.

In figure 3 the value $U = 3\,eV$ corresponds to the symmetric location of the Hubbard levels with regard to conduction band. In this case the curve $\Delta\Omega^2/\omega^2$ *vs* μ is also symmetric, see thick broken line in figure 3. The increase of U in figure 3 introduces moderate asymmetry for the location of the Hubbard levels. As a result the strong dependence $\Delta\Omega^2/\omega^2$ *vs* μ appears in the center of conduction band ($\mu = 0$) related to the presence of van Hove singularity accompanied by the asymmetry of the function $\Delta\Omega^2(\mu)$ in the domains $\mu < 0$ and $\mu > 0$. One can observe in figure 3 that $\Delta\Omega^2(\mu) > 0$, i.e. superconductivity induces here the hardening of phonon dynamics which is, however, suppressed in the region $\mu > 0$ by the increase of U. If the chemical potential approaches the edges of conduction band, the effect of hardening increases.

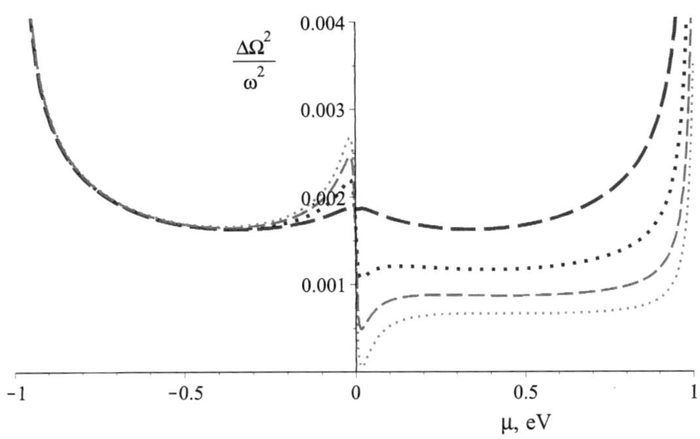

Figure 3. The dependence of $\Delta\Omega^2/\omega^2$ on μ for $U = 3\,eV$ (thick broken line), $3.25\,eV$ (thick dotted line), $3.5\,eV$ (thin broken line), $3.75\,eV$ (thin dotted line). Parameters: $\chi = 1$, $t = 0.25\,eV$, $\varepsilon_d = -1.5\,eV$, $|\Delta_{\mathrm{sc}}| = 0.02\,eV$.

The dependence of $\Delta\Omega^2/\omega^2$ on the position of chemical potential in the case of the larger values of U is depicted in figure 4. Here the rapid fall of $\Delta\Omega^2/\omega^2$ if the increasing μ passes through the van Hove singularity leads to the softening of phonon dynamics in the domain

$\mu > 0$, see especially thin full line and thin chain line. The softening may appear also if the chemical potential approaches the top of the band of itinerant electrons (thin full line and thin chain line in figure 4).

One can follow the tendencies demonstrated in figures 3,4 also in the functions $\Delta\Omega^2/\omega^2$ *vs* U for various values of μ sown in figure 5. In particular, one can distinguish the regions of the values of μ where $\Delta\Omega^2/\omega^2$ increases or decreases with the increase of U.

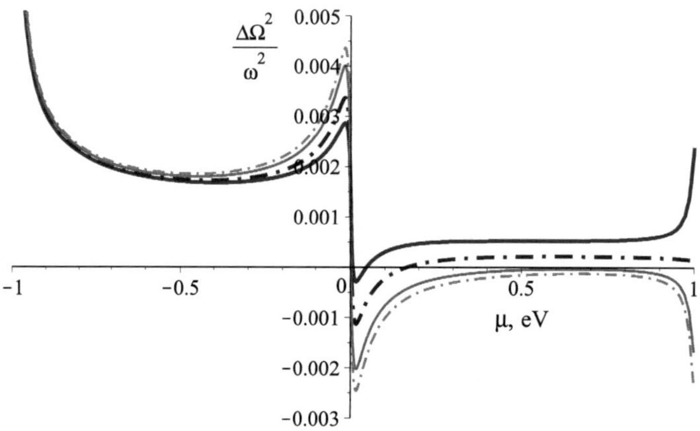

Figure 4. The dependence of $\Delta\Omega^2/\omega^2$ on μ for $U = 4\,eV$ (thick full line), $5\,eV$ (thick chain line), $8\,eV$ (thin full line), $13\,eV$ (thin chain line). For the rest of the parameters see figure 3

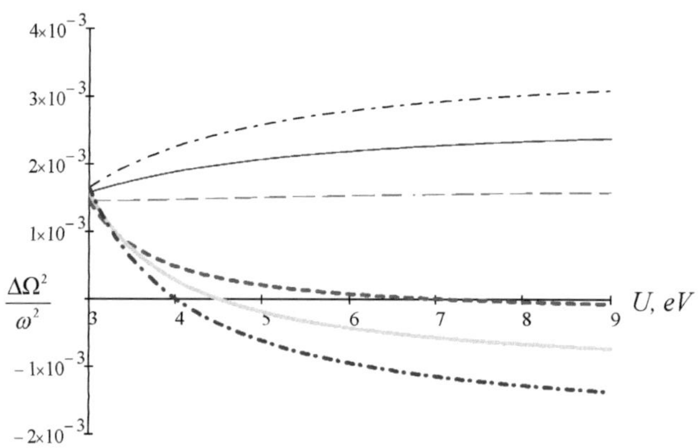

Figure 5. The dependence of $\Delta\Omega^2/\omega^2$ on U for $\mu = -0.4\,eV$ (thin broken line), $-0.1\,eV$ (thin full line), $-0.05\,eV$ (thin chain line), $0.05\,eV$ (thick chain line), $0.1\,eV$ (thick full line), $0.4\,eV$ (thick broken line). For the rest of the parameters see figure 3

4.2. Constant density of electron states

To illustrate more precisely the role of van Hove singularity in the superconductivity-induced shift of phonon frequency, we will consider for the comparison a model where the density of itinerant electron states $\rho = \text{const}$. In this case Eq. (10) must be replaced by the expression

$$\frac{\Omega^2}{\omega^2} = 1 - \kappa \int\limits_{\varepsilon_{\min}}^{\varepsilon_{\max}} \left\{ \frac{1}{2}\left[1 + \frac{\tilde{\varepsilon}(\varepsilon)}{E(\varepsilon)}\right] \frac{1}{E(\varepsilon) - \varepsilon_1} + \frac{1}{2}\left[1 - \frac{\tilde{\varepsilon}(\varepsilon)}{E(\varepsilon)}\right] \frac{1}{\varepsilon_2 - \varepsilon_1 + E(\varepsilon)} \right\} d\varepsilon \tag{12}$$

with $\kappa = 2|V|^2\rho/N_0\omega^2$.

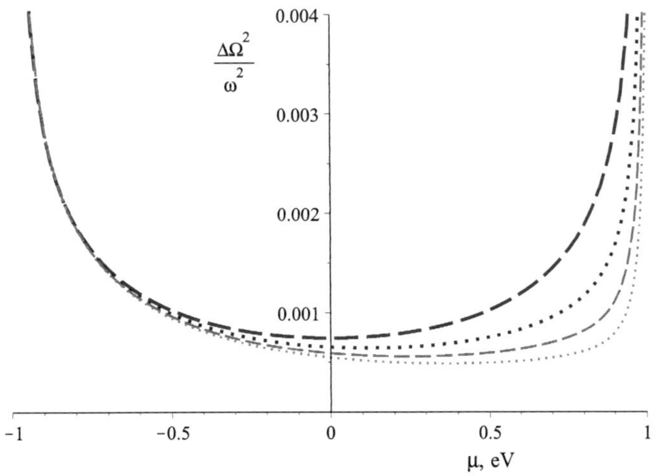

Figure 6. The dependence of $\Delta\Omega^2/\omega^2$ on μ for $U = 3\,eV$ (thick broken line), $3.25\,eV$ (thick dotted line), $3.5\,eV$ (thin broken line), $3.75\,eV$ (thin dotted line) in the case of constant density of states. Parameters: $\kappa = 1$, $\varepsilon_{\min} = -1\,eV$, $\varepsilon_{\max} = 1\,eV$, $\varepsilon_d = -1.5\,eV$, $|\Delta_{\rm sc}| = 0.02\,eV$.

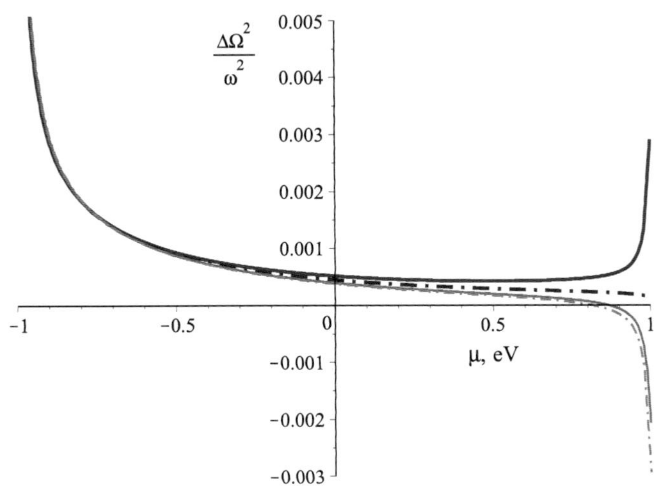

Figure 7. The dependence of $\Delta\Omega^2/\omega^2$ on μ for $U = 4\,eV$ (thick full line), $5\,eV$ (thick chain line), $8\,eV$ (thin full line), $13\,eV$ (thin chain line) in the case of constant density of states. For the rest of the parameters see figure 6

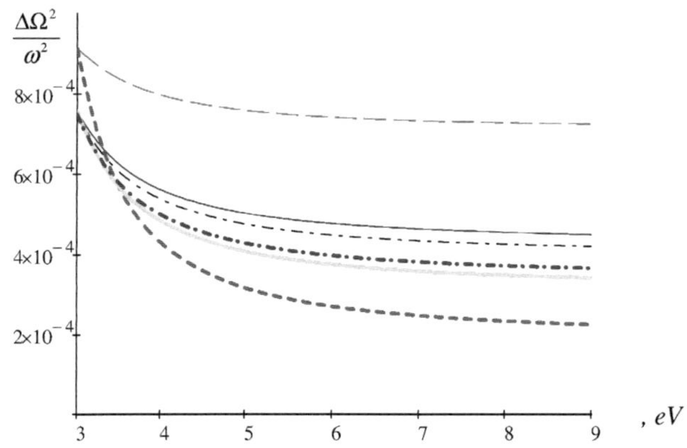

Figure 8. The dependence of $\Delta\Omega^2/\omega^2$ on U for $\mu = -0.4\,eV$ (thin broken line), $-0.1\,eV$ (thin full line), $-0.05\,eV$ (thin chain line), $0.05\,eV$ (thick chain line), $0.1\,eV$ (thick solid line), $0.4\,eV$ (thick broken line) in the case of constant density of states. For the rest of the parameters see figure 3

For the constant density of itinerant electron states, the superconductivity-induced shifts $\Delta\Omega^2/\omega^2$ *vs* chemical potential position calculated on the basis of Eqs. (12) and (10) are shown in figures 6, 7 for various values of U. As expected, the rapid change of $\Delta\Omega^2/\omega^2$ if chemical potential passes through the middle of the band of itinerant electrons is absent in the present case. The hardening of phonon frequency induced by superconductivity is suppressed by the increase of U more substantially in the domain $\mu > 0$. If chemical potential approaches the top of the band of itinerant electrons the effect of softening appears for sufficiently large values of U (thin full line and thin chain line in figure 7). Figure 8 demonstrates the decreasing dependence of $\Delta\Omega^2/\omega^2$ *vs* U for the various positions of chemical potential (c.f. figure 5).

5. Summary

- The impact of electron energy gap formation in the superconducting phase on phonon frequency renormalized by the vibronic mixing of itinerant and localized electron states has been studied.
- If the location of the lower and upper Hubbard levels with regard to the band of itinerant electrons is symmetric or weakly asymmetric, superconductivity induces the hardening of phonon dynamics. The dependence of the effect on μ is stronger near the edges and near the center of conduction band.
- If the location of the Hubbard levels is sufficiently asymmetric, the softening of phonon dynamics appears in certain regions of the chemical potential position.
- For the constant density of electron states the strong dependence of superconductivity-induced shift of phonon frequency on the chemical potential position near the center of conduction band disappears because in this case the van Hove singularity is absent in the center of conduction band.

Acknowledgement

The research was supported by the Estonian Research Council through the Institutional Research Funding IUT2-27 and by the Estonian Science Foundation, Grant No 8991.

References

[1] Testardi L R 1975 *Rev. Mod. Phys.* **47** 637
[2] Plakida 2009 *High Temperature Superconductors* (Heidelberg: Springer)

IOP Conf. Series: Journal of Physics: Conf. Series **833** (2017) 012018 doi:10.1088/1742-6596/833/1/012018

[3] Vojta M 2009 *Adv. Phys.* **58** 699
[4] Zeyher R and Zwicknagel G 1988 *Solid State Commun.* **66** 617
[5] Zeyher R and Zwicknagel G 1990 *Z. Phys. B - Condensed Matter* **78** 175
[6] Nicol E J, Jiang C and Carbotte J P 1993 *Phys. Rev. B* **47** 8131
[7] Bill A, Hizhnyakov V and Sigmund E 1995 *Phys. Rev. B* **52** 7637
[8] Örd T and Kristoffel N 1997 *Z. Physik. Chemie* **201** 167
[9] Kristoffel N 1990 *phys. stat. sol. (b)* **160** K93
[10] Kristoffel N and Örd T 1991 *Fiz. Nizk. Temp.* **17** 1218; 1991 *Sov. J. Low Temp. Phys.* **17** 641
[11] Bersuker I B 2006 *The Jahn-Teller Effect* (Cambridge: Cambridge University Press)
[12] Kristoffel N and Konsin P 1988 *phys. stat. sol. (b)* **149** 11
[13] Schrieffer J R and Wolff P A 1966 *Phys. Rev.* **149** 491
[14] Zaanen J and Oleś A M 1988 *Phys. Rev. B* **37** 9423
[15] Mattheiss L F and Hamann D R 1989 *Phys. Rev. B* **40** 2217
[16] Emery V J 1987 *Phys. Rev. Lett.* **58** 2794
[17] Hirsch J E 1987 *Phys. Rev. Lett* **59** 228

The spin-1 J_1-J_3 Heisenberg model on a triangular lattice

P Rubin and A Sherman

Institute of Physics, University of Tartu, W. Ostwaldi 1, 50411 Tartu, Estonia

E-mail: rubin@ut.ee

Abstract. Motivated by the experimental data for $NiGa_2S_4$, the spin-1 Heisenberg model on a triangular lattice with the ferromagnetic nearest- and antiferromagnetic third-nearest-neighbor exchange interactions, $J_1 = -(1-p)J$ and $J_3 = pJ$, $J > 0$, is studied in the range $0 \leq p \leq 1$. Mori's projection operator technique and the Lanczos exact diagonalization are used. Mori's method retains the rotation symmetry of spin components and does not anticipate any magnetic ordering. For zero temperature several phase transitions are observed. At $p_{cr} \approx 0.2$ the ground state is transformed from the ferromagnetic spin structure into a disordered state, which in its turn is changed to an antiferromagnetic long-range ordered state with the incommensurate ordering vector $\mathbf{Q}' \approx (1.16, 0)$ at $p \approx 0.31$. With growing p the ordering vector moves along the X axis to the commensurate point $\mathbf{Q_c} = (2\pi/3, 0)$ which is reached at $p = 1$. The final state with an antiferromagnetic long-range order can be conceived as four interpenetrating sublattices with the 120° spin structure on each of them. The model is able to describe the state with the incommensurate short-range order observed in $NiGa_2S_4$. To verify the used approach the ground state energy and corresponding spin-spin correlations are compared with exact-diagonalization results obtained with the SPINPACK code (the Lanczos exact diagonalization). Results of the two methods are in qualitative agreement.

1. Introduction

Investigation of the spin-1 Heisenberg model on a two-dimensional (2D) triangular lattice is of interest in connection with recently synthesized compound $NiGa_2S_4$ [1]. Magnetism in this system is mainly connected with Ni^{2+} ($S = 1$) ions, which form a 2D triangular lattice. Experiments on neutron scattering reveal a short-range antiferromagnetic order, being characterized by an incommensurate ordering vector. For the description of the observed order the classical $J_1 - J_3$ 2D Heisenberg model on a triangular lattice was proposed [1]. The dominating antiferromagnetic interaction J_3 between the third nearest neighbors (TNN) and the weak ferromagnetic interaction J_1 between the nearest neighbors (NN) were taken into account. This model agrees with first-principle calculations of the exchange parameters [2]. In this work the corresponding quantum model with $S = 1$ is considered with the aim to describe all phases of the model.

2. Ground state properties in Mori's approach

The Hamiltonian of the model reads

$$H = \frac{1}{2} \sum_{\mathbf{nm}} J_{\mathbf{nm}} \left(s_{\mathbf{n}}^z s_{\mathbf{m}}^z + s_{\mathbf{n}}^{+1} s_{\mathbf{m}}^{-1} \right), \tag{1}$$

where $s^z_{\mathbf{n}}$ and $s^{\pm 1}_{\mathbf{m}}$ are components of the spin-1 operators, **n** and **m** label sites of the triangular lattice. Only the NN and TNN interactions are taken into account,

$$J_{\mathbf{nm}} = \sum_{\mathbf{a}} (J_1 \delta_{\mathbf{n},\mathbf{m}+\mathbf{a}} + J_3 \delta_{\mathbf{n},\mathbf{m}+2\mathbf{a}}),$$

with the vectors **a** and 2**a** connecting the NN and TNN sites. We use the lattice spacing $a = |\mathbf{a}|$ as the unit of length. The exchange interactions, $J_1 = -(1-p)J$ and $J_3 = pJ$, $J > 0$, are expressed through the frustration parameter p, which varies in the full range from 0 to 1 (J is set as the unit of energy). Mori's projection operator technique [3] gives the following expression for the spin Green's function

$$D(\mathbf{k}\omega) = \frac{6J[(1-p)[1-\gamma(\mathbf{k})]C_1 - p[1-\gamma(2\mathbf{k})]\, C_{2a}]}{\omega^2 - \omega^2_{\mathbf{k}}}, \qquad (2)$$

where $\gamma(\mathbf{k}) = \frac{1}{3}\cos(k_x) + \frac{2}{3}\cos\left(\frac{k_x}{2}\right)\cos\left(\frac{k_y\sqrt{3}}{2}\right)$, the frequency of spin excitations $\omega_{\mathbf{k}}$ contains products of $[1-\gamma(\mathbf{k})]$, $[1-\gamma(2\mathbf{k})]$ and spin correlations such as $C_1 = \langle s^{+1}_{\mathbf{n}} s^{-1}_{\mathbf{n}+\mathbf{a}}\rangle$ and $C_{2a} = \langle s^{+1}_{\mathbf{n}} s^{-1}_{\mathbf{n}+2\mathbf{a}}\rangle$. These correlations can be expressed through $D(\mathbf{k}\omega)$ and calculated self-consistently. Details of our approach can be found in [4, 5]. If the system has a long-range order at $T = 0$ the summation over the wave vector in an expression for the spin correlation function can be divided into the contribution yielded at the ordering vector $\mathbf{k} = \mathbf{Q}$, which is proportional to the condensation part C, and the fluctuation contribution given by other wave vectors [6]. C plays the role of the order parameter. In the case $|J_1| \gg J_3$ the ground state is ferromagnetic and the ordering vector $\mathbf{Q} = (0, 0)$. We have found that the long-range ferromagnetic order exists for $0 \le p \le p_{cr}$, $p_{cr} \approx 0.2$. In this range of p the ferromagnetic condensation part is practically constant, and it vanishes abruptly at $p \approx p_{cr}$. At $T = 0$ the spectra of magnetic excitations for different values of p are shown in figure 1. As it can be seen, in the ferromagnetic range the dispersion $\omega_{\mathbf{k}}$ is parabolic near the Γ-point, which is typical for a ferromagnet. With the growth of the antiferromagnetic exchange J_3 the system passes into a disordered state, where the condensation part is equal to zero. When the value $p \approx 0.31$ is reached the system goes into the state with the long-range antiferromagnetic order with the incommensurate ordering vector $\mathbf{Q}' \approx (1.16, 0)$. The frequency of magnetic excitations vanishes at this wave vector (see figure 1) and a corresponding condensation part is finite. With further growth of the frustration parameter p the ordering vector moves along the X axis from the point $\mathbf{Q}'$ to $\mathbf{Q_c} = (2\pi/3, 0)$, which is reached at $p = 1$. For this p the incommensurate antiferromagnetic order transforms to a commensurate phase, which can be conceived as four interpenetrating 120° spin structures on sublattices with twice as large lattice spacing. Spin orientations in these sublattices are independent of one another.

At $p \approx 0.82$ the calculated ordering vector is close to that observed in $NiGa_2S_4$ [1]. For the finite temperature of the experiment, in accord with the Mermin-Wagner theorem, our long-range order passes into the short-range one. However, its calculated correlation length is an order of magnitude larger than the experimental value. This discrepancy can be related to the influence of impurities.

3. Lanczos exact diagonalization

Applying Mori's method one has to use a number of approximations. Therefore, it is of interest to study the same model with another method and compare obtained results [7]. To do this we employ Schulenburg's SPINPACK code [8]. This package is intended for exact diagonalization (ED) of finite spin systems using the Lanczos algorithm. For $0 \le p \le 1$ we found energies of the ground, low-lying excited states and spin-spin correlation functions for lattices containing

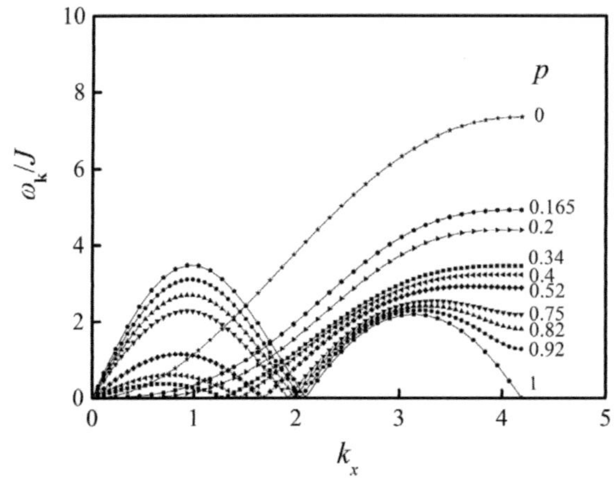

Figure 1. The dispersion of spin excitations $\omega_{\mathbf{k}}$ along the X axis for different values of the frustration parameter p at T=0.

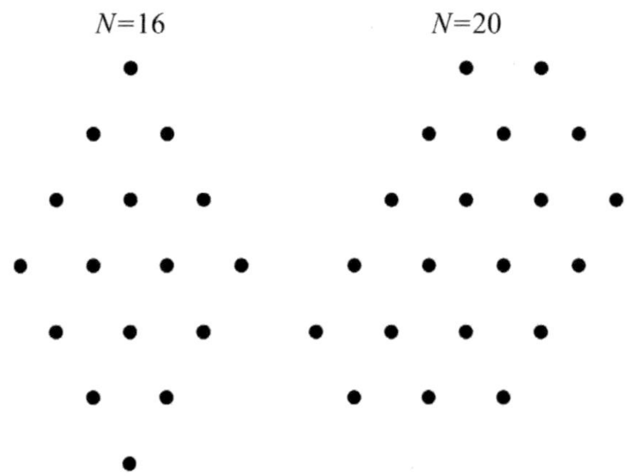

Figure 2. The triangular lattices with $N = 16$ and $N = 20$ sites studied in this work.

$N = 16$ and 20 sites. These lattices are shown in figure 2. The periodic boundary conditions were used.

Dependencies of the ground and low-lying excited state energies of the $N = 20$ lattice on the frustration parameter p are shown in figure 3 (corresponding dependencies for the $N = 16$ lattice are, in general, analogous). As seen, the ground state (GS) of the N-site lattice is transformed from the classic ferromagnet ($S_{\text{tot}} = N$) to the singlet state at some critical value of p. For the lattices with $N = 16$ and $N = 20$ these critical values are $p_{16} \approx 0.45$ and $p_{20} \approx 0.28$, respectively. As mentioned above, in Mori's approach we found that the transition from the ferromagnetically ordered state to a spin disorder occurs at the value of the frustration parameter $p_{cr} \approx 0.2$. The condensation part vanishes at this value of the frustration parameter. Thus p_N and p_{cr} indicate a border of the ferromagnetic region. It can be supposed that with the rise of the lattice size p_N will tend to the value p_{cr} obtained in Mori's approach because p_{cr} was obtained in a 216×216 lattice [5]. In figure 3a the dependencies of the GS energy (E_{GS}, $S_{\text{tot}} = N = 20$) and the first excited state (E_1, $S_{\text{tot}} = 19$) on p are shown. These dependencies are linear. Differences between the GS and excited-state energies disappear at the critical value p_{20}. For $p > p_{20}$ the energies of the low-lying states are shown in figure 3b. The GS is a singlet, $S_{\rm tot} = 0$. The lowest excited states are characterized by $S_{\rm tot} = 0$ (E_2) and $S_{\rm tot} = 1$ (E_3).

In figure 4 we compare the dependencies of the GS energy per site on the frustration parameter p, obtained by ED in the $N = 20$, $N = 16$ lattices and obtained by Mori's projection operator technique in a 216×216 lattice. As a whole these dependencies are similar. As one can see, the dependencies are linear in the ferromagnetic region ($p < p_{16}$, p_{20} or p_{cr}). Besides, $p_{cr} < p_{20} < p_{16}$. This sequence of the critical values of p seems reasonable taking into account the sizes of the lattices under consideration. For p larger than these critical values all curves have maxima. However, positions of these maxima are different – the dependence is peaked at $p \approx 0.4$ in the results obtained with Mori's technique, at $p \approx 0.7$ in the $N = 16$ and at $p \approx 0.9$ in the $N = 20$ ED results.

Spin correlation functions for nearest and third-nearest neighbors are shown in figure 5. The data were obtained by the exact diagonalization in the $N = 20$ lattice and by Mori's technique in a 216×216 lattice. In panels a) and b) ED correlations are constant and equal to unity in the ferromagnetic phase for $p < p_{20}$. In this region correlation functions obtained by Mori's method are also constant. However, they are somewhat smaller than one due to approximations made for the considered $S = 1$ Heisenberg model [5]. The interaction between nearest neighbor spins vanishes at $p = 1$, which manifests itself in the vanishing correlation $\langle \mathbf{S_0 S_a} \rangle$. The correlation $\langle \mathbf{S_0 S_{2a}} \rangle$ depends only weakly on p in the range $p > p_{20}$ and $p > p_{cr}$. Notice that the phase transition from the spin disordered state to the state with the incommensurate antiferromagnetic long-range order at $p \approx 0.31$ does not reveal itself in figure 5. As seen from the figure, curves obtained in the larger lattice are more smooth than those in the $N = 20$ lattice, which at least

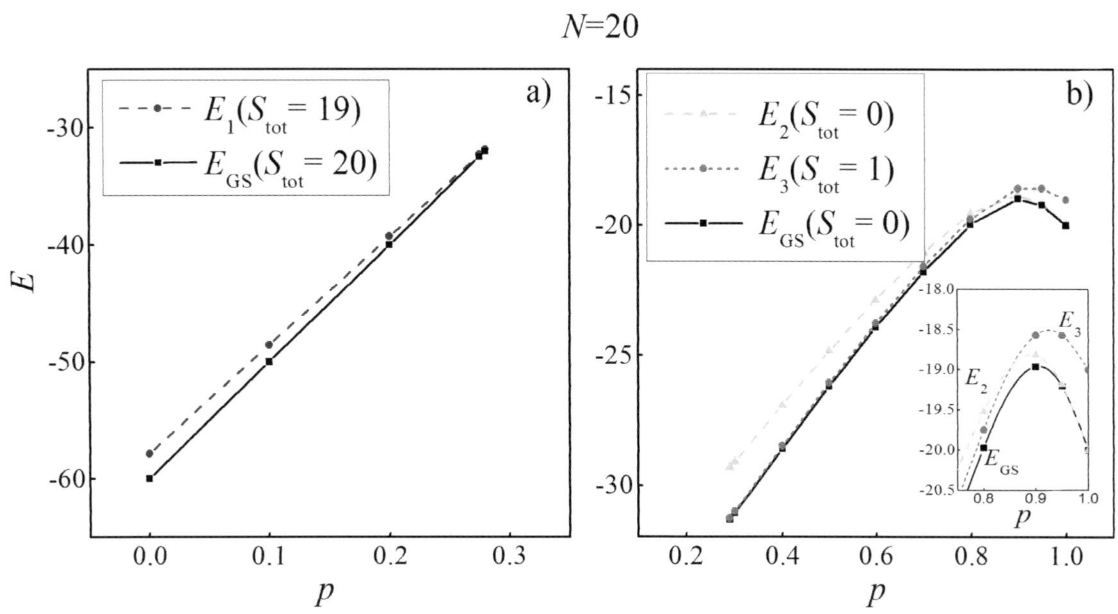

Figure 3. a) The dependencies of the ground state energy ($E_{\rm GS}$, $S_{\rm tot} = N = 20$, black squares and solid line) and the energy of the lowest excited state with $S_{\rm tot} = N - 1 = 19$ (E_1, blue circles and dashed line) on the frustration parameter $p \leq 0.28$ for the $N = 20$ lattice. In this interval of p the system is characterized by the ferromagnetic order. b) The dependencies of the ground state energy ($E_{\rm GS}$, $S_{\rm tot} = 0$, black squares and solid line), the energy of the lowest singlet (E_2, green triangles and short dashed line) and triplet (E_3, red circles and dashed line) excited states on the frustration parameter $p > 0.28$.

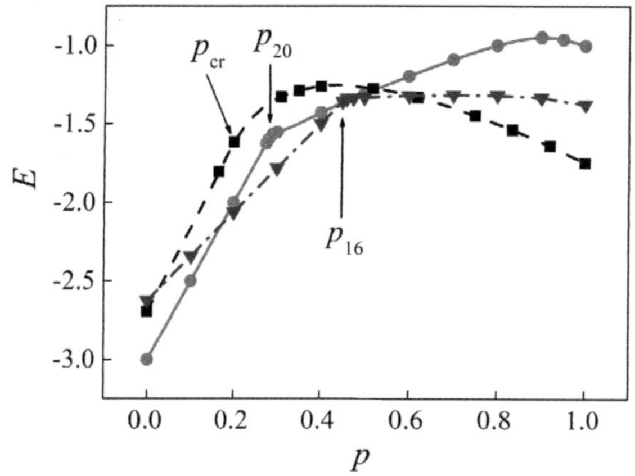

Figure 4. The dependencies of the ground state energy per site (E_{GS}/N) on the frustration parameter p obtained by ED in the $N = 20$ lattice (red circles and solid line), $N = 16$ lattice (blue triangles and dash-dotted line), and obtained by Mori's approach in a 216×216 lattice (black squares and dashed line). The parameters p_{20} and p_{16} are values of p, at which the transition from the ferromagnetic to the singlet GS occurs. p_{cr} corresponds to the transition to a disordered state.

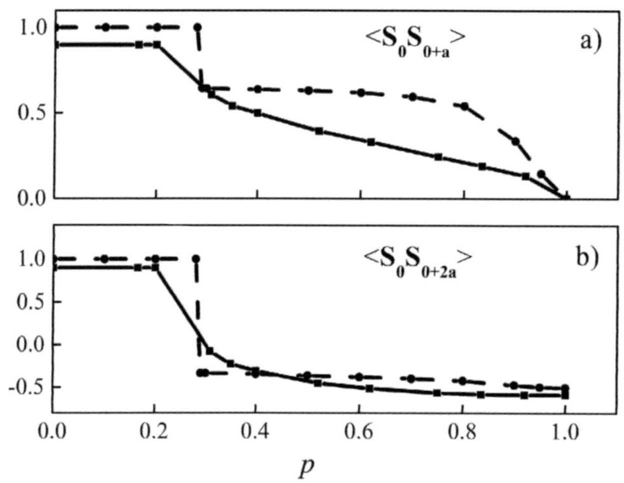

Figure 5. The nearest neighbor (a) and third-nearest neighbor spin correlations obtained by ED in the $N = 20$ lattice (circles and dashed lines) and by Mori's technique in a 216×216 lattice at $T = 0$ (squares and solid lines).

partly may be connected with finite-size effects. However, spin correlations calculated in the two lattices by two different methods are in general close and behave similarly with changing the frustration parameter. We can conclude that results obtained with the approximate approach based on Mori's projection operator technique are in reasonable agreement with the exact-diagonalization data.

4. Concluding remarks

In this work, Mori's projection operator technique and the Lanczos exact diagonalization have been applied for investigating magnetic properties of the quantum J_1-J_3 Heisenberg model on a triangular lattice in the entire range of the frustration parameter $0 \leq p \leq 1$, where $p = J_3/(J_3 - J_1)$, J_1 and J_3 are the nearest and third-nearest exchange constants. With the use of Mori's method we have shown that at zero temperature for different values of p the system is characterized by the ferromagnetic ordering, spin disorder, incommensurate and commensurate antiferromagnetic ordering. At $p \approx 0.82$ the model describes the key feature observed in $NiGa_2S_4$ [1] – the incommensurate short-range antiferromagnetic order at finite temperature. The results were verified using the exact diagonalization of small lattices with periodic boundary conditions. The SPINPACK package implementing the Lanczos algorithm was applied to find energies of the ground and low-lying exited states. Besides, spin-spin correlation functions between nearest- and third-nearest spins were calculated. We found qualitative and in some cases quantitative agreement between results obtained by exact diagonalization of small clusters and by Mori's projection technique in larger lattices.

Acknowledgments
This work was supported by the research project IUT2-27 and the ESF Grant No. 9371.

References
[1] Nakatsuji S, Nambu Y, Tonomura H, Sakai O, Jonas S, Broholm C, Tsunetsugu H, Qiu Y and Maeno Y 2005 *Science* **309** 1697-1700
[2] Mazin I I 2007 *Phys. Rev.* B **76** 40406
[3] Mori H 1965 *Prog. Theor. Phys.* **34** 399-416
[4] Sherman A and Schreiber M 2002 *Phys. Rev.* B **65** 134520
[5] Rubin P, Sherman A and Schreiber M 2012 *Phys. Lett.* A **376** 1062-66
[6] Shimahara H, Takada S 1991 *J. Phys. Soc. Jpn.* **60** 2394-2405
[7] Rubin P, Sherman A and Schreiber M 2014 *Phys. Lett.* A **378** 3572-74
[8] Schulenburg J *Program package SPINPACK* https://www-e.uni-magdeburg.de/jschulen/spin/.

IOP Conf. Series: Journal of Physics: Conf. Series **833** (2017) 012020 doi:10.1088/1742-6596/833/1/012020

Efficient emitting molecules in organic light-emitting diodes on the basis of the control of vibronic couplings

Tohru Sato[a,b]

[a]Department of Molecular Engineering, Graduate School of Engineering, Kyoto University, Nishikyo-ku, Kyoto 615-8510, Japan

[b]Unit of Elements Strategy Initiative for Catalysts & Batteries, Kyoto University, Nishikyo-ku, Kyoto 615-8510, Japan

E-mail: tsato@moleng.kyoto-u.ac.jp

Abstract. A high external quantum efficiency observed in organic light-emitting diodes using PTZ-BZP (PTZ: 10-hexyl-phenothiazin, and BZP: 4-phenyl-2,1,3-benzothiadiazole) as an emitting molecule is studied based on the vibronic coupling density analyses employing the time-dependent density functional theory. The high efficiency can be attributed to a high efficiency of exciton generation which originates from reverse intersystem crossings from the higher T_3 and T_2 states than T_1 to the emitting S_1 state. The nonradiative transitions from T_3 and T_2 to the lower triplet states are suppressed because of small offdiagonal vibronic coupling constants. The small vibronic couplings originate from the multi-configurationality of these triplet states. A molecular design principle for efficient emitting molecules is discussed from the view of the control of vibronic couplings.

1. Introduction

Vibronic couplings play important roles in light emission from a molecule. Off-diagonal vibronic couplings give rise to nonradiative transitions, and diagonal vibronic couplings are related to the geometrical deviation of the minimum of the ground and excited state[1]. A large vibrational relaxation results in small Franck-Condon factors which give rise to a small rate constant of the radiative transition. Therefore, small diagonal and off-diagonal vibronic couplings are crucial in designing of an efficient emitting molecule.

We have proposed concepts of vibronic coupling densities (VCD) to analyze and control diagonal and off-diagonal vibronic couplings[2, 3]. On the basis of vibronic coupling density analysis, we designed efficient emitting molecules by the suppression of vibronic couplings in anthracene[4] and triphenylamine[5], and we experimentally observed the increases of quantum yields of the designed molecules[6, 7]. It should be noted that triphenylamine is known as a non-emitting molecule. We can make a non-emitting molecule fluorescent by a molecular design on the basis of a vibronic coupling density analysis.

Thermally-activated delayed fluorescence (TADF) has attracted much attention as a highly-efficient emitting mechanism for organic light-emitting diodes (OLED) [8, 9, 10, 11]. Thermally excited T_1 excitons are converted to S_1 excitons via reverse intersystem crossing. While a fluorescent OLED and a phosphorescent OLED utilize 25% singlet and 75% triplet excitons at best, respectively, in TADF OLED, both T_1 and S_1 excitons which are generated by current excitations are utilized for light emission. The external quantum efficiency (EQE) of a fluorescent, phosphorescent, and TADF OLED is 0.05 of photoluminescence quantum yield (PLQY), 0.15 of PLQY, and 0.20 of PLQY, respectively.

We have proposed a fluorescence via higher triplets mechanism to increase the efficiency of exciton generation[12, 13, 14]. Fluorescence via reverse intersystem crossing from T_n states higher than T_1 to S_m is possible by suppressing non-radiative transitions from T_n based on the concept of vibronic coupling density as long as the energy level of T_n is close to S_m.

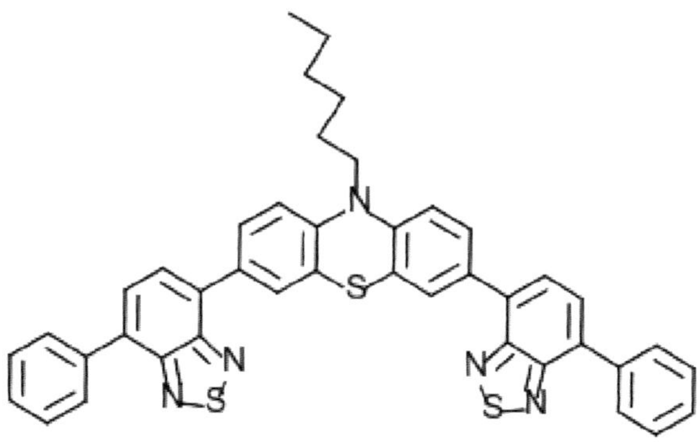

Figure 1. Molecular structure of PTZ-BZP. PTZ-BZP consists of the central 10-hexyl-phenothiazin (PTZ) moiety and the two 4-phenyl-2,1,3-benzothiadiazole (BZP) moieties.

PTZ-BZP (see figure 1) consists of the central 10-hexyl-phenothiazin (PTZ) moiety and the two 4-phenyl-2,1,3-benzothiadiazole (BZP) moieties. Yao *et al.* have reported that they observed the electroluminescence (EL) of a device using PTZ-BZP as an emitting molecule with EQE=1.54 %[15]. Since the observed PLQY was 16%, we cannot explain the observed EQE based on the fluorescent EL mechanism. In other word, triplet excitons as well as singlet ones should be utilized in this device. In this study, we analyze the EL mechanism on the basis of VCD to elucidate the high efficiency of exciton generation in the OLED of PTZ-BZP employing the time-dependent density functional theory (TD-DFT). We also propose a molecular design principle for the fluorescence via higher triplets mechanism.

2. Vibronic coupling constants and vibronic coupling density

The definitions of vibronic coupling constants and vibronic coupling densities are briefly summarized here. See details in the literature[2, 3]. Vibronic coupling operator is written as

$$\hat{V}Q_\alpha = \left(\frac{\partial \hat{H}}{\partial Q_\alpha}\right)_{\boldsymbol{R}_0} Q_\alpha, \tag{1}$$

where $\hat{H}$ is a molecular Hamiltonian, Q_α is the normal coordinate of mode α, and $\boldsymbol{R}_0$ is a reference molecular geometry.

Diagonal vibronic coupling constant (VCC) of state n is defined by

$$V_n^\alpha := \left\langle \Psi_n(\boldsymbol{R}_0, \boldsymbol{r}) \left| \hat{V}_\alpha \right| \Psi_n(\boldsymbol{R}_0, \boldsymbol{r}) \right\rangle + \left(\frac{\partial U_{nn}}{\partial Q_\alpha}\right)_{\boldsymbol{R}_0}, \tag{2}$$

where $\Psi_n(\boldsymbol{R}_0, \boldsymbol{r})$ is the electronic wavefunction of state n, and U_{nn} is the nuclear-nuclear repulsion. $\boldsymbol{r}$ denotes the coordinates of electrons. A diagonal VCC is the driving force of a vibrational relaxation. The reorganization energy can be expressed in terms of diagonal VCCs,

$$E^n_{\rm reorg} = \sum_\alpha \frac{(V^n_\alpha)^2}{2\omega_\alpha^2}. \tag{3}$$

Offdiagonal VCC between state m and n is defined by

$$V^\alpha_{mn} := \left\langle \Psi_m(\boldsymbol{R}_0, \boldsymbol{r}) \left| \hat{V}_\alpha \right| \Psi_n(\boldsymbol{R}_0, \boldsymbol{r}) \right\rangle \tag{4}$$

A offdiagonal VCC is a driving force of a nonradiative transition. Small offdiagonal VCCs result in the suppression of a nonradiative transition.

Offdiagonal vibronic coupling density (VCD) is defined by

$$\eta^\alpha_{mn}(\boldsymbol{r}_1) := \rho_{mn}(\boldsymbol{r}_1) v^\alpha(\boldsymbol{r}_1). \tag{5}$$

VCD describes a local picture of a vibronic coupling (see Appendix). Potential derivative $v^\alpha(\boldsymbol{r}_1)$ is defined in Appendix, and overlap density $\rho_{mn}(\boldsymbol{r}_1)$ is defined by

$$\rho_{mn}(\boldsymbol{r}_1) := N \int \cdots \int d\sigma_1 d\boldsymbol{x}_2 \cdots d\boldsymbol{x}_N \Psi^*_m(\boldsymbol{R}_0, \boldsymbol{x}_1, \cdots, \boldsymbol{x}_N) \Psi_n(\boldsymbol{R}_0, \boldsymbol{x}_1, \cdots, \boldsymbol{x}_N), \tag{6}$$

where $\boldsymbol{x}_i = (\boldsymbol{r}_i, \sigma_i)$ with the space coordinates $\boldsymbol{r}_i$ and spin coordinate σ_i) of electron i. Offdiagonal VCC V^α_{mn} is the integral of offdiagonal VCD $\eta^\alpha_{mn}(\boldsymbol{r}_1)$:

$$V^\alpha_{mn} = \int \eta^\alpha_{mn}(\boldsymbol{r}_1) d\boldsymbol{r}_1. \tag{7}$$

3. Method of calculation

The ground state was optimized employing M06-2X/6-31G(d,p). The semi-empirical functional M06-2X[16] has been employed in theoretical studies on TADF molecules[9]. The optimized structure of S_0 was confirmed by performing vibrational analysis. The excited states, S_1, T_1, T_2, and T_3 were optimized by using time-dependent density functional theory (TD-DFT) at the M06-2X/6-31G(d,p) level of theory. In our preliminary investigation at the B3LYP/6-31G(d,p) level of theory[13], the calculated absorption wavelength is too long compared with the experimental value[15].

The offdiagonal vibronic coupling constants and vibronic coupling densities were calculated from the excited electronic wavefunction for the optimized geometries. These electronic and vibrational structures were obtained by Gaussian09[17]. The VCC calculations and VCD analyses were performed by using our in-house codes. The vibrational vectors employed for the VC calculations and analyses were obtained from the vibrational analysis for the S_0 states at the optimized geometries of the excited states, since only a numerical frequency analysis is available for an excited state in Gaussian09. A numerical frequency analysis is computationally very demanding.

4. Results and discussion

4.1. Franck-Condon S_1 state

Table 1 summarizes the Franck-Condon excited sates S_n@S_0 and T_n@S_0, where @S_n means that the excited states are calculated for the S_0 optimized geometry. The calculated excitation energy of S_1 reproduces the experimental absorption wavelength well.

The calculated diagonal VCCs are shown in figure 2 for the vibrational modes. A large reorganization energy, or vibrational relaxation of S_1 suppresses the rate constant of a fluorescent process[1]. The total reorganization energy was estimated to be 372 meV by using Eq.(3), which is not so large as an emitting molecule[1].

Table 1. Franck-Condon excited states calculated at the M06-2X/6-31G(d,p) level of theory.

State	Excitation energy eV	nm	f	Major configurations (CI coefficient)
T_1 (^3A)	2.3225	533.85	0.0000	NHO → NLU (0.43658)
				HO → LU (0.35711)
T_2 (^3A)	2.3308	531.94	0.0000	NHO → LU (0.41920)
				HO → NLU (0.33955)
S_1 (^1A)	3.2051	386.83	0.7867	HO → LU (0.59349)
T_3 (^3A)	3.2794	378.07	0.0000	HO → LU (0.36497)
				HO → LU+2 (0.32673)
S_2 (^1A)	3.3470	370.43	0.1971	NHO → LU (0.30039)
				HO → NLU (0.58363)
Exp[15]	2.77	447		

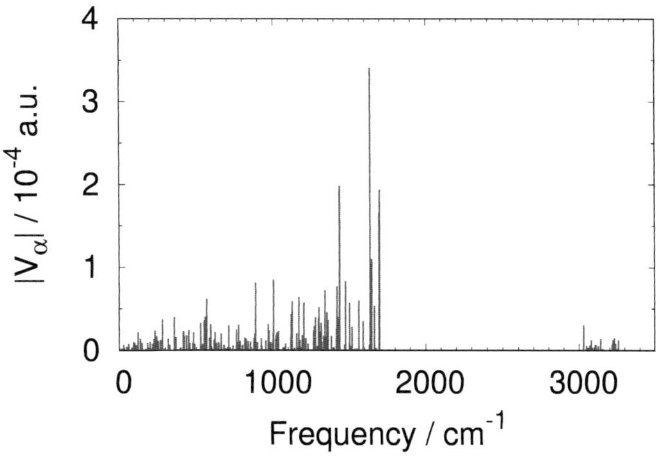

Figure 2. Diagonal vibronic coupling constants for the Franck-Condon S_1 state at the M06-2X/6-31G(d,p) level of theory.

4.2. S_1 adiabatic state

The excited states for the optimized structure of the S_1 state are tabulated in table 2. The energy levels of T_2 and T_3 are close to that of S_1. This suggests that the reverse intersystem crossing from T_2 or T_3 to S_1 could be possible if the energy levels of T_2@T_2 and S_1@T_2, or those of T_1@T_1 and S_1@T_1 are close each other with the offdiagonal VCCs between the higher triplet states and the lower ones small.

Offdiagonal vibronic couplings between S_1 and S_0 should be small as a fluorescent molecule. The offdiagonal VCCs between the S_1@S_1 and S_0@S_1 are shown in figure 3. They are less than 1×10^{-3} a.u. which is typical in a fluorescent molecule. The strong coupling modes are 7.29×10^{-4} a.u. of mode 111 (960 cm^{-1}) and 7.14×10^{-4} a.u. of mode 45 (403 cm^{-1}). We performed the VCD analyses for these modes.

Figure 4 shows the VCD analyses of the strong coupling modes, mode 111 and mode 45, for the S_1 adiabatic state. The overlap density between S_1@S_1 and S_0@S_1 is localized on the one

Table 2. S_1 adiabatic excited states calculated at the M06-2X/6-31G(d,p) level of theory.

State	Excitation energy eV	nm	f	Major configurations (CI coefficient)
T_1 (^3A)	1.5619	793.81	0.0000	NHO → LU (0.42818)
				HO → LU (0.51651)
T_2 (^3A)	2.2316	555.58	0.0000	HO-2 → NLU (0.40509)
				NHO → NLU (-0.30471)
				HO → NLU (0.38302)
S_1 (^1A)	2.4851	498.92	0.6998	HO → LU (0.66212)
T_3 (^3A)	2.6813	462.40	0.0000	NHO → LU (0.34910)
				HO → LU (-0.34410)
S_2 (^1A)	3.0121	411.61	0.2710	HO → NLU (0.62714)
Exp[15]	2.10	590		

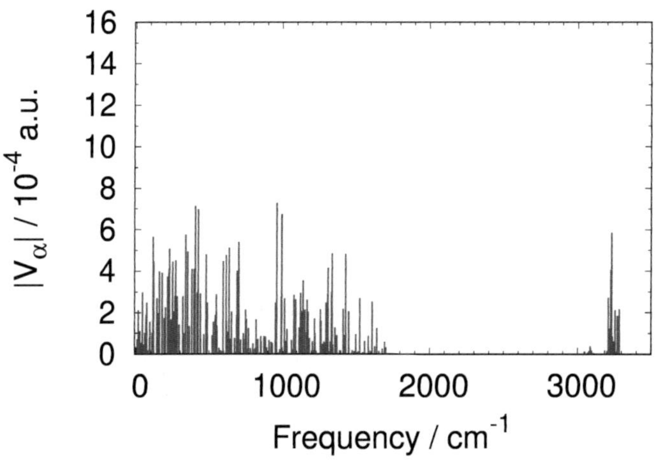

Figure 3. Offdiagonal vibronic coupling constants between S_1 and S_0 states for the adiabatic S_1 state at the M06-2X/6-31G(d,p) level of theory.

of the BZP moieties. As shown in figure 4 (c) and (e), therefore, the VCs originate from the overlap density on the BZP moiety. Accordingly, in order to enhance the PLQY, the overlap density on the BZP moiety should be reduced.

4.3. T_3 adiabatic state

The excited states for the optimized geometry of T_3 is summarized in table 3. The energy level of T_3@T_3 is close to that of S_1@T_3. The energy gap between S_1@T_3 and T_3@T_3, $\Delta E_{S_1@T_3-T_3@T_3} = 25$ meV. This small energy gap makes the thermal activation from T_3 to S_1 possible if the offdiagonal vibronic couplings are small.

Figures 5 and 7 show the offdiagonal VCCs between the T_3 and T_2 states, and the T_3 and T_1 states, respectively. The strong coupling modes between T_3 and T_2 are 3.20×10^{-4} a.u. of mode 201 (1629 cm^{-1}), and -2.87×10^{-4} a.u. of mode 177 (1420 cm^{-1}). The strong coupling modes between T_3 and T_1 are 3.62×10^{-4} a.u. of mode 78 (697 cm^{-1}), and 3.11×10^{-4} a.u.

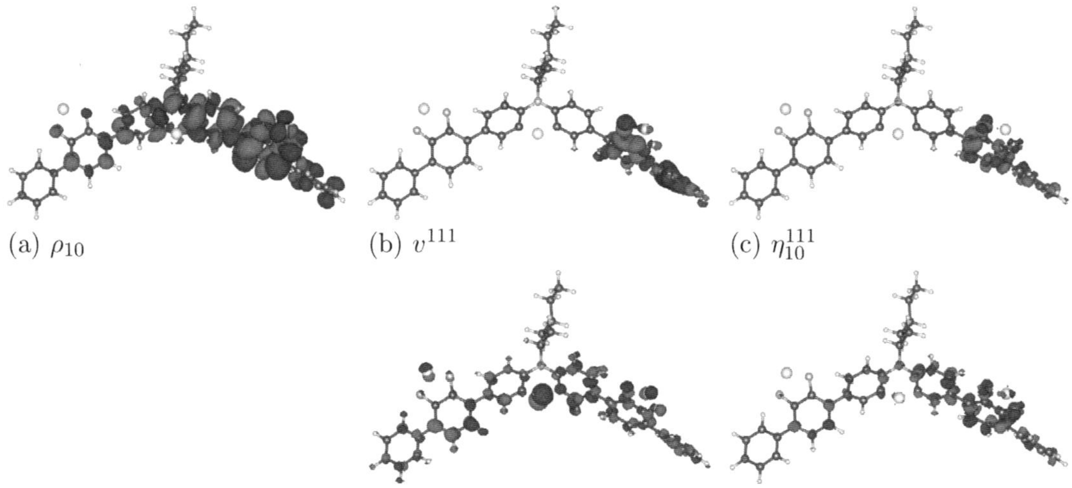

(a) ρ_{10} (b) v^{111} (c) η_{10}^{111}

(d) v^{45} (e)η_{10}^{45}

Figure 4. The VCD analyses for the S_1 adiabatic state at the M06-2X/6-31G(d,p) level of theory. Mode 111 and mode 45 indicate strong couplings. (a) Overlap density between S_1@S_0 and S_0@S_1, ρ_{10}, (b) potential derivative of mode 111, v^{111}, (c) VCD of mode 111, η_{10}^{111}, (d) potential derivative of mode 45, v^{45}, and (e) VCD of mode 45, η_{10}^{45}. The isosurface values are 5.0×10^{-4} for ρ, 0.004 for v, and 5.0×10^{-6} for η, respectively.

Table 3. T_3 adiabatic excited states calculated at the M06-2X/6-31G(d,p) level of theory.

State	Excitation energy		f	Major configurations (CI coefficient)
	eV	nm		
T_1 (^3A)	2.0115	616.39	0.0000	NHO → NLU (-0.36213)
				HO → LU (0.47322)
				HO-2 → LU (0.24834)
T_2 (^3A)	2.0799	596.09	0.0000	NHO → LU (0.41808)
				HO → NLU (-0.40890)
				HO-2 → NLU (-0.29704)
T_3 (^3A)	2.6324	471.00	0.0000	HO → LU+2 (0.39798)
				HO-2 → LU (0.28310)
				NHO → NLU (-0.25387)
				HO → LU (-0.28352)
S_1 (^1A)	2.6574	466.56	0.6657	HO → LU (0.65139)
S_2 (^1A)	2.9119	425.79	0.1057	HO → NLU (0.66925)

of mode 202 (1630 cm^{-1}). These coupling constants are rather small, which suggests that the nonradiative transitions from the T_3 state to the T_2 and T_1 states are suppressed.

4.4. T_2 adiabatic state

Table 4 summarizes the excited states for the T_2 optimized geometry. The energy gap between the T_2@T_2 and S_1@T_2 states, $\Delta E_{S_1@T_2-T_2@T_2}$ is equal to -14.9 meV. If the offdiagonal vibronic couplings between T_2 and T_1 are small, the reverse intersystem crossing from T_2 to S_1 is possible.

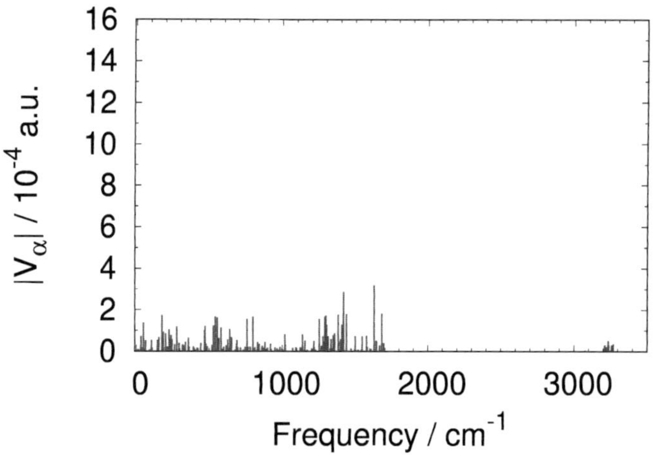

Figure 5. Offdiagonal vibronic coupling constants between T_3 and T_2 states for the adiabatic T_3 state at the M06-2X/6-31G(d,p) level of theory.

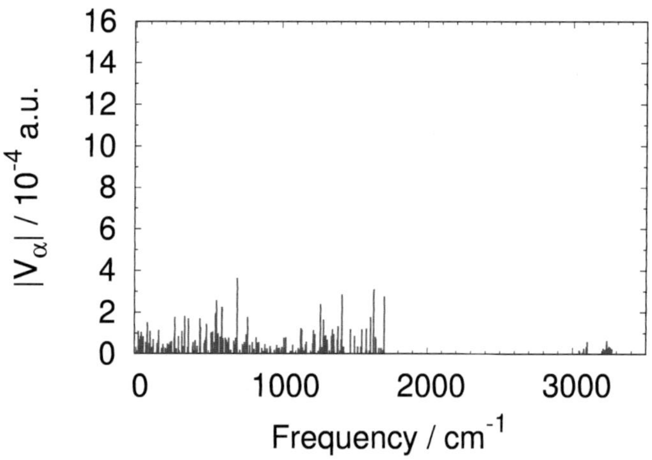

Figure 6. Offdiagonal vibronic coupling constants between T_3 and T_1 states for the adiabatic T_3 state at the M06-2X/6-31G(d,p) level of theory.

The offdiagonal VCCs between the T_2 and T_1 states are shown in figure 7. It should be noted that the order of the coupling constants is of 10^{-5} a.u. The extremely small coupling constants originate from the small overlap density discussed in the following subsection.

4.5. Reductions of the overlap densities

Figure 8 shows the overlap densities between the T_3 and T_2 states, the T_3 and T_1 states, and the T_2 and T_1 states. Note that the isosurface value (0.0005) of the overlap density between S_1 and S_0 shown in figure 4 (a) is larger than the isosurface values (0.0002) of figure 8. The overlap densities between the T_3 and T_2 states (figure 8 (a)), and the T_3 and T_1 states (figure

Table 4. T_2 adiabatic excited states calculated at the M06-2X/6-31G(d,p) level of theory.

State	Excitation energy eV	nm	f	Major configurations (CI coefficient)
T_1 (^3A)	1.0565	1173.56	0.0000	NHO → LU (0.35875)
				HO → LU (0.60451)
S_1 (^1A)	2.3072	537.38	0.6700	HO → LU (0.67479)
T_2 (^3A)	2.3221	533.94	0.0000	HO-2 → NLU (0.42305)
				NHO → NLU (-0.39498)
T_3 (^3A)	2.7331	453.64	0.0000	NHO → LU (0.42675)
				HO → LU (-0.31247)
T_4 (^3A)	3.1743	390.58	0.0000	HO-10 → LU (0.50716)
				HO-9 → LU (0.34630)
S_2 (^1A)	3.1799	389.90	0.2617	NHO → LU (0.42893)
				HO → NLU (-0.40933)

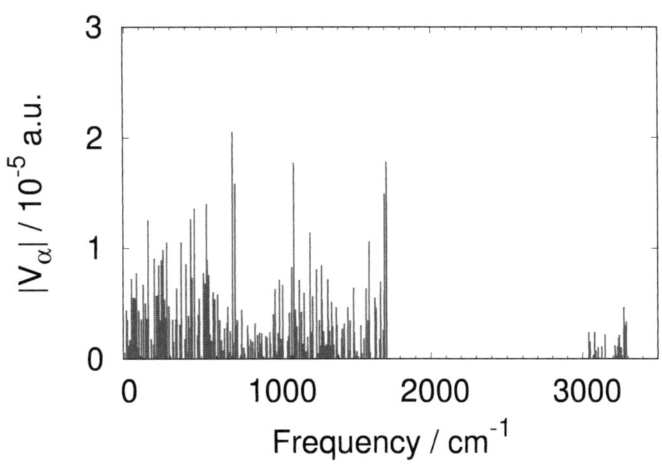

Figure 7. Offdiagonal vibronic coupling constants between T_2 and T_1 states for the adiabatic T_2 state at the M06-2X/6-31G(d,p) level of theory.

8 (b)) are reduced compared with that between the S_1 and S_0 states (figure 4 (a)). The overlap density between the T_2 and T_1 states (figure 8 (c)) are much reduced.

Compared with offdiagonal vibronic couplings between S_1 and S_0, which is a driving force of the nonradiative transition that competes with the fluorescent process, those between the higher triplet states and lower triplet ones are small in PTZ-BZP. This is one of the necessary condition for the fluorescence via higher triplets mechanism. We discuss this reductions from the view of the structures of the excited wave functions.

Overlap density between states $\Psi_1 = c_{1a}\Phi_a + c_{1b}\Phi_b$ and $\Psi_2 = c_{2c}\Phi_c + c_{2d}\Phi_d$, $\rho(\Psi_1, \Psi_2)$, can be expanded as a sum of the overlap densities between electronic configurations (Slater determinants):

$$\rho(\Psi_1, \Psi_2) = c_{1a}^* c_{2c}\rho(\Phi_a, \Phi_c) + c_{1a}^* c_{2d}\rho(\Phi_a, \Phi_d) + c_{1a}^* c_{2c}\rho(\Phi_a, \Phi_c) + c_{1a}^* c_{2d}\rho(\Phi_a, \Phi_d), \quad (8)$$

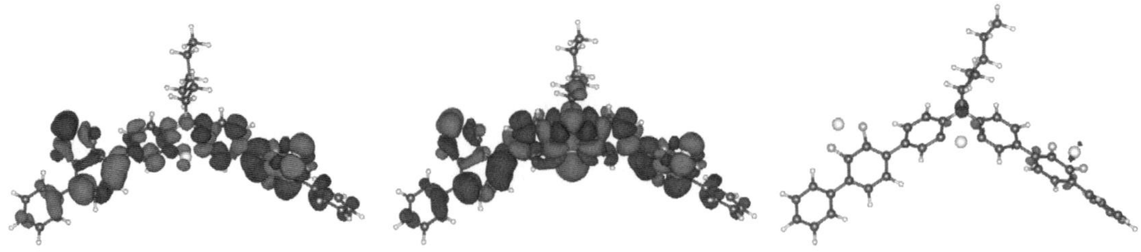

(a) ρ_{32} (b) ρ_{31} (c) ρ_{21}

Figure 8. Overlap densities between (a) $T_3@T_3$ and $T_2@T_3$, (b) $T_3@T_3$ and $T_1@T_3$, and (c) $T_2@T_2$ and $T_1@T_2$. All the isosurface values are 2×10^{-4}.

where $\Phi_i \quad (i = a, b, c, d)$ is an electronic configuration, and $c_{ki} \quad (k = 1, 2)$ is a configuration interaction (CI) coefficient.

A TD-DFT wavefunction consists of one-electron excited electronic configurations, Φ_i^r, where i denotes an occupied molecular orbital ψ_i, and r is an unoccupied molecular orbital ψ_r:

$$\Psi = \sum c_i^r \Phi_i^r. \tag{9}$$

Overlap density between the electronic configurations satisfies the following:

$$\begin{aligned} \rho(\Phi_i^r, \Phi_i^r) &= \rho_0 - \rho_i + \rho_r, & (10)\\ \rho(\Phi_i^r, \Phi_i^s) &= \psi_r^* \psi_s, & (11)\\ \rho(\Phi_i^r, \Phi_j^r) &= \psi_i^* \psi_j, & (12)\\ \rho(\Phi_i^r, \Phi_j^s) &= 0 & (13) \end{aligned}$$

where ρ_0 is the electron density of the ground electronic configuration Φ_0, and $\rho_a = \psi_a^* \psi_a \quad (a = i, r)$.

From table 4, the wavefunctions of triplet excited states $T_2@T_2$ and $T_1@T_2$ can be approximately expressed by

$$\begin{aligned} \Psi_{T_2@T_2} &= a\Phi_{\mathrm{HO}-2}^{\mathrm{NLU}} - a\Phi_{\mathrm{NHO}}^{\mathrm{NLU}}, & (14)\\ \Psi_{T_1@T_2} &= b\Phi_{\mathrm{NHO}}^{\mathrm{LU}} + c\Phi_{\mathrm{HO}}^{\mathrm{LU}}, & (15) \end{aligned}$$

respectively. Therefore, the overlap density can be written as

$$\rho(\Psi_{T_2@T_2}, \Psi_{T_1@T_2}) = -ab\psi_{\mathrm{NLU}}^* \psi_{\mathrm{LU}}. \tag{16}$$

It should be noted that $\rho(\Phi_{\mathrm{HO}-2}^{\mathrm{NLU}}, \Phi_{\mathrm{NHO}}^{\mathrm{LU}})$, $\rho(\Phi_{\mathrm{HO}-2}^{\mathrm{NLU}}, \Phi_{\mathrm{HO}}^{\mathrm{LU}})$, and $\rho(\Phi_{\mathrm{NHO}}^{\mathrm{NLU}}, \Phi_{\mathrm{HO}}^{\mathrm{LU}})$ are vanishing. In other words, the overlap density between $T_2@T_2$ and $T_1@T_2$ are reduced because of the multiconfigurational structure of the TD-DFT wavefunctions. Accordingly, the nonradiative transition from T_2 to T_1 is suppressed.

Similarly in the T_3 adiabatic state, from table 3, the wavefunctions of triplet excited states $T_3@T_3$, $T_2@T_3$, and $T_1@T_3$ can be approximately expressed by

$$\begin{aligned} \Psi_{T_3@T_3} &= p\Phi_{\mathrm{HO}}^{\mathrm{LU}+2} + q\Phi_{\mathrm{HO}-2}^{\mathrm{LU}} - q\Phi_{\mathrm{NHO}}^{\mathrm{NLU}} - q\Phi_{\mathrm{HO}}^{\mathrm{LU}}, & (17)\\ \Psi_{T_2@T_3} &= p\Phi_{\mathrm{NHO}}^{\mathrm{LU}} - p\Phi_{\mathrm{HO}}^{\mathrm{NLU}} - q\Phi_{\mathrm{HO}-2}^{\mathrm{NLU}}, & (18)\\ \Psi_{T_1@T_3} &= -p\Phi_{\mathrm{NHO}}^{\mathrm{NLU}} + r\Phi_{\mathrm{HO}}^{\mathrm{LU}} + s\Phi_{\mathrm{HO}-2}^{\mathrm{LU}}, & (19) \end{aligned}$$

IOP Conf. Series: Journal of Physics: Conf. Series **833** (2017) 012020 doi:10.1088/1742-6596/833/1/012020

where $p \approx 0.4$, $q \approx 0.3$, $r \approx 0.5$, and $s \approx 0.2$. The overlap densities are derived as follows:

$$\rho(\Psi_{T_3@T_3}, \Psi_{T_2@T_3}) = -p^2\psi^*_{\text{LU+2}}\psi_{\text{NLU}} + p(p+q)\psi^*_{\text{HO}-2}\psi_{\text{NHO}} - q^2\psi^*_{\text{LU}}\psi_{\text{NLU}}. \quad (20)$$

$$\begin{aligned}\rho(\Psi_{T_3@T_3}, \Psi_{T_1@T_3}) = &\ pr\psi^*_{\text{LU+2}}\psi_{\text{LU}} + q(r-s)\psi^*_{\text{HO}-2}\psi_{\text{HO}} \\ &+qs(\rho_0 - \rho_{\text{HO}-2} + \rho_{\text{LU}}) + qp(\rho_0 - \rho_{\text{NHO}} + \rho_{\text{NLU}}) \\ &-qr(\rho_0 - \rho_{\text{HO}} + \rho_{\text{LU}})\end{aligned} \quad (21)$$

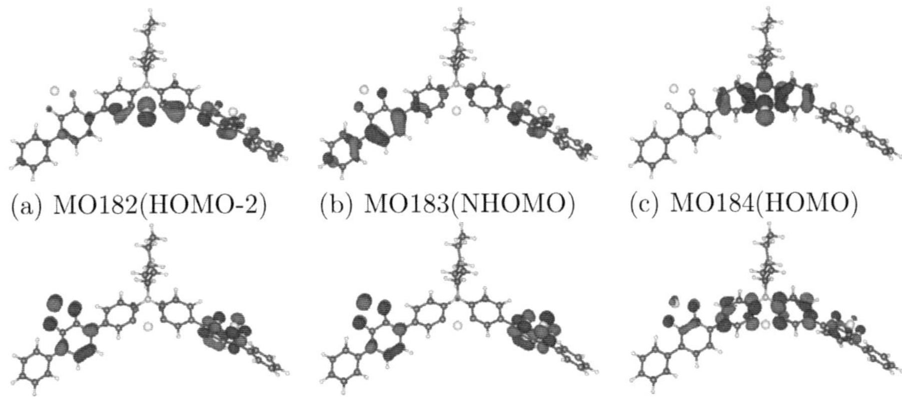

(a) MO182(HOMO-2) (b) MO183(NHOMO) (c) MO184(HOMO)

(d) MO185(LUMO) (e) MO186(NLUMO) (f) MO187(LUMO+2)

Figure 9. Frontier orbitals for the $T_3@T_3$ states at the M06-2X/6-31G(d,p) levels of theory.

As shown in figure 9, $\psi^*_{\text{LU+2}}\psi_{\text{NLU}} \approx 0$, $\psi^*_{\text{LU+2}}\psi_{\text{LU}} \approx 0$, $\psi^*_{\text{HO}-2}\psi_{\text{NHO}} \approx 0$, and $\psi^*_{\text{HO}-2}\psi_{\text{HO}} \approx 0$,

$$\rho(\Psi_{T_3@T_3}, \Psi_{T_2@T_3}) = -q^2\psi^*_{\text{LU}}\psi_{\text{NLU}}, \quad (22)$$

$$\begin{aligned}\rho(\Psi_{T_3@T_3}, \Psi_{T_1@T_3}) = &+qs(\rho_0 - \rho_{\text{HO}-2} + \rho_{\text{LU}}) + qp(\rho_0 - \rho_{\text{NHO}} + \rho_{\text{NLU}}) \\ &-qr(\rho_0 - \rho_{\text{HO}} + \rho_{\text{LU}}).\end{aligned} \quad (23)$$

The orbital localizations give rise to the reduction of the overlap densities.

5. Conclusion

The pseudodegenerate electronic structure can give rise to the excited state wavefunctions with the multi-configurationality. Offdiagonal VCCs between such excited state wavefunctions can be small because of the reductions of the overlap densities between them. Small VCCs make nonradiative transitions from the higher triplet states to the lower triplet states suppressed. If the energy levels of these excited state, T_2 and T_3 in PTZ-BZP, are close to the emitting state, S_1 in PTZ-BZP, the reverse intersystem crossings from the higher triplet states to the emitting state is possible (fluorescence via higher triplets mechanism). The efficiency of exciton generation in such a system can be high because both single and triplet excitons can be utilized under current excitation as is the same in TADF OLED.

For PTZ-BZP, the energy gap between S_1 and T_3 is 25 meV, and that between S_1 and T_2 is -15 meV. The offdiagonal VCCs between the T_3 and T_2 states and those between the T_3 and T_1 states are less than 3.7×10^{-4} a.u., and those between the T_2 and T_1 states are less than 2×10^{-5} a.u. Therefore, the high EQE in PTZ-BZP can be attributed to fluorescence via higher triplet states because of the suppression of the nonradiative transitions from the T_3 and

T_2 states. Since the triplet excitons can be converted to the S_1 state via reverse intersystem crossing from these higher triplet states, the efficiency of exciton generation is increased.

To increase EQE, not only the efficiency of exciton generation should be increased, but also the PLQY should be enhanced. From the offdiagonal VCD analysis for the S_1 state, the overlap density on the benzothiadiazole moieties should be reduced by some chemical modification to increase the PLQY of PTZ-BZP.

One class of the candidates of the fluorescence via higher triplets mechanism is a set of molecules of X-Y-X type. For PTZ-BZP, BZP is X, and PTZ is Y. Two same moieties of X can yield a pseudodegenerate electronic structure which can give rise to multiconfigurational excited states.

Acknowledgments
Numerical calculations were partly performed at the Supercomputer Laboratory of Kyoto University. This study was also supported by a Grant-in-Aid for Scientific Research (C) (15K05607) from the Japan Society for the Promotion of Science (JSPS).

Appendix

Vibronic coupling density operator

The field operators of an electron are defined by

$$\hat{\psi}(\boldsymbol{r}_1) := \sum_b \psi_b(\boldsymbol{r}_1)\hat{c}_b, \tag{24}$$

$$\hat{\psi}^\dagger(\boldsymbol{r}_1) := \sum_a \psi_a^*(\boldsymbol{r}_1)\hat{c}_a^\dagger, \tag{25}$$

where $\psi_b(\boldsymbol{r}_1)$ denotes a molecular orbital, and $\hat{c}_b$ and $\hat{c}_a^\dagger$ are a annihilation and creation operator, respectively.

In second-quantized form, vibronic coupling density operator[18] is written as

$$\hat{\eta}^\alpha(\boldsymbol{r}_1) := \hat{\psi}^\dagger(\boldsymbol{r}_1)v^\alpha(\boldsymbol{r}_1)\hat{\psi}(\boldsymbol{r}_1) = \sum_{a,b} \eta_\alpha^{ab}(\boldsymbol{r}_1)\hat{c}_a^\dagger\hat{c}_b, \tag{26}$$

where potential derivative $v^\alpha(\boldsymbol{r}_1)$ is defined by

$$v^\alpha(\boldsymbol{r}_1) := \left(\frac{\partial u(\boldsymbol{r}_1)}{\partial Q_\alpha}\right)_{\mathbf{R}_0}, \text{with} \quad u(\boldsymbol{r}_1) = -\sum_A \frac{Z_A e^2}{4\pi\epsilon_0 |\boldsymbol{r}_1 - \boldsymbol{R}_A|}. \tag{27}$$

Orbital vibronic coupling density (OVCD) $\eta_{ab}^\alpha(\boldsymbol{r}_1)$ is defined by

$$\eta_{ab}^\alpha(\boldsymbol{r}_1) := \psi_a^*(\boldsymbol{r}_1)v^\alpha(\boldsymbol{r}_1)\psi_b(\boldsymbol{r}_1). \tag{28}$$

Offdiagonal VCC V_{mn}^α can be written by using VCD operator as

$$V_{mn}^\alpha = \langle \Psi_m | \hat{\eta}^\alpha | \Psi_n \rangle = \int \eta_{mn}^\alpha(\boldsymbol{r}_1) d\boldsymbol{r}_1. \tag{29}$$

VCD $\eta_{mn}^\alpha(\boldsymbol{r}_1)$ and OVCD $\eta_{ab}^\alpha(\boldsymbol{r}_1)$ describe a local picture of a vibronic coupling of mode α.

IOP Conf. Series: Journal of Physics: Conf. Series **833** (2017) 012020 doi:10.1088/1742-6596/833/1/012020

References

[1] Uejima M, Sato T, Yokoyama D, Tanaka K and Park J -W 2014 *Phys. Chem. Chem. Phys.* **16** 14244
[2] Sato T, Tokunaga K and Tanaka K 2008 *J. Phys. Chem. A* **112** 758
[3] Sato T, Uejima M, Iwahara N, Haruta N, Shizu K and Tanaka K 2013 *J. Phys. Conf. Ser.* **428** 012010
[4] Uejima M, Sato T, Tanaka K and Kaji H 2014 *Chem. Phys.* **430** 47
[5] Kameoka Y, Uebe M, Ito A, Sato T and Tanaka K 2014 *Chem. Phys. Lett.* **615** 44
[6] Uejima M, Sato T, Detani M, Wakamiya A, Suzuki F, Suzuki H, Fukushima T, Tanaka K, Murata Y, Adachi C and Kaji H 2014 *Chem. Phys. Lett.* **602** 80
[7] Uebe M, Ito A, Kameoka Y, Sato T and Tanaka K 2015 *Chem. Phys. Lett.* **633** 190
[8] Endo A, Ogasawara M, Takahashi A, Yokoyama D, Kato Y and Adachi C 2009 *Adv. Mater.* **21** 4802
[9] Uoyama H, Goushi K, Shizu K, Nomura H and Adachi C 2012 *Nature* **492** 234
[10] Kaji H, Suzuki H, Fukushima T, Shizu K, Suzuki K, Kubo S, Komino T, Suzuki F, Wakamiya A, Murata Y, Adachi C 2015 *Nature Commun.* **6** 8476
[11] As a review article, Adachi C 2014 *Jpn. J. Appl. Phys.* **53** 060101
[12] Sato T, Uejima M, Tanaka K, Kaji H and Adachi C 2015 *J. Mater. Chem. C* **3** 870
[13] Sato T 2015 *J. Comput. Chem., Jpn.* **14** 189
[14] Sato T, Hayashi R, Haruta N and Pu Y -J arXiv:1609.06122 [cond-mat.mtrl-sci], 20 Sep 2016.
[15] Yao L, Zhang S, Wang R, Li W, Shen F, Yang B and Ma Y 2014 *Angew. Chem.* **126** 2151
[16] Zhao Y and Truhlar D 2008 *Theor. Chem. Acc.* **120** 215
[17] Frisch M J, Trucks G W, Schlegel H B, Scuseria G E, Robb M A, Cheeseman J R, Scalmani G, Barone V, Mennucci B, Petersson G A, Nakatsuji H, Caricato M, Li X, Hratchian H P, Izmaylov A F, Bloino J, Zheng G, Sonnenberg J L, Hada M, Ehara M, Toyota K, Fukuda R, Hasegawa J, Ishida M, Nakajima T, Honda Y, Kitao O, Nakai H, Vreven T, Montgomery Jr. J A, Peralta J E, Ogliaro F, Bearpark M, Heyd J J, Brothers E, Kudin K N, Staroverov V N, Kobayashi R, Normand J, Raghavachari K, Rendell A, Burant J C, Iyengar S S, Tomasi J, Cossi M, Rega N, Millam J M, Klene M, Knox J E, Cross J B, Bakken V, Adamo C, Jaramillo J, Gomperts R, Stratmann R E, Yazyev O, Austin A J, Cammi R, Pomelli C, Ochterski J W, Martin R L, Morokuma K, Zakrzewski V G, Voth G A, Salvador P, Dannenberg J J, Dapprich S, Daniels A D, Farkas O, Foresman J B, Ortiz J V, Cioslowski J and Fox D J 2009 *Gaussian 09 revision D.01* Gaussian Inc. Wallingford CT
[18] Uejima 2014 PhD Thesis Kyoto University

Charged ultradiscrete supersonic kinks and discrete breathers in nonlinear molecular chains with realistic interatomic potentials and electron-phonon interactions

Yuriy A. Kosevich
Semenov Institute of Chemical Physics, Russian Academy of Sciences, 4 Kosygin str., 119991 Moscow, Russia
E-mail: yukosevich@gmail.com

Abstract.
The problem of electron or hole trapping by supersonic lattice kink is revisited. Supersonic kinks in molecular chains with realistic interatomic potential produce local compression of the lattice. Lattice compression enhances electron Fermi energy and therefore produces for the electron a local potential hill, rather than a potential well, through the deformation potential of the proper sign. Here we discuss the possibility of electron trapping above the top of its tight-binding conduction band, where it possesses negative effective mass, by supersonic kink in a molecular chain with realistic interatomic potentials and electron-phonon interactions. The localization length of the electron wave function is much larger than lattice period in the case of adiabatic electron dynamics and decreases with the velocity of the ultradiscrete supersonic kink with the approximately sinusoidal envelope with the "magic" wave number. Such kinks were revealed in lattices with different interatomic potentials with hardening anharmonicity. Electron or hole can also be trapped by discrete breather (intrinsic localized mode) in the lattice with realistic asymmetric anharmonic potential. The local quasi-static strain, produced by the stationary or slowly-moving discrete breather in the lattice, can trap the electron (or hole) with its localization below the lower edge of the conduction (or above the upper edge of the valence) band.

1. Introduction

Plane one-dimensional shock waves can be considered as an example of supersonic propagation of locally enhanced density in macroscopic condensed matter systems [1, 2]. One the other hand, macromolecules in polymers present an example of nanoscale quasi-one-dimensional systems. Electron-phonon interaction in quasi-one-dimensional macromolecules like polyacetylene can result in the Pierles lattice dimerization and the formation of the gap at the Fermi level, which strongly affect transport properties of such low-dimensional materials [3]. Essentially the Pierles lattice dimerization in a quasi-one-dimensional system can be considered as the *band Jahn-Teller effect* [4].

Supersonic kinks with local lattice compression can also propagate in quasi-one-dimensional macromolecules of the polyacetilene type [5]. In this paper we discuss the scenario of electron trapping above the top of its tight-binding conduction band, where it possesses negative effective mass, by supersonic kink in a molecular chain with realistic interatomic potentials and electron-phonon interactions. The reason of such kind of electron trapping is related with the property that lattice compression enhances electron Fermi energy and therefore produces for the electron a local potential hill, rather than a potential well, through the deformation potential of the proper sign. Electron trapping by the potential hill is different from the previously considered electron trapping by the effective potential well, produced by supersonic kink [6, 7, 8, 9, 10]. We consider electron trapping by ultradiscrete supersonic kinks with the approximately sinusoidal envelope with the "magic" wave number, which were revealed in lattices with different interatomic potentials with hardening anharmonicity [11, 12, 13]. It is worth mentioning that the trapping by *potential hill* of the quasiparticle with negative effective mass is similar to the formation of intrinsic localized mode (discrete breather) above the phonon band in a crystal with *repulsive* anharmonicity [14].

Electron, exciton or hole can also be localized by the discrete breather in the lattice with realistic asymmetric anharmonic potential. In the case of stationary or slowly moving discrete breather (DB), there is a kink-like distribution of static or quasi-static lattice displacements, caused by the asymmetry of the interparticle potential [15] (see also [16]). In the lattice with realistic interatomic potentials, these displacements cause local lattice stretching and the formation of local potential well for the electron (or exciton), produced by the discrete breather. This in turn results in the trapping of slowly-moving electron (or exciton) by DB with its localization below the lower edge of the conduction (or exciton) band. The hole can be trapped by the stationary or slowly-moving DB above the valence band, below the Fermi level. The papers [17, 18] consider different mechanisms of electron localization by DB, not related with static stretching of the anharmonic lattice.

2. The model

We consider the following semiclassical Hamiltonian, which describes the interaction of classical longitudinal phonons with a quantum quasiparticle (electron, exciton or hole) in 1D lattice with nearest-neighbor interactions between atoms (nuclei) and between corresponding quasiparticles (the tight-binding model):

$$H = H_{lat} + H_{el} + H_{el-ph}, \tag{1}$$

$$H_{lat} = \sum_n [\frac{1}{2M} p_n^2 + U(u_n - u_{n-1})], \tag{2}$$

$$H_{el} = \sum_n [E_0 a_n^* a_n - J a_n^* (a_{n+1} + a_{n-1})], \tag{3}$$

$$H_{el-ph} = -\sum_n (u_n - u_{n-1})[g_1 a_n^* a_n - g_2 a_n^* (a_{n+1} + a_{n-1})], \tag{4}$$

where u_n is atomic displacement from the equilibrium position in the lattice site n, M is atomic mass, p_n is the momentum conjugate to u_n, $U(r)$ is the (nonlinear in general) interparticle potential, where $r = u_n - u_{n-1}$ is relative particles displacement, a_n is the complex amplitude of the quasiparticle in the lattice site n such that $|a_n|^2$ gives the probability to find the quasiparticle at this site, E_0 is energy in the center of the conduction band, $4J$ gives the width of the conduction band, g_1 and g_2 are (positive) constants of the electron-phonon interaction, which determine the shift of the center and the change of the width of the conduction band caused by lattice strain (1D deformation) $u_n - u_{n-1} = d\partial u/\partial x$, d is lattice period. The model of the electron-phonon interaction with the two parameters g_1 and g_2, eq. (4), combines the deformation-potential model for nonpolar semiconductors [19] with the Su-Schriffer-Heeger (SSH) model for quasi-1D conducting systems [20] (with the parameters g_1 and g_2, respectively).

From equations (1)-(4) we obtain the corresponding equations of motion for a_n and u_n:

$$\begin{aligned} i\hbar\dot{a}_n &= E_0 a_n - J(a_{n+1}+a_{n-1}) - (u_n - u_{n-1})[g_1 a_n - g_2(a_{n+1}+a_{n-1})], \qquad (5) \\ M\ddot{u}_n &= -\frac{\partial}{\partial u_n}[U(u_n-u_{n-1}) - U(u_{n+1}-u_n)] + [g_1(|a_n|^2 - |a_{n-1}|)^2 \\ &\quad - g_2(a_n^*(a_{n+1}+a_{n-1}) - a_{n-1}^*(a_n + a_{n-2}))]. \qquad (6) \end{aligned}$$

Both at the lower and upper edges of the quasiparticle band, one has $|a_n| \approx |a_{n+1}|$ and the amplitudes a_n at the nearest-neighbour sites are in phase at the lower edge and are anti-phase at the upper edge of the band. Therefore at both edges the effective Hamiltonian of the electron-phonon interaction H^{eff}_{el-ph} can be written in the following universal form (cf. eq. (4)):

$$H^{eff}_{el-ph} = -g^{eff}_{\pm}\sum_n (u_n - u_{n-1}) a_n^* a_n, \qquad (7)$$

where $g^{eff}_{\pm} = g_1 \pm 2g_2$, and sign plus (minus) corresponds to the upper (lower) edge of the quasiparticle band. The corresponding equations of motion for a_n and u_n take the following form:

$$\begin{aligned} i\hbar\dot{a}_n &= E_0 a_n - J(a_{n+1}+a_{n-1}) - g^{eff}_{\pm}(u_n - u_{n-1}) a_n, \qquad (8) \\ M\ddot{u}_n &= -\frac{\partial}{\partial u_n}[U(u_n-u_{n-1}) - U(u_{n+1}-u_n)] + g^{eff}_{\pm}(|a_n|^2 - |a_{n-1}|)^2. \qquad (9) \end{aligned}$$

In the continuum approximation with respect to the quasiparticle wave function $\psi = \psi(x,t)$, equations (8) and (9) can be written as:

$$\begin{aligned} i\hbar\dot{\psi} &= E_{\pm}\psi \pm Jd^2\frac{\partial^2\psi}{\partial x^2} - g^{eff}_{\pm}(u_n - u_{n-1})\psi, \qquad (10) \\ M\ddot{u}_n &= -\frac{\partial}{\partial u_n}[U(u_n-u_{n-1}) - U(u_{n+1}-u_n)] + g^{eff}_{\pm} d\frac{\partial|\psi|^2}{\partial x}, \qquad (11) \end{aligned}$$

where $\psi(x) = a_n$ and $\psi(x) = (-1)^n a_n$ at the lower and upper edges of the quasiparticle band, respectively, $x = nd$, $E_{\pm} = E_0 \pm 2J$. Equation (10) for the quasiparticle wave function close to the upper edge of the band corresponds to the anti-continuum limit, which is known in the envelope-function approach to the short-wavelength lattice and spin dynamics, both linear and nonlinear, see, e.g., Refs. [21, 22].

3. Results and Discussion

Below we list the realistic nonlinear interatomic potentials $U(r)$, which are known to support the propagation of supersonic kinks:

$$\begin{aligned} U(r) &= \frac{1}{2}K_2 r^2 - \frac{1}{3}K_3 r^3, \quad K_3 > 0, \qquad (12) \\ U(r) &= \frac{1}{2}K_2 r^2 - \frac{1}{3}K_3 r^3 + \frac{1}{4}K_4 r^4, \quad K_3 \geq 0, \quad K_4 > 0, \qquad (13) \\ U(r) &= \epsilon\left(\frac{1}{(1+r)^{12}} - \frac{2}{(1+r)^6}\right), \qquad (14) \\ U(r) &= \frac{1}{1+r}, \qquad (15) \end{aligned}$$

which are the α-Fermi-Pasta-Ulam (α-FPU), α-β-FPU, Lennard-Jones (LJ) and repulsive Coulomb (RC) potentials, respectively. In all these interparticle potentials, the part of the potential, which describes *contraction* of the lattice, is always more steep than the part, describing *stretching* of the lattice, which ensures the possibility of propagation of the *compression* supersonic kinks. It is worth mentioning that the α-β-FPU potential is also realized for the intermediate relative displacements r in the Lennard-Jones and repulsive Coulomb potentials.

3.1. Trapping of quasiparticle by supersonic kink

To describe a supersonic kink traveling to the right in the one-dimensional chain with the α-β-FPU interatomic potential, we use the following *ansatz* with the "magic" wave number $2\pi/3d$, which was introduced in Refs. [11, 12] and was successfully applied for the description of ultradiscrete supersonic kinks in repulsive Coulomb lattice in Ref. [13]:

$$u_n - u_{n-1} = -\frac{A}{2}(1+\cos(\frac{2\pi}{3d}(nd - Vt)) \tag{16}$$

for $-\pi \leq (2\pi/3d)(n - Vt) < \pi$, and $u_n - u_{n-1} = 0$ otherwise, where $A > 0$ is the kink amplitude of relative particle displacements, $V > \sqrt{K_2/M}d$ is supersonic velocity of the kink, $\sqrt{K_2/M}d$ is a speed of sound in the system. Localization of such nonlinear excitations increases with its velocity V, when the inverse localization length κ_{lat} of the lattice excitation is determined by its velocity and follows the equation [11]:

$$MV^2\kappa_{lat}^2 = 4K_2 \sinh^2(\kappa_{lat}d/2). \tag{17}$$

The kink velocity V grows in the large-amplitude limit (in β-FPU lattice) as [12]

$$V = \frac{3d}{2\pi}\sqrt{\frac{3K_2 + (45/16)K_4A^2}{M}}. \tag{18}$$

Supersonic kink with the pattern of relative displacements (16) can be considered as a discrete limit of the kink in the Korteweg - de Vries or modified Korteweg - de Vries equation.

Following the known solution of Schrödinger equation for localization in the 1D potential well which can be considered as perturbation [23], with the use of equation (10) we obtain the quasiparticle inverse localization length $\kappa_{qp} > 0$, given by the integral over the discrete breather width of the deformation potential (7), in which we substitute $u_n - u_{n-1}$=$d\partial u/\partial x$:

$$\kappa_{qp} = \frac{|m_{qp}|}{\hbar}\int_{-\infty}^{+\infty} g_+^{eff}(d\partial u/\partial x)dx = \frac{|m_{qp}|}{\hbar}g_+^{eff}dA \tag{19}$$

for $g_+^{eff} > 0$, where $|m_{qp}| = \hbar/(2d^2J)$ is the modulus of the negative quasiparticle effective mass at the top of the band. From equation (19) follows that quasiparticle localization length decreases with the increase of kink amplitude and velocity, and the localization length is much larger than lattice period in the case of adiabatic electron dynamics, for $V \ll 2Jd/\hbar$

With the use of equations (10) and (19), we obtain the energy E of the moving with group velocity V quasiparticle, localized above the upper edge of the band E_+:

$$E = E_+ + \frac{|m_{qp}|}{2\hbar}(g_+^{eff}dA)^2 - \frac{|m_{qp}|}{2}V^2 , \tag{20}$$

where $V = \partial E/(\hbar\partial q) = 2Jd\sin(qd)/\hbar = 2Jd(\pi - qd)/\hbar$ is the group velocity of the quasiparticle with the wavenumber q close to the Brillouin zone edge π/d, which coincides with the supersonic kink velocity V in eq. (16). It is worth mentioning that the shift $-|m_{qp}|V^2/2$ of the energy of the moving quasiparticle in equation (20) is consistent with the transformation of wave function of the particle with a given mass, moving with respect to laboratory frame [23]: the mass is effective and negative in equation (20). According to equation (19), the finite quasiparticle velocity V does not affect the localization condition, and the energy of the moving localized quasiparticle can enter the band, for $\Delta E < 0$ and $E < E_+$, since the latter is determined in the laboratory frame.

We can compare the inverse localization length (19) and energy (20) of the moving localized quasiparticle with the inverse localization length $\kappa_{DB} > 0$ and energy $\hbar\omega_{DB}$ of the moving discrete breather, localized above the phonon band in the α-β-FPU lattice, see, e.g., [14, 16, 24]:

$$\kappa_{DB} = \sqrt{6K_4/K_2 - 8(K_3/K_2)^2} A_{DB}/d, \tag{21}$$

$$\hbar\omega_{DB} = \hbar\omega_{max}(1 + (\frac{3K_4}{4K_2} - \frac{K_3^2}{K_2^2})A_{DB}^2) - \frac{|m_{ph}|}{2}V^2, \tag{22}$$

$$m_{ph} = -\frac{2\hbar}{d^2}\sqrt{\frac{M}{K_2}}, \tag{23}$$

$$\omega_{max} = 2\sqrt{\frac{K_2}{M}}, \tag{24}$$

where A_{DB} is displacement oscillation amplitude in the discrete breather, m_{ph} is effective "phonon mass" at the top of phonon band with the dispersion $\omega = 2\sqrt{K_2/M}\sin(qd/2)$, $1/m_{ph} = \partial^2\omega/\partial q^2/\hbar$ and $V = \partial\omega/\partial q$. Similar to the case of moving quasiparticle, equation (20), frequency of the moving discrete breather can, in principle, enter phonon band, when $\omega_{DB} < \omega_{max}$, see equation (22).

3.2. Trapping of quasiparticle by discrete breather

Discrete breather in the lattice with realistic asymmetric interparticle potential $U(r)$ causes local static deformation (stretching) of the lattice [15]. For instance, in the α-β FPU chain with $3K_4K_2 > 4K_3^2$, we have the following expression for the local static stretching deformation $(\partial u/\partial x)_{st}$ [16]:

$$(\partial u/\partial x)_{st} = \langle\frac{1}{2}\partial(u_n + u_{n-1})/\partial x\rangle = \frac{K_3}{dK_2}\langle(u_n - u_{n-1})_{DB}^2\rangle > 0\,, \tag{25}$$

where $(u_n - u_{n-1})_{DB}^2$ denotes square of relative particle displacements in the discrete breather and angle brackets denote time averaging (one has $\langle(u_n - u_{n-1})_{DB}^2\rangle \sim A_{DB}^2$, cf. equations (21) and (22)). This relation shows that discrete breather produces local *energetic expansion* of the lattice with realistic asymmetric interatomic potential, which is caused by atoms with locally enhanced vibrational energy, forming discrete breather [16].

With the use of equations (10) and (25) with substitution $u_n - u_{n-1}$=$d(\partial u/\partial x)_{st}$, we obtain the inverse localization length $\kappa_{qp} > 0$,

$$\kappa_{qp} = \frac{2m_{qp}}{\hbar}\frac{K_3}{K_2}g_-^{eff}\langle(u_n - u_{n-1})_{DB}^2\rangle\lambda_{DB}, \tag{26}$$

and energy E of the quasiparticle trapped below its conduction band where is possesses positive effective mass m_{qp}:

$$E = E_- - \frac{2m_{qp}}{\hbar}(\frac{K_3}{K_2}g_-^{eff}\langle(u_n - u_{n-1})_{DB}^2\rangle\lambda_{DB})^2 + \frac{m_{qp}}{2}V^2\,, \tag{27}$$

where $\lambda_{DB} = 1/\kappa_{DB}$ is localization length of the discrete breather, see equation (21). The trapping can be realized for g_-^{eff}=$g_1 - 2g_2$> 0. Since discrete breathers are stationary or slowly-moving nonlinear lattice excitations, kinetic energy of the moving quasiparticle cannot significantly affect the trapping condition (26), but the energy of the moving quasiparticle can, in principle, enter the band, for $E > E_-$, cf. equations (20) and (22).

In p-doped semiconductor, a hole can be trapped by the stationary or slowly-moving discrete breather. Since the sign of the deformation potential (7) for holes is opposite to that for electrons [19], the hole can be trapped above the upper edge of the valence band. This property reflects the particle-hole symmetry in the quasiparticle trapping.

4. Conclusions

We describe the possibility of electron trapping above the top of its tight-binding conduction band, where it possesses negative effective mass, by supersonic kink in a molecular chain with realistic interatomic potentials and electron-phonon interactions. The localization length of the electron wave function is in general much larger than lattice period for adiabatic electron dynamics and decreases with the velocity of the ultradiscrete supersonic kink. We consider electron trapping by ultradiscrete supersonic kinks with the approximately sinusoidal envelope with the "magic" wave number, which were revealed in lattices with different interatomic potentials with hardening anharmonicity. Electron or hole can also be trapped by discrete breather (intrinsic localized mode) in the lattice with realistic asymmetric anharmonic potential. The local quasi-static strain, produced by the stationary or slowly-moving discrete breather in the lattice, can trap the electron (or hole) with its localization below the lower edge of the conduction (or above the upper edge of the valence) band.

5. References

[1] Zel'dovich Ya B and Raizer Yu P 1966 *Physics of Shock Waves and High-Temperature Hydrodynamic Phenomena* vol 1 (New York: Academic Press)
[2] Forbes J W 2012 *Shock Wave Compression of Condensed Matter* (Berlin: Springer-Verlag)
[3] Heeger A J, Kivelson S, Schrieffer J R and Su W P 1988 *Rev. Mod. Phys.* **60** 781
[4] Bersuker I B and Polinger V Z 1989 *Vibronic Interactions in Moleclules and Crystals* (Berlin: Springer-Verlag) p 365
[5] Savin A V, Manevich L I, Christiansen P L and Zolotaryuk A V 1999 *Physics-Uspekhi* **42** 245
[6] Zmuidzinas J S 1978 *Phys. Rev. B* **10** 3919
[7] Davydov A S 1985 *Solitons in Molecular Systems* (Dordrecht: Springer Science+Business Media)
[8] Zolotaryuk A V, Spatschek K H and Savin A V 1996 *Phys. Rev. B* **24** 266
[9] Hennig D, Chetverikov A, Velarde M G and Ebeling W 2007 *Phys. Rev. E* **76** 046602
[10] Velarde M G 2010 *J. Comput. and Appl. Math.* **233** 1432
[11] Kosevich Yu A 1993 *Phys. Rev. Lett.* **71** 2058
[12] Kosevich Yu A, Khomeriki R and Ruffo S 2004 *Europhys. Lett.* **66** 21
[13] Archilla J F R, Kosevich Yu A, Jiménez N, Sánchez-Morcillo V J and García-Raffi L M 2015 *Phys. Rev. E* **91** 022912
[14] Kosevich Yu A and Lepri S 2000 *Phys. Rev. B* **61** 299
[15] Kosevich A M and Kovalev A 1974 *Sov. Phys. JETP* **40** 891
[16] Kosevich Yu A, Manevitch L I and Savin A 2008 *Phys. Rev. E* **77** 046603
[17] Vekhter B G and Ratner M A 1995 *Phys. Rev. B* **51** 3469
[18] Flach S and Kladko K 1996 *Phys. Rev. B* **53** 11531
[19] Bardeen J and Shockley W 1950 *Phys. Rev.* **80** 72
[20] Su W P, Schrieffer J R and Heeger A J 1979 *Phys. Rev. Lett.* **42** 1698
[21] Kosevich Yu A 1993 *Phys. Rev. B* **47** 3138
[22] Kosevich Yu A and Gann V V 2013 *J. Phys.: Cond. Mat.* **25** 246002
[23] Landau L D and Lifshitz E M 1991 *Quantum Mechanics. Non-Relativistic Theory* (Oxford: Pergamon)
[24] Bickham S R, Kiselev S A and Sievers A J 1993 *Phys. Rev. B* **47** 14206

Understanding the electron-phonon interaction in polar crystals: Perspective presented by the vibronic theory

A. Pishtshev

Institute of Physics, University of Tartu, W. Ostwaldi 1, 50411 Tartu, Estonia

E-mail: aleksandr.pishtshev@ut.ee

N. Kristoffel

Institute of Physics, University of Tartu, W. Ostwaldi 1, 50411 Tartu, Estonia

Abstract. We outline our novel results relating to the physics of the electron-TO-phonon (el-TO-ph) interaction in a polar crystal. We explained why the el-TO-ph interaction becomes effectively strong in a ferroelectric, and showed how the electron density redistribution establishes favorable conditions for soft-behavior of the long-wavelength branch of the active TO vibration. In the context of the vibronic theory it has been demonstrated that at the macroscopic level the interaction of electrons with the polar zone-centre TO phonons can be associated with the internal long-range dipole forces. Also we elucidated a methodological issue of how local field effects are incorporated within the vibronic theory. These result provided not only substantial support for the vibronic mechanism of ferroelectricity but also presented direct evidence of equivalence between vibronic and the other lattice dynamics models. The corresponding comparison allowed us to introduce the original parametrization for constants of the vibronic interaction in terms of key material constants. The applicability of the suggested formula has been tested for a wide class of polar materials.

IOP Conf. Series: Journal of Physics: Conf. Series **833** (2017) 012022 doi:10.1088/1742-6596/833/1/012022

1. Introduction

In a lattice dynamics model of a polar crystalline material the expression for the squared dipole-active vibration in the Γ point has an effective representation in terms of two contributions:

$$\Omega_{0j}^2 = \widetilde{\omega}_{0j}^2 + \Delta\omega_{0j}^2 \, . \tag{1}$$

The first one $\widetilde{\omega}_{0j}^2$ describes an overall positive contribution of the short range forces acting between ions. Since the second contribution $\Delta\omega_{0j}^2$ is negative, comparison of $\widetilde{\omega}_{0j}^2$ and $\Delta\omega_{0j}^2$ indicates that the lattice becomes dynamically unstable through the imaginary transverse mode. Microscopic understanding of $\Delta\omega_{0j}^2$, was given by the vibronic theory (e.g., [1, 2, 3, 4, 5, 6, 7]). According to the vibronic mechanism the interband mixing of valence electrons by the zone-center polar TO lattice vibrations is the leading driving force of displacive structural instability. Outstanding success has been established in an explanation of structural changes in the perovskite oxides for which the lattice instability was associated with the effect of dynamical hybridization. The hybridization involves the mixing the filled O 2p and the empty Ti d^0 electronic states by the IR-active TO F_{1u} soft vibrations and gives rise to an additional chemical interaction, known as dynamic covalency, that governs a structural distortion of a parent high-symmetry cubic configuration.

On the other hand, as it has been previously observed in a number of classic works on lattice dynamics [8, 9, 10, 11], just the presence of long-range forces predetermines the polar properties of crystals. Furthermore, in the context of Born and Huang theory of lattice dynamics [8], the specific contribution represented by the second term of (1) can be expressed via the regular part of the Coulomb contribution to the force matrix of the system[12, 13]:

$$\Delta\omega_{0j}^2 = -\frac{4\pi e^2}{v} \sum_{\alpha,\beta} \frac{\delta_{\alpha\beta}}{(\epsilon_\infty + 2)} \left(\sum_s Z_s^* \frac{w_\alpha(s, 0j)}{\sqrt{M_s}} \right) \left(\sum_t Z_t^* \frac{w_\beta(t, 0j)}{\sqrt{M_t}} \right), \tag{2}$$

where v is the volume of the unit cell, Z_s^* are the Born dynamical charges, and ϵ_∞ is the electronic (static-high-frequency) dielectric constant. Comparison of (1) and (2) allows to interpret the stability issue of a polar material as a delicate balance of the short-range and long-range internal forces.

The vibronic theory has been the subject of intense development over the past several decades. The results were reported in a number of books and review papers (e.g., [1, 2, 3, 4, 5, 6, 7] and references therein). One of its fundamental results is that the theory provided a faithfully microscopic insight into the origin of displacive structural transitions in crystalline solids. As it follows from the theory, the magnitude of vibronic interaction is particularly important to promote and trigger displacive structural instability. However, the problem of accurate characterizing the interaction of valence electrons with vibrations of the transverse optical branch received much less attention and had remained a challenge for a substantially long time. Furthermore, the

IOP Conf. Series: Journal of Physics: Conf. Series **833** (2017) 012022 doi:10.1088/1742-6596/833/1/012022

important part of the challenge has been the methodological difference positioned as the difference between the underlying mechanisms of the structural instability, one of which is based on the vibronic interaction and the other one relies upon the long-wavelength dynamics of force balances. Thus, the main purpose of the present paper was (i) to clarify why the e-TO-ph interaction is strong enough in polar materials, and (ii) to find a proper way according which the vibronic interaction and mechanism relating structure and forces could be entirely linked at the level of the specific contribution $\Delta\omega_{0j}^2$ to the dipole-active vibrational mode.

2. Results

We first present here the vibronic Hamiltonian in terms of the second-quantization formalism:

$$H_{el-ph} = N^{-1/2} \sum_{\sigma,\sigma'} \sum_{j} \sum_{\mathbf{k},\mathbf{q}} g_{\sigma\sigma'}(\mathbf{k},\mathbf{q}j)\, a^{+}_{\sigma\mathbf{k}} a_{\sigma'\mathbf{k}-\mathbf{q}}\, u_{\mathbf{q}j}\,, \tag{3}$$

where $g_{\sigma\sigma'}(\mathbf{k},\mathbf{q}j)$ are the matrix elements of the electron-TO-phonon interaction (the vibronic coupling) calculated on the Bloch wave functions $\psi_{\sigma\mathbf{k}}$. As usual, $\psi_{\sigma\mathbf{k}}$ labeled by the band index σ represent the basis of the one-particle electronic states with energies $E_\sigma(\mathbf{k})$ and Fermi occupation factors $f_\sigma(\mathbf{k}) \equiv f(E_\sigma(\mathbf{k}))$. Note that the detailed evaluation of how the electron-TO-phonon interaction contributes into the total energy of the system has been performed in [14].

2.1. Why the e-TO-ph interaction is strong enough in polar materials

As we consider a model of the polar crystal at the atomistic level, it is important to emphasize that the microscopic expression for the quantities $g_{\sigma\sigma'}(\mathbf{k},\mathbf{q}j)$ involves the dipole polarization (dipole moment per unit volume) $\mathbf{P}(\mathbf{q}j)$ (defined in [8]) associated with the polar TO mode. Let us introduce in the long-wavelength limit $\mathbf{q} \to 0$ the amplitude of the interband el-TO-ph interaction $|\, g_{\sigma\sigma'}(\mathbf{k})| = (\lim_{\mathbf{q}\to 0} |g_{\sigma\sigma'}(\mathbf{k},\mathbf{q}j)|^2)^{1/2}$. Due to the dependence of $|\, g_{\sigma\sigma'}(\mathbf{k})|$ on the polarization $P(0j)$ one can represent the amplitude in terms of macroscopic quantities by using results of [8]. Details of calculations can be found in [14]. We demonstrate here the final expression:

$$|\, g_{\sigma\sigma'}(\mathbf{k})| \cong \sqrt{\frac{v}{4\pi}\,\epsilon_\infty\,(\Delta\Omega_{LT})^2}\, \left| < \sigma\mathbf{k}|\, \nabla\phi_e(\mathbf{r})\, |\sigma'\mathbf{k} > \right| . \tag{4}$$

The most important feature of equation (4) is that it displays the straightforward interplay of two macroscopic effects:
(i) it depends directly on the LO-TO splitting $\Delta\Omega_{LT} = (\Omega_{LO}^2 - \Omega_{TO}^2)^{1/2}$, the magnitude of which is determined by the difference between the longitudinal (Ω_{LO}) and the transverse (Ω_{TO}) vibrational modes, and
(ii) it is determined by the spatial distribution of gradients of the overall Coulomb

electrostatic potential $\phi_e(\mathbf{r})$

$$\phi_e(\mathbf{r}) = \sum_{\mathbf{l}} \frac{e}{|\mathbf{r}-\mathbf{l}|}\,. \tag{5}$$

represented as the sum over the lattice site $\mathbf{l}$. The potential is associated with an internal electric field which via macroscopic changes in valence electron charge densities $\delta\rho_e(\mathbf{r}) = e\sum_{\mathbf{l}}\delta(\mathbf{r}-\mathbf{l})$ is generated by the actual long-wavelength distortions [14].

Another feature of equation (4) is determined by the product of two macroscopic parameters: $\epsilon_\infty\,(\Delta\Omega_{LT})^2 \sim |Z^*|^2$. Thereby, equation (4) provides the necessary link connecting theoretical result (4) with the corresponding results of studies [16] on Born's dynamical effective charges in ferroelectrics. The comparison shows that the presence of $\epsilon_\infty\,(\Delta\Omega_{LT})^2$ guarantees the significantly large values of the el-TO-ph interaction in ferroelectrics [15]. We can thus conclude that the vibronic coupling of valence band electrons with dipole-active zone-centre TO phonons has a long-range character, is governed by the internal electric field, and is essentially sensitive to a key macroscopic parameter of a polar material – Born's dynamical effective charge.

2.2. Vibronic mechanism and long-range forces

In previous subsection we demonstrated a proof of the canonical view of the vibronic theory that in a polar crystalline material the dipole-active vibrational modes may strongly interact with valence band electrons. We now show how vibronic interaction channel (3) can be related to lattice dynamics models of a polar insulator considered in [8, 9, 10, 11]. The idea is to reconcile the microscopic results by mapping theirs to the macroscopic scale determined by equation (2) of the Born and Huang theory [8]. Thus, our task is to check whether calculations of the softening vibronic contribution to the TO vibrational mode can reveal a solution that corresponds exactly the formulation of equation (2). In a general RPA approach, such examination was performed in [14]. In order to draw comparisons with the previous theoretical studies we here report a special case of the zone-centre TO vibrational modes. At the Γ point, the vibronic renormalization of the TO phonon frequency is given by

$$\begin{cases} \Delta\omega_{0j}^2 = 4\pi v \sum\limits_{\alpha,\beta} B_{\alpha\beta}(0)\,P_\alpha(0j)\,P_\beta(0j)\,, \\ B_{\alpha\beta} = \dfrac{v}{4\pi e^2} \sum\limits_{\mathbf{G},\mathbf{G}'\neq 0} [\,G_\alpha\, v_c(\mathbf{G})\,\chi_{el}(\mathbf{G},\mathbf{G}')v_c(\mathbf{G}')\,G'_\beta\,]\,, \\ v_c(\mathbf{G}) = \dfrac{4\pi e^2}{v\,|\mathbf{G}|^2}\,, \quad \hat{\chi}_{el} = \dfrac{\hat{\Pi}_0}{\hat{I}-\hat{v}_c\hat{\Pi}_0}\,, \\ \Pi_0(\mathbf{G},\mathbf{G}') = -\dfrac{1}{N}\sum\limits_{\sigma,\sigma'}\sum\limits_{\mathbf{k}} \dfrac{f_\sigma(\mathbf{k})-f_{\sigma'}(\mathbf{k})}{E_{\sigma'}(\mathbf{k})-E_\sigma(\mathbf{k})} \\ \times \int \psi^*_{\sigma\mathbf{k}}\, e^{i\mathbf{G}\,\mathbf{r}}\psi_{\sigma'\mathbf{k}-\mathbf{q}}\,d\tau \int \psi^*_{\sigma'\mathbf{k}-\mathbf{q}}\, e^{-i\mathbf{G}'\,\mathbf{r}'}\psi_{\sigma\mathbf{k}}\,d\tau'\,, \end{cases} \tag{6}$$

where the microscopical dielectric susceptibility of the electron subsystem $\chi_{el}(\mathbf{G},\mathbf{G}')$, written in the operator form, is expressed via the polarization operator $\hat{\Pi}_0$, the identity matrix $\hat{I}$, the bare Coulomb interaction $\hat{v}_c$. In the context of summation over the reciprocal lattice vectors $\mathbf{G}$ and $\mathbf{G}'$, it is important to emphasize that thanks to a fractional form of $\hat{\chi}_{el}$ the system of equations (6) accounts for the many-body polarization effects. As shown in [14], comparison with previously reported studies on the "macroscopic" counterpart of $\Delta\omega_{0j}^2$ [17] allowed one to obtain a result of summation over all $\mathbf{G}$, $\mathbf{G}'$ in the equation for the tensor $B_{\alpha\beta}(0)$. The final expression contains the local-field factor:

$$B_{\alpha\beta}(0) = \frac{v}{4\pi e^2} \sum_{\mathbf{G},\mathbf{G}'\neq 0} G_\alpha v_c(\mathbf{G})\, \chi_{el}(\mathbf{G},\mathbf{G}')\, v_c(\mathbf{G}')\, G'_\beta = -\frac{\delta_{\alpha\beta}}{(\epsilon_\infty + 2)}. \quad (7)$$

The minus sign displayed on the right hand of (7) implies that the elements of $B_{\alpha\beta}(0)$ are allowed to take only negative values, and, correspondingly, $\Delta\omega_{0j}^2$ describes the destabilizing contribution. Result (7) leads to the following expression for $\Delta\omega_{0j}^2$ which is represented in terms of the local field factor and the dipole oscillator strengths $S_{\alpha\beta}(j)$:

$$\begin{cases} \Delta\omega_{0j}^2 = \sum_{\alpha,\beta} B_{\alpha\beta}(0)\, S_{\alpha\beta}(j)\,; \\ S_{\alpha\beta}(j) = \dfrac{4\pi e^2}{v} \left(\sum_s Z_s^* \dfrac{w_\alpha(s,0j)}{\sqrt{M_s}} \right) \left(\sum_t Z_t^* \dfrac{w_\beta(t,0j)}{\sqrt{M_t}} \right). \end{cases} \quad (8)$$

Comparison of equation (2) with its vibronic counterpart given at the macroscopic level by equation (8) recognizes the typical patterns corresponding to the contribution of the long-range dipole forces. It is clearly seen that both equations display a full identity. In this context, we can thus conclude that we achieved the link between two different descriptions of the dynamics of TO vibrations: with respect to final results the description based on the vibronic mechanism appears entirely identical with that [8, 12, 13] exploiting the mechanism based on the dipole-dipole interactions.

2.3. El-TO-ph coupling via macroscopic parameters

The results of previous subsection offer an interesting possibility to connect (conceptually) the microscopic nature of the interaction of the valence band electrons with the polar long-wavelength TO modes to the regular part of the long-range dipole-dipole interaction. A comparative analysis revealed that the quantities $S_{\alpha\beta}(j)$ that serve as a direct mediator of the long-range dipole forces can be used for the parametrization of the el-TO-ph interaction in terms of material parameters. As it was reported in [14, 15], for an insulator with the electron band gap E_g the square of the effective $\mathbf{k}$-independent vibronic interaction renormalized (due to screening) by the local field factor can be represented as follows

$$\bar{g}_j^2 = 3g_{0j}^2/(\epsilon_\infty + 2)\,, \; g_{0j}^2 = \frac{1}{12} E_g\, M_j\, S(j)\,. \quad (9)$$

Here g_{0j}^2 are the squared el-TO-ph matrix elements averaged over the one-particle electronic states:

$$g_{0j}^2 = \sum_{\sigma' > \sigma} \frac{1}{2N} \sum_{\mathbf{k}} |\, g_{\sigma\sigma'}^{j}(\mathbf{k})|^2 \frac{E_g}{|\, E_{\sigma'}(\mathbf{k}) - E_{\sigma}(\mathbf{k})|} \, ;$$

$S(j) = \sum_{\alpha} S_{\alpha\alpha}(j)$, and the dipole oscillator strengths $S_{\alpha\alpha}(j)$ are associated with the dipole-active vibrational modes of the branch j and of the reduced mass M_j; the band gap value E_g plays the role of a material parameter. In analyzing relation (9), we should note three significant points. First, with respect to direct dependence on Born's dynamic charges it is fully consistent with microscopic equation (4). Likewise, the presence of Born's dynamic charges represents here a key signature indicating that the el-TO-ph coupling is particularly strong in a ferroelectric system. Second, the useful prognostic potential of (9), i. e. it allows directly evaluate the el-TO-ph coupling constants from the experimental/computational data. Third, very important is the observation that the electron-phonon Fröhlich Hamiltonian can be employed as the model Hamiltonian of the vibronic theory under the following rigorous requirement: the model constants of el-TO-ph interaction should be taken renormalized by the local field effect.

2.4. Tailoring of macroscopic measure of ferroelectricity

Based on the results of our comparative analysis, one can demonstrate that the vibronic theory describes the lattice instability for a wide number of ferroelectric materials. We start our rationale from the following arguments. On a microscopic scale, the vibronic theory rationalizes the underlying physics of the ferroelectric instability in terms of the specific matching of structural and electronic parameters. On the other hand, on a macroscopic scale, the dipole oscillator strengths provide reliable information on the properties of dipole-active lattice vibrational modes. Taking into account that the dipole oscillator strengths contain information on the strength of vibronic interactions in the system, we can quantify the interplay of both scales by introducing parameter τ written in terms of material constants Ω_{0j}, $S(j)$, and ϵ_∞:

$$\tau_j = \frac{\widetilde{S}(j)}{\widetilde{S}(j) + \epsilon_\infty + 2}\,, \quad \text{where} \quad \widetilde{S}(j) = S(j)/\Omega_{0j}^2 \qquad (10)$$

In ferroelectrics, $\widetilde{S}(j) >> \epsilon_\infty + 2$ for the soft-mode, so that $\tau_j \rightarrow 1$. Accordingly, the parameter τ_j represents a measure of the polarity. In fact, τ integrates the polar features of the system via the presence of the material constants in equation (10). In this sense one can say that (10) indicates how close the system under consideration resides to the point of ferroelectric instability.

The numerical investigation of (10) was performed in our recent work [18] for several representative polar compounds such as typical insulating perovskites, and the $A^{IV}B^{VI}$-type narrow-gap semiconductors. Estimates were made for the case of the most low-lying zone-center polar TO vibrational mode. Calculations demonstrated that in accordance with out theoretical description the values of the parameter τ fall for all the materials considered within the range: 0.89 (SnTe) – 0.99 ($BaTiO_3$).

3. Conclusion

In conclusion, we showed the full correspondence between two different lattice dynamics models on description of displacive structural instabilities in polar crystals. The obtained result is of principal importance because it confirms that the vibronic theory a priori accounts for the effects of the local field. We suggested a representation of the vibronic coupling via a combination of dipole oscillator strengths of the zone-center TO vibrational mode, electronic forbidden gap, and the dielectric constant. This approach provided a novel relationship that by using infrared spectroscopy and electron structure data allows one to estimate the strength of the el-TO-ph interaction in any polar crystalline material.

References

[1] Bersuker I B and Polinger V Z 1989 Vibronic interactions in molecules and crystals (New York: Springer)
[2] Bersuker I B 2006 The Jahn-Teller effect (Cambridge: Cambridge University Press)
[3] Bersuker I B 1995 *Ferroelectrics* **164** 75
[4] Bersuker I B 2013 *Chem. Rev.* **113** 1351
[5] Kristoffel N N 1984 *Czech. J. Phys.* B **34** 1253
[6] Kristoffel N and Konsin P 1988 *Phys. Status Solidi* (b) **149** 11
[7] Konsin P and Kristoffel N 1999 Ferroelectrics **226** 95
[8] Born M and Huang K 1954 Dynamical Theory of Crystal Lattices (Oxford: Oxford Univ. Press)
[9] Keating P N 1968 *Phys. Rev.* **175** 1171.
[10] Sham L J 1969 *Phys. Rev.* **188** 1431
[11] Pick R M, Cohen M H and Martin R M 1970 *Phys. Rev.* B **1** 910
[12] Kvyatkovskii O E 1985 *Sov. Phys. Solid State* **27** 1603
[13] Kvyatkovskii O E 1993 *Phys. Solid State* **35** 1071; 1997 *Phys. Solid State* **39** 602
[14] Pishtshev A 2011 *Physica* B **406** 1586
[15] Pishtshev A 2010 *Physica* B **405** 4128
[16] Zhong W, King-Smith R D and Vanderbilt D 1994 *Phys. Rev. Lett.* **72** 3618
[17] Vinetskii V L, Itskovskii M A and Kukushkin L S 1971 *Fiz. Tverd. Tela* **13** 76
[18] Kristoffel N, Pishtshev A, Rubin P and Klopov M 2016 *J. Supercond. Nov. Magn.* (doi:10.1007/s10948-016-3765-y)

Theory of three-step absorption of three light pulses

I Rebane

Institute of Physics, University of Tartu, W. Ostwaldi Str 1, 50411 Tartu, Estonia

inna.rebane@ut.ee

Abstract. The theory of the absorption of three light pulses with an arbitrary duration in four-level systems is proposed. The pulses are in resonance with the corresponding transitions. By using the time-dependent perturbation theory the probability that the fourth level is excited at the moment t is found. The possible absorption lines and parameters on which their widths depend on are analysed.

1. Introduction

Multistep absorption/excitation has been investigated in many ways and also both monochromatic light and light pulses have been used [1-7]. In [7] the possible time dependent two-step absorption spectra of two light pulses are theoretically investigated by using the simplest model for matter and in the case of an arbitrary duration of absorbed light pulses from monochromatic light to ultrashort pulses.

Thereby, we note that if at the two-step absorption in the three-level system the monochromatic light with the frequency ω_1 is absorbed at the first step $0 \to 1$, then in the spectrum where the frequency ω_1 is fixed and the frequency ω_2 of the maximum of the second light pulse (which is absorbed at the second step $1 \to 2$) varies, there exists one line whose width is determined by the rate of energy relaxation of the second excited level 2 and by the spectral width of the second pulse. In our opinion, this is exactly the same situation as in [6]. We note that if at the first step of the absorption the light pulse is absorbed and/or the pure phase relaxation of the first excited level exists, the second line is added to the aforementioned spectrum. The width of this line depends on the rates of the energy relaxations of both excited levels 1 and 2, the rate of the pure phase relaxation and on the spectral width of the second pulse, but not on the spectral width of the first pulse. On the other hand, the width of the initial line, in addition to the aforementioned parameters, depends on the spectral width of the first pulse. This line corresponds to the coherent contribution of the spectrum [3]. We note that the linewidths depend on the timeframes between the pulses and the time t as well.

In this paper the three-step absorption of three light pulses in a four-level system (the pulses are in resonance with the corresponding transitions) is observed to analyse which absorption lines are possible at all and which parameters their widths depend on. The interaction with the environment (phonons) is taken into account phenomenologically via the relaxation constants of the electronic levels. The presented model holds for impurity centers at low temperatures with weak electron-phonon coupling.

IOP Conf. Series: Journal of Physics: Conf. Series **833** (2017) 012023 doi:10.1088/1742-6596/833/1/012023

2. Starting formulas and model

The process starts from ground level 0. The resonance conditions are $\omega_1 \approx \Omega_{01}$, $\omega_2 \approx \Omega_{12}$ and $\omega_3 \approx \Omega_{23}$, where ω_1, ω_2 and ω_3 are the frequencies of the maximums of the pulses, Ω_{01}, Ω_{12} and Ω_{13} are the frequencies of the transitions between levels $0 \rightarrow 1$, $1 \rightarrow 2$ and $2 \rightarrow 3$.

By using the time-dependent perturbation theory [8, 9] analogically with the theory of the absorption of two light pulses in three-level systems [7], the probability that at the time moment t all three pulses are absorbed and the fourth level 3 is excited is the following:

$$W(t)=\int_{-\infty}^{t} dt_1 \int_{-\infty}^{t} dt'_1 S_3(t_1,t'_1) \int_{-\infty}^{t_1} dt_2 \int_{-\infty}^{t'_1} dt'_2 S_2(t_2,t'_2) \int_{-\infty}^{t_2} dt_3 \int_{-\infty}^{t'_2} dt'_3 S_1(t_3,t'_3) F(t,t_1,t'_1,t_2,t'_2,t_3,t'_3), \quad (1)$$

where F is the correlation function of the four-level system and S_1, S_2 and S_3 are the correlation functions of the first, second and third pulses. This probability $W(t)$ decreases to zero with the increase of time t because of the energy relaxation of the fourth level.

To simplify the calculations, the light pulses are taken coherent and of a single-sided exponential shape. The corresponding correlation functions are

$$\begin{aligned} S_1(t_3,t'_3) &= \theta(t_3-\tau_1)\theta(t'_3-\tau_1)\Delta_1 \exp[i\omega_1(t_3-t'_3)-\Delta_1(t_3+t'_3-2\tau_1)/2], \\ S_2(t_2,t'_2) &= \theta(t_2-\tau_2)\theta(t'_2-\tau_2)\Delta_2 \exp[i\omega_1(t_2-t'_2)-\Delta_2(t_2+t'_2-2\tau_2)/2], \\ S_3(t_1,t'_1) &= \theta(t_1-\tau_3)\theta(t'_1-\tau_3)\Delta_3 \exp[i\omega_2(t_1-t'_1)-\Delta_3(t_1+t'_1-2\tau_3)/2], \end{aligned} \quad (2)$$

where $\theta(x)$ is Heaviside step function, τ_1, τ_2 and τ_3 are the time moments when the pulses begin to pass through the impurity centre, Δ_1, Δ_2 and Δ_3 are the FWHM spectral widths of the pulses.

The correlation function of the four-level system is [9]

$$\begin{aligned} F(t,t_1,t'_1,t_2,t'_2,t_3,t'_3) = \Big\langle & v^+_{\omega_1} \exp\left[i\left(\hat{H}+\frac{i}{2}\hat{\gamma}\right)(t_2-t_3)\right] v^+_{\omega_2} \exp\left[i\left(\hat{H}+\frac{i}{2}\hat{\gamma}\right)(t_1-t_2)\right] v^+_{\omega_3} \\ & \times \exp\left[i\hat{H}(t'_1-t_1)-\frac{1}{2}\hat{\gamma}(2t-t_1-t'_1)\right] v_{\omega_3} \exp\left[-i\left(\hat{H}-\frac{i}{2}\hat{\gamma}\right)(t'_1-t'_2)\right] \\ & \times v_{\omega_2} \exp\left[-i\left(\hat{H}-\frac{i}{2}\hat{\gamma}\right)(t'_2-t'_3)\right] v_{\omega_1} \exp\left[-i\hat{H}(t'_3-t_3)\right] \Big\rangle. \end{aligned} \quad (3)$$

In (3) $\hat{H}$ is Hamiltonian and $\hat{\gamma}$ is the operator of the radiation decay of the four-level system, $\langle \ldots \rangle$ is the symbol of averaging over the ensemble of vibrations on the initial electronic level, and the single-photon matrix elements of the interaction Hamilton $\hat{V}$ of the field and matter are

$$\begin{aligned} v_{\omega_1} &= \left\langle N-1_{\omega_1} \middle| \hat{V} \middle| N \right\rangle, v_{\omega_2} = \left\langle N-1_{\omega_1}-1_{\omega_2} \middle| \hat{V} \middle| N-1_{\omega_1} \right\rangle, \\ v_{\omega_3} &= \left\langle N-1_{\omega_1}-1_{\omega_2}-1_{\omega_3} \middle| \hat{V} \middle| N-1_{\omega_1}-1_{\omega_2} \right\rangle. \end{aligned} \quad (4)$$

They describe the annihilation of the photons of the frequencies ω_1, ω_2 and ω_3. In (4) $|N\rangle$ is the initial state of the electromagnetic field. In our elementary model the phase relaxation and the phonon wings

are not taken into account, the relaxation processes of the excited levels 1, 2 and 3 are described by the rates of energy relaxation γ_1, γ_2 and γ_3. Thus the interaction with phonons is taken into account phenomenologically via the relaxation constants of the electronic levels.
Then the correlation function of the four-level system is

$$F(t,t_1,t'_1,t_2,t'_2 t_3,t'_3) = C\ \exp[-\gamma_3(2t - t_1 - t'_1)/2 - i\Omega_{23}(t_1 - t'_1) - \gamma_2(t_1 + t'_1 - t_2 - t'_2)/2 \\ - i\Omega_{12}(t_2 - t'_2) - \gamma_1(t_2 + t'_2 - t_3 - t'_3)/2 - i\Omega_{01}(t_3 - t'_3)], \quad (5)$$

where C is a constant.

In summary we have twelve parameters/variables which have influence on the spectra (Δ_1, Δ_2, Δ_3, $\tau_2 - \tau_1$, $\tau_3 - \tau_2$, $t - \tau_3$, $\omega_1 - \Omega_{01}$, $\omega_2 - \Omega_{12}$, $\omega_3 - \Omega_{23}$, γ_1, γ_2 and γ_3).

3. Results

Subsequently, $\tau_1 = 0$, $T_1 \equiv \tau_2 - \tau_1$, $T_2 \equiv \tau_3 - \tau_2$, $x \equiv \Omega_{01} - \omega_1$, $y \equiv \Omega_{12} - \omega_2$, and $z \equiv \Omega_{21} - \omega_3$. In figures 1 – 3 the spectral widths of light pulses Δ_1, Δ_3, the time intervals T_1, T_2, the time t and the relaxation constants γ_1, γ_2 and γ_3 are the same.

In figure 1 the dependence of the probability $W(t)$ on the frequency of the maximum of the first pulse ω_1 is calculated, the other frequencies ω_2, ω_3 are fixed and the quantities $\omega_2 - \Omega_{12}$, $\omega_3 - \Omega_{23}$ are chosen different from zero to separate possible lines spectroscopically.

In figure 2 the dependence of the probability $W(t)$ on the frequency of the maximum of the second pulse ω_2 is presented, the other frequencies ω_1, ω_3 are fixed.

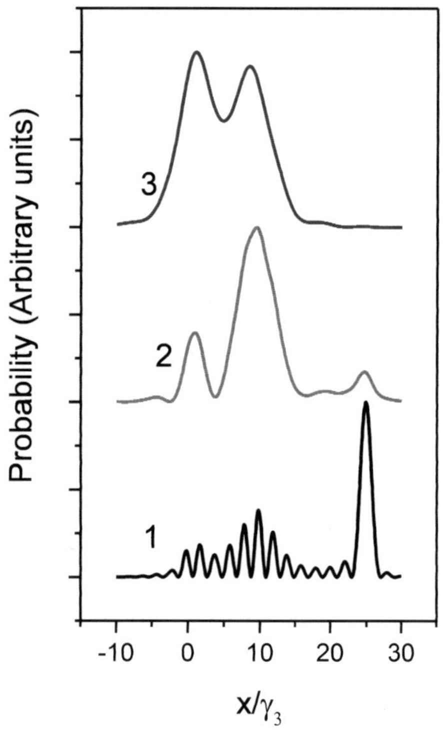

Figure 1. Probability $W(t)$ for different values of Δ_2, $x \equiv \Omega_{01} - \omega_1$ is variable. $y = -10\gamma_3$, $z = -15\gamma_3$. Curve 1 – $\Delta_2 = 0$, curve 2 – $\Delta_2 = 3\gamma_3$, curve 3 – $\Delta_2 = 10\gamma_3$. $\Delta_1 = 0.1\gamma_3$, $\Delta_3 = \gamma_3$, $\gamma_1 = \gamma_2 = 5\gamma_3$, $T_1 = T_2 = \gamma_3^{-1}$, $t = 5.1\gamma_3^{-1}$. Curves are normalized

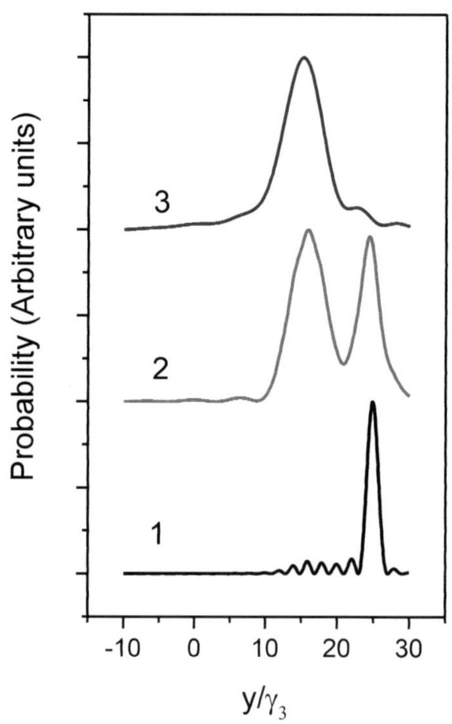

Figure 2. Probability $W(t)$ for different values of Δ_2, $y \equiv \Omega_{12} - \omega_2$ is variable. $x = -15\gamma_3$, $z = -10\gamma_3$. Curve 1 – $\Delta_2 = 0$, curve 2 – $\Delta_2 = 3\gamma_3$, curve 3 – $\Delta_2 = 10\gamma_3$. The other parameters are the same as in figure 1. Curves are normalized to 1.

to 1.

In figure 3 the dependence of the probability $W(t)$ on the frequency of the maximum of the third pulse ω_3 is calculated, the other frequencies ω_1, ω_2 are fixed. At different values of the spectral width of the second pulse Δ_2 different lines exist. When the frequency ω_1 is variable, then the possible maximums of the lines are at $x = 0$, $x = -y$ (in figure 1 $x = 10\gamma_3$) and $x = -y - z$ (in figure 1 $x = 25\gamma_3$). When the frequency ω_2 is variable, the possible maximums of the lines are at $y = 0$, $y = -x$ (in figure 2 $y = 15\gamma_3$), $y = -z$ and $y = -x - z$ (in figure 2 $y = 25\gamma_3$). Finally, when the frequency ω_3 is variable, the possible maximums of the lines are at $z = 0$, $z = -y$ and $z = -x - y$ (in figure 3 $z = 25\gamma_3$).

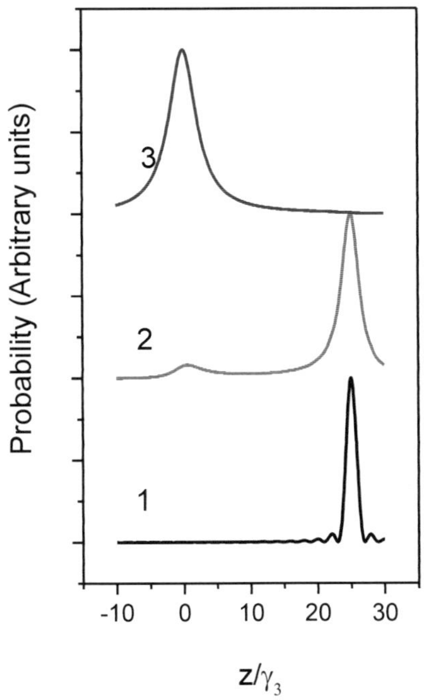

Figure 3. Probability $W(t)$ for different values of Δ_2, $z \equiv \Omega_{23} - \omega_3$ is variable. $x = -15\gamma_3$, $y = -10\gamma_3$. Curve 1 – $\Delta_2 = 0$, curve 2 – $\Delta_2 = 3\gamma_3$, curve 3 – $\Delta_2 = 10\gamma_3$. The other parameters are the same as in figure 1. Curves are normalized to 1.

Figure 4 shows the case where $y \equiv \Omega_{12} - \omega_2$ varies, in which all four possible lines are visible.

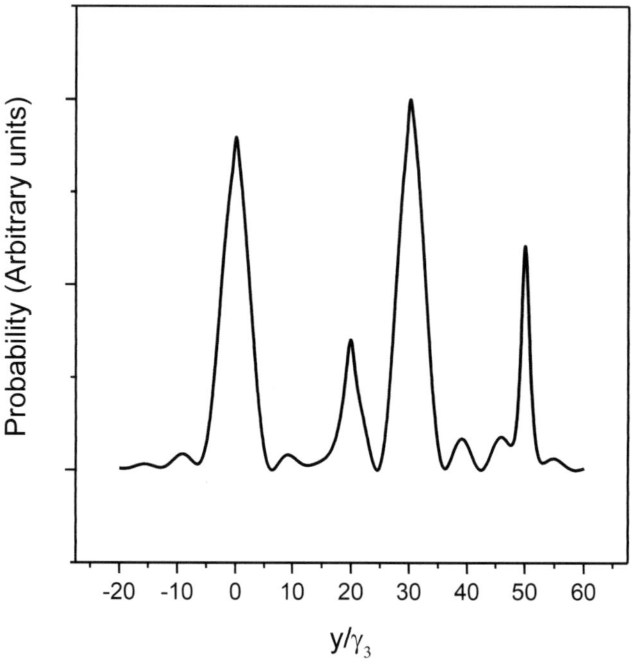

Figure 4. Probability $W(t)$ in the case where all four lines exist, $y \equiv \Omega_{12} - \omega_2$ is variable. $x = -30\gamma_3$, $z = -20\gamma_3$, $\Delta_1 = \Delta_2 = \Delta_3 = \gamma_1 = \gamma_2 = \gamma_3$, $T_1 = 2\gamma_3^{-1}$, $T_2 = \gamma_3^{-1}$, $t = 10.1\gamma_3^{-1}$. Curve is normalized to 1.

The widths of the different lines depend on different parameters (see table). It can be seen from the Table that if z varies, the width of the line with the maximum at $z = 0$ depends on the parameters γ_2, γ_3 and Δ_3; the width of the line with the maximum at $z = -y$ depends on the parameters γ_1, γ_3, Δ_2 and Δ_3; the width of the line with the maximum at $z = -x - y$ depends on the parameters γ_3, Δ_1, Δ_2 and Δ_3.

In the case, where the monochromatic light ($\Delta_1 = 0$) is used at the first step of the absorption ($0 \rightarrow 1$) in the spectrum where x varies, all three lines remain, in the spectrum where y varies the lines with the maximums at $y = -x$ and $y = -x - z$ remain, and in the spectrum where z varies the lines with the maximums at $z = 0$ and $z = -x - y$ remain.

If, additionally, the monochromatic light is used at the second step of the absorption ($1 \rightarrow 2$) in the spectrum where z is a variable, the line with the maximum at $z = 0$ disappears as well.

Note, that the lines for which the equation $x + y + z = 0$ applies are the coherent parts of the spectrums. Furthermore, the widths of all lines depend on the timespans T_1, T_2 and the time t. In general, with the increase of T_1, T_2 and t the lines become narrower.

Table. The dependence of the widths of different lines on different parameters

Locations of the maximums of the lines	$\Delta_1 \neq 0$, $\Delta_2 \neq 0$, $\Delta_3 \neq 0$	$\Delta_1 = 0$, $\Delta_2 \neq 0$, $\Delta_3 \neq 0$	$\Delta_1 = \Delta_2 = 0$, $\Delta_3 \neq 0$	$\Delta_1 = \Delta_2 = \Delta_3 = 0$
***x* varies**				
$x = 0$	γ_1, Δ_1	γ_1	γ_1	γ_1
$x = -y$	γ_2, Δ_1, Δ_2	γ_2, Δ_2	γ_2	γ_2
$x = -y - z$	γ_3, Δ_1, Δ_2, Δ_3	γ_3, Δ_2, Δ_3	γ_3, Δ_3	γ_3
***y* varies**				
$y = 0$	γ_1, γ_2, Δ_2	-	-	-
$y = -x$	γ_2, Δ_1, Δ_2	γ_2, Δ_2	γ_2	γ_2
$y = -z$	γ_1, γ_3, Δ_2, Δ_3	-	-	-
$y = -x - z$	γ_3, Δ_1, Δ_2, Δ_3	γ_3, Δ_2, Δ_3	γ_3, Δ_3	γ_3
***z* varies**				
$z = 0$	γ_2, γ_3, Δ_3	γ_2, γ_3, Δ_3	-	-

$z = -y$	$\gamma_1, \gamma_3, \Delta_2, \Delta_3$	-	-	-
$z = -x - y$	$\gamma_3, \Delta_1, \Delta_2, \Delta_3$	$\gamma_3, \Delta_2, \Delta_3$	γ_3, Δ_3	γ_3

4. Conclusions

The time-dependence theory of three-step absorption of three different light pulses with an arbitrary duration in the electronic four-level model is proposed. The probability that the fourth level is excited at the time moment t is found to be depending on the time delays between the pulses T_1 and T_2, the spectral widths of the pulses Δ_1, Δ_2 and Δ_3 and the energy relaxation constants γ_1, γ_2 and γ_3 of the excited electronic levels 1, 2 and 3. The time dependent perturbation theory is applied.

In the calculations the pulses are taken as coherent and of a single-sided exponential shape, ω_1, ω_2 and ω_3 are the frequencies of the maximums, 0, T_1 and $T_1 + T_2$ are the time moments when the pulses begin to pass through the impurity centre. The resonance conditions are $\omega_1 \approx \Omega_{01}$, $\omega_2 \approx \Omega_{12}$ and $\omega_3 \approx \Omega_{23}$ where Ω_{01}, Ω_{12} and Ω_{13} are the frequencies of the transitions $0 \rightarrow 1$, $1 \rightarrow 2$ and $2 \rightarrow 3$.

In the general case, where the spectral widths of the pulses Δ_1, Δ_2 and Δ_3 are comparable with the energy relaxation constants γ_1, γ_2 and γ_3, three lines may exist in the spectra in the cases where ω_1 or ω_3 are variable (corresponding ω_2 and ω_3 or ω_1 and ω_2 are fixed, figures 1 and 3). In the case where ω_2, is variable there may exist four lines (ω_1 and ω_3 are fixed, figure 4). An analysis shows that the widths of the possible lines depend on different parameters (Table).

This work was supported by the European Union through the European Regional Development Fund (Centre of Excellence "Mesosystems: Theory and Applications", TK114) and by institutional research funding (IUT2-27„Nonlinear theory of solids and fundamental fields“) of the Estonian Ministry of Education and Research.

References

[1] Yan Yi J, Fried L E and Mukamel S 1989 *J. Phys. Chem.* **93** 8149
[2] Yan Yi J and Mukamel S 1990 *Phys. Rev.* A **41** 6485
[3] Gelin M F, Egorova D, Pisliakov A V and Domcke W 2004 *Chem. Phys. Letters* **391** 234
[4] Dobryakov A L, Kovalenko S A and Ernsting N P 2005 *J. Chem. Phys.* **123** 044502
[5] Dobryakov A L, Perez Lustres J L, Kovalenko S A and Ernsting N P 2008 *Chem. Phys.* **347** 127
[6] Tanasittikosol M, Carr C, Adams C S and Wetherill K J 2012 *Phys. Rev.A* **85** 033830
[7] Rebane I 2015 *Physics Letters A* **379** 2135
[8] Hau-Riege S P 2015 *Nonrelativistic Quantum X-Ray Physics* (Weinheim: Wiley-VCH) pp 133-158
[9] Rebane I 2017 *Proc. Estonian Acad. Sci.* **66** 1